Gene Therapy for Rare Diseases

Gene Therapy for Rare Diseases

Guest Editor

Cord Brakebusch

MDPI

Basel • Beijing • Wuhan • Barcelona • Belgrade • Novi Sad • Cluj • Manchester

Guest Editor
Cord Brakebusch
Biotech Research &
Innovation Centre
The University of
Copenhagen
Copenhagen
Denmark

Editorial Office
MDPI AG
Grosspeteranlage 5
4052 Basel, Switzerland

This is a reprint of the Special Issue, published open access by the journal *Cells* (ISSN 2073-4409), freely accessible at: https://www.mdpi.com/journal/cells/special_issues/01PXQ7M9X8.

For citation purposes, cite each article independently as indicated on the article page online and as indicated below:

Lastname, A.A.; Lastname, B.B. Article Title. *Journal Name* **Year**, *Volume Number*, Page Range.

ISBN 978-3-7258-7446-0 (Hbk)
ISBN 978-3-7258-7447-7 (PDF)
https://doi.org/10.3390/books978-3-7258-7447-7

Contents

About the Editor

Cord Brakebusch

Cord Brakebusch is working at the Biotech Research and Innovation Center (BRIC) at the University of Copenhagen in Denmark. He received his M.SC. in Biochemistry in 1989 at the University of Hannover in Germany and performed his Ph.D. work at the Weizmann Institute of Science, Rehovot, Israel. Following postdocs at the Max Planck Institute of Biochemistry, Martinsried, Germany, he became Assistant Professor at Lund University, Lund, Sweden, in 1998 and a Heisenberg group leader at the Max Planck Institute of Biochemistry, Martinsried, Germany, in 2002. Cord Brakebusch joined the University of Copenhagen as a Professor in 2006. He also assumed the Directorship of the University's Transgenic Animal Core Facility in 2006, and since 2007 he has been a group leader of Biotech Research and Innovation Centre (BRIC) at the University of Copenhagen. Dr. Brakebusch investigates the role Rho GTPase signaling in development and disease using genetically modified mice. Rho GTPase signaling is regulating multiple important cellular functions including cell migration, cell polarity, and cell proliferation and work by the Brakebusch group revealed novel roles, for example, in skin inflammation and skin tumor formation.

cells

Editorial

Editorial for Special Issue on Gene Therapy of Rare Diseases

Cord Brakebusch

Biotech Research and Innovation Center (BRIC), University of Copenhagen, Ole Maaløes Vej 5, 2200 Copenhagen, Denmark; cord.brakebusch@bric.ku.dk

1. What Are Rare Diseases?

A rare disease is a condition that affects only a small portion of the population. In Europe, rare diseases are defined as diseases with a frequency of less than 1 in 2000. However, more than 7000 rare diseases have been described to date, and although each single disease is rare, the total number of patients suffering from rare diseases is high. It is estimated that 6% of the world population (around 450 million people) are affected by rare diseases [1]. Most rare diseases have a genetic background. Importantly, nearly all rare diseases lack an approved treatment; hence, the development of novel therapies for rare diseases is a major strategic research goal in the European Union, the US, and other countries.

2. Challenges for the Treatment of Rare Diseases

Developing therapies for rare diseases is challenging [2]. Gene therapy is the only curative option for the treatment of most rare diseases, but here, many obstacles need to be overcome in the development of an effective drug [3]. Firstly, many different genes and within each gene often different disease-causing mutations need to be targeted. This requires the development of a large number of therapies. Second, since each new therapy addresses only a very small number of patients, the costs per treatment are very high, often exceeding the financial capacities of individual healthcare systems. Gene therapy products such as Lenmeldy, Hemgenix, Skysona, Zynteglo, or Zolgensma, developed for the metachromatic leukodystrophy, hemophilia B, β-thalassemia, cerebral adrenoleukodystrophy, and spinal muscular atrophy, respectively, show clinical efficacy, but with prices of up to $4.3 million per patient, sales are still slow [4]. Moreover, unexpected and expensive failures at stage 3 clinical trials can occur despite convincing animal studies [5]. Finally, gene therapy is associated with several safety issues, including unwanted genetic alterations and immune reactions against the viral vectors used to introduce gene therapy tools into the target cells. More research is therefore required to reduce costs and minimize the adverse effects of gene therapy.

3. Content of This Special Issue

This Special Issue on "Gene Therapy of Rare Disease" collects several reviews and one original article covering the ongoing research into increasing the viability of gene therapy as a clinical treatment option.

For recessive rare diseases, where a functional gene is lacking, gene addition therapy—that is, is the introduction of a normal version of the diseased gene into the target cells—is possible. Lin et al. [Contribution 1], in an original article, report the successful gene therapy of a lysosomal storage disorder involving the intracranial viral transduction of the galactocerebrosidase gene into Twitcher mice, a mouse model for globoid cell leukodystrophy.

 https://doi.org/10.3390/cells15060504

The transduced gene was shown to be widely expressed in both the central and peripheral nervous systems and to greatly improve the survival of the treated Twitcher mice.

Next, Giometti and Papanikolaou [Contribution 2] provide an overview of the currently used viral transduction systems for hematopoietic stem cell gene therapy. This includes improvements to the viral vectors, small molecules enhancing transduction efficiency, safety aspects, ongoing clinical trials, and corresponding gene therapies already on the market.

Crucial for the success of gene addition therapies is the expression of the added gene being sufficiently high and restricted to the target cells. The choice of the promoter directing the expression level and cell type specificity of the added gene is therefore of the utmost importance. Artemyev et al. [Contribution 3] provide a detailed introduction to liver-specific promoters for application in the treatment of rare diseases of the liver, including α1-antitrypsin disease, Wilson disease, and hemophilia A and B. They describe the advantages and shortcomings of all the promoters currently used in liver-directed gene therapy in great detail. In an additional review, Artemyev et al. provide an overview of the attempts to generate synthetic promoters combining high tissue-specificity with high expression levels, thus widening the repertoire of natural promoters currently being used [Contribution 4].

More elegant than therapy by gene addition is therapy by gene repair, which, ideally, changes the defective gene back to the normal variant of the gene. In recent years, CRISPR technologies, including classical CRISPR/Cas9 genome editing, base editing, and prime editing, have given this field a significant boost. In 2025, Casgevy, a CRISPR-based cure for sickle cell anemia and transfusion-dependent β-thalassemia, was approved by the FDA, and more than 100 clinical trials for CRISPR-based medicines are ongoing. Mentani et al. [Contribution 5] describe the molecular mechanisms underlying the repair of DNA double-strand breaks, as well as the prime editing technology, which requires only a single-strand DNA break, resulting in less off-target effects.

Improving the efficiency and specificity of gene editing is a major research goal in CRISPR gene editing. He and Brakebusch [Contribution 6] discuss the question of whether modulation of nuclear actin polymerization could be used to increase the efficiency of precise CRISPR/Cas9 genome editing by homologous recombination.

Finally, two reviews present the current state of the art in gene therapy for specific rare diseases. Leal et al. describe the challenges and opportunities encountered in treating lysosomal storage disorders via CRISPR/Cas technologies [Contribution 7], and Vale et al. demonstrate the gene therapy options for the treatment of Diamond-Blackfan anemia [Contribution 8].

4. Conclusions

Taken together, gene therapy for rare diseases is a vibrant research field that, increasingly, is producing drugs approved for clinical application. However, the costs of these curative therapies need to be reduced in order to promote their wider application.

Funding: This study was supported by the European Union (Project: 101,072,427—GetRadi—HORIZON-MSCA-DN-2021).

Data Availability Statement: No new data were created or analyzed in this study. Data sharing is not applicable to this article.

Acknowledgments: The views and opinions expressed above are those of the authors only and do not necessarily reflect those of the European Union or the European Research Executive Agency. Neither the European Union nor the European Research Executive Agency can be held responsible for them.

https://doi.org/10.3390/cells15060504

Conflicts of Interest: The author declares no conflicts of interest.

List of Contributions:

1. Lin, D.-S.; Ho, C.-S.; Huang, Y.-W.; Lee, T.-H.; Huang, Z.-D.; Wang, T.-J.; Cheng, W.-C.; Huang, S.-F. Durable Global Correction of CNS and PNS and Lifespan Rescue in Murine Globoid Cell Leukodystrophy via AAV9-Mediated Monotherapy. *Cells* **2025**, *14*, 1942. https://doi.org/10.3390/cells14241942.
2. Giommetti, A.; Papanikolaou, E. Advancements in Hematopoietic Stem Cell Gene Therapy: A Journey of Progress for Viral Transduction. *Cells* **2024**, *13*, 1039. https://doi.org/10.3390/cells13121039.
3. Artemyev, V.; Paremskaia, A.I.; Dzhioeva, A.A.; Mishina, D.; Bogdanov, V.; Krupinova, J.; Mazloum, A.; Feoktistova, S.G.; Mityaeva, O.N.; Volchkov, P.Y. Engineering Liver-Specific Promoters: A Comprehensive Review of Design, Mechanisms, and Clinical Applications in Gene Therapy. *Cells* **2026**, *15*, 14. https://doi.org/10.3390/cells15010014.
4. Artemyev, V.; Gubaeva, A.; Paremskaia, A.I.; Dzhioeva, A.A.; Deviatkin, A.; Feoktistova, S.G.; Mityaeva, O.; Volchkov, P.Y. Synthetic Promoters in Gene Therapy: Design Approaches, Features and Applications. *Cells* **2024**, *13*, 1963. https://doi.org/10.3390/cells13231963.
5. Mentani, A.; Maresca, M.; Shiriaeva, A. Prime Editing: Mechanistic Insights and DNA Repair Modulation. *Cells* **2025**, *14*, 277. https://doi.org/10.3390/cells14040277.
6. He, X.; Brakebusch, C. Regulation of Precise DNA Repair by Nuclear Actin Polymerization: A Chance for Improving Gene Therapy? *Cells* **2024**, *13*, 1093. https://doi.org/10.3390/cells13131093.
7. Leal, A.F.; Prieto, L.E.; Pachajoa, H. CRISPR/Cas-Based Ex Vivo Gene Therapy and Lysosomal Storage Disorders: A Perspective Beyond Cas9. *Cells* **2025**, *14*, 1147. https://doi.org/10.3390/cells14151147.
8. Vale, M.; Prochazka, J.; Sedlacek, R. Towards a Cure for Diamond–Blackfan Anemia: Views on Gene Therapy. *Cells* **2024**, *13*, 920. https://doi.org/10.3390/cells13110920.

References

1. Taruscio, D.; Gahl, W.A. Rare diseases: Challenges and opportunities for research and public health. *Nat. Rev. Dis. Primers* **2024**, *10*, 13. [CrossRef] [PubMed]
2. Tambuyzer, E.; Vandendriessche, B.; Austin, C.P.; Brooks, P.J.; Larsson, K.; Miller Needleman, K.I.; Valentine, J.; Davies, K.; Groft, S.C.; Preti, R.; et al. Therapies for rare diseases: Therapeutic modalities, progress and challenges ahead. *Nat. Rev. Drug Discov.* **2020**, *19*, 93. [CrossRef] [PubMed]
3. Fox, T.A.; Booth, C. Improving access to gene therapy for rare diseases. *Dis. Models Mech.* **2024**, *17*, dmm050623. [CrossRef] [PubMed]
4. Drugs.com. Available online: https://www.drugs.com/article/top-10-most-expensive-drugs.html (accessed on 5 March 2026).
5. Servais, L.; Oskoui, M. Broad lessons from negative trials in rare diseases. *Lancet Neurol.* **2026**, *25*, 215. [CrossRef]

https://doi.org/10.3390/cells15060504

cells

MDPI

Article

Durable Global Correction of CNS and PNS and Lifespan Rescue in Murine Globoid Cell Leukodystrophy via AAV9-Mediated Monotherapy

Dar-Shong Lin [1,2,3,*], Che-Sheng Ho [2,4], Yu-Wen Huang [5], Tsung-Han Lee [5], Zo-Darr Huang [5], Tuan-Jen Wang [6], Wern-Cherng Cheng [6,7] and Sung-Fu Huang [6]

1 Department of Translational Medicine, MacKay Memorial Hospital, Taipei 104217, Taiwan
2 Department of Medicine, MacKay Medical University, New Taipei 252005, Taiwan
3 Department of Pediatrics, MacKay Memorial Hospital, Taipei 104217, Taiwan
4 Department of Neurology, MacKay Children's Hospital, Taipei 104217, Taiwan
5 Department of Medical Research, MacKay Memorial Hospital, Taipei 104217, Taiwan
6 Department of Laboratory Medicine, MacKay Memorial Hospital, Taipei 104217, Taiwan
7 Department of Laboratory Medicine, School of Medicine, College of Medicine, National Taiwan University, Taipei 106319, Taiwan
* Correspondence: dslin@mmh.org.tw

Highlights:

What are the main findings?

- A single, region-specific intracranial dose of AAV9-GALC achieves widespread CNS-PNS biodistribution, sustained supraphysiological GALC activity, and complete lifelong psychosine normalization in Twitcher mice.
- This streamlined monotherapy preserves myelin integrity, proteostasis, and motor function, leading to near–wild-type lifespan without the need for HSCT, repeated dosing, multi-route administration, or high systemic AAV exposure.

What are the implications of the main findings?

- Demonstrating durable metabolic and structural correction after a single intracranial dose provides strong rationale for translational intracerebral AAV9 strategies that minimize procedural burden and systemic risks.
- These results establish a practical therapeutic framework for GLD, highlighting the feasibility of achieving long-term CNS-PNS wide correction through targeted, low-exposure gene delivery.

Abstract: Globoid cell leukodystrophy (GLD) is a devastating lysosomal storage disorder caused by galactocerebrosidase (GALC) deficiency, leading to cytotoxic psychosine accumulation, broad neuroinflammation, dysfunction of autophagy and ubiquitin-proteasome system, progressive demyelination in both the central (CNS) and peripheral nervous systems (PNS), and premature death. Curative treatments are lacking, highlighting the urgent need for transformative approaches. Existing therapies have failed to achieve durable metabolic correction across neural compartments or sustained functional recovery. Here, we demonstrate that a single intracranial administration of high-titer AAV9-GALC targeting the thalamus and deep cerebellar nuclei achieves unprecedented and lifelong therapeutic efficacy in the Twitcher mouse model of GLD. This region-specific monotherapy achieved broad neuronal and glial transduction throughout the CNS and PNS, resulting in sustained supraphysiological GALC activity and complete normalization of psychosine levels. Treated mice exhibited preserved proteostasis, axonal architecture, and myelin integrity, inhibition of neuroinflammation, alongside restored motor function. Remarkably,

https://doi.org/10.3390/cells14241942

treated mice attain lifespans approaching wild-type levels, far surpassing all previously reported interventions in this model, indicating a durable, possibly lifelong therapeutic effect. By achieving durable and comprehensive metabolic and structural correction across neural systems without repeated dosing, multi-route delivery, combinational therapy, hematopoietic stem cell transplantation, or high-dose systemic delivery, this study establishes CNS-directed AAV9 monotherapy as a clinically translatable and potentially lifelong therapeutic paradigm for GLD.

Keywords: AAV; globoid cell leukodystrophy; GALC; gene therapy; psychosine; demyelination; neuroinflammation; autophagy

1. Introduction

Globoid cell leukodystrophy (GLD), or Krabbe disease, is caused by lysosomal storage disease. The infantile form of GLD accounts for approximately 90% of cases and represents the most severe phenotype. Affected infants typically present with extreme irritability, spasticity, and feeding difficulties as early as two months of age [1]. These symptoms rapidly progress to include vision loss, seizures, developmental regression, and profound neurodegeneration, culminating in early mortality, often by the age of two years [1]. The sole established therapeutic intervention for infantile GLD is hematopoietic stem cells transplantation (HSCT) for presymptomatic patients, extending median survival to 15.5 years but failing to halt neurological decline [2].

GLD is caused by autosomal recessive mutations in the GALC gene, which encodes lysosomal galactocerebrosidase, an enzyme essential for galactosylceramide degradation, a key sphingolipid component of myelin [1]. GALC deficiency results in impaired catabolism of both galactosylceramide and galactosylsphingosine (psychosine), the latter being a cytotoxic metabolite that accumulates in the central nervous system (CNS) and peripheral nervous system (PNS). Psychosine integrates into lipid rafts, disrupting membrane architecture and intracellular signaling critical for oligodendrocyte survival [3]. Additionally, psychosine interferes with autophagy and the ubiquitin-proteasome system (UPS), leading to cytoplasmic aggregate formation, mitochondrial dysfunction, increased reactive oxygen species, and apoptotic death of myelin-forming cells [4,5].

The Twitcher (Twi) mouse, an authentic murine model of infantile GLD, harbors a spontaneous nonsense mutation in Galc [6]. These mice exhibit progressive neurological decline that includes tremors, ataxia, paralysis of the hindlimb, reduced food intake, and weight loss, eventually succumbing around postnatal day 40 (P40). Pathologically, Twi mice exhibit absent GALC activity, elevated psychosine levels, widespread demyelination, and astrogliosis in the CNS and PNS, closely recapitulating human disease [7]. This model is extensively used to investigate disease mechanisms and evaluate emerging therapeutic strategies.

Several therapeutic approaches have been explored in Twi mice, including HSCT, enzyme replacement therapy (ERT), substrate reduction therapy (SRT), and gene therapy. HSCT has demonstrated benefits in reducing neuroinflammation and partially delaying disease progression, particularly when administered pre-symptomatically [7]. ERT has demonstrated limited effectiveness due to challenges in achieving sufficient enzyme delivery across the blood-brain barrier [8]. SRT achieves modest reductions in psychosine accumulation, though it has limited impact on survival [9]. Gene therapy using adeno-associated viral (AAV) vectors has shown the most promising approach to date, especially when administered neonatally, significantly extending lifespan and preserving myelin integrity [10]. While no single therapy fully reverses Krabbe disease, combinatorial strategies addressing enzyme deficiency, substrate toxicity, and neuroinflammation show the most promise in pre-clinical studies [10]. Yet, effective interventions restoring long-term central and peripheral myelin integrity, metabolism, and survival remain limited.

In this study, we demonstrate comprehensive correction of pathological, metabolic, and functional deficits in Twitcher mice through a single-dose, CNS-directed AAV9-GALC monotherapy. This targeted approach achieves sustained supraphysiological GALC activity, normalization of psychosine levels across both the CNS and PNS, and long-term preservation of proteostasis and myelin integrity. Notably, this treatment extends lifespan to near wild-type levels, surpassing all previously reported outcomes from either monotherapy or combinatorial gene therapy strategies. By achieving durable correction without repeated administration, combination therapy, HSCT, or high-dose systemic delivery, this study establishes CNS-directed AAV9 monotherapy as a clinically translatable and potentially lifelong therapeutic paradigm for GLD.

2. Materials and Methods

2.1. Animals and Therapy

All animal procedures were conducted in accordance with the guidelines for animal care sanctioned by the Animal Care and Use Committee of MacKay Memorial Hospital. The colonies of heterozygous (twi/+) twitcher mic, possessing a genetic background of C57BL/6J, were maintained through inbreeding under pathogen-free conditions in the animal research facility at MacKay Memorial Hospital. Genotyping of twitcher mice was performed at three days of age using a molecular PCR assay, as previously described [11]. Experimental subjects, including wild-type (+/+) and homozygous (twi/twi) mice, were allowed to live freely until reaching the moribund stage.

Newborn Twi mice were injected with AAV9-GALC at P3. The viral suspension was slowly injected into thalamus (1.5 mm rostral and 1 mm lateral from the lambda and 3 mm deep), and cerebellum (2 mm caudal and 1.5 mm lateral from the lumbda and 30° medial oblique 3 mm deep) at both hemispheres with Hamilton syringe (Hamilton Company, Reno, NV, USA). The animals received a total of 1.2×10^{12} AAV9 particles in a volume of 5 µL and 10 µL at each thalamus and deep cerebellum injection site, respectively. Untreated Twi and normal wild-type (WT) mice were included as controls.

2.2. Vector Production

The preparation of recombinant adeno-associated virus (rAAV) vectors was performed using a cotransfection method at the AAV Core Facility of Academia Sinica. The AAV9-GALC vector was designed to incorporate murine GALC cDNA under the regulation of the human cytomegalovirus enhancer and chicken-β-actin promoter. The recombinant AAV9-GALC vector was packed through a cotransfection procedure, involving vector plasmids and helper plasmids. The purification of the viral vector was performed by means of cesium chloride gradient centrifugation followed by extensive dialysis. The AAV titter was determined by droplet digital PCR.

2.3. GALC Activity

GALC activity was assessed using a previously described modified method [12]. Fresh tissue samples were homogenized in 10 mM sodium phosphate buffer (pH 6.0) containing 0.1% (*v*/*v*) NP-40, followed by centrifugation. An aliquot of 50 µL supernatant (5 µg protein) was incubated at 37 °C for 1.5 h in 100 µL reaction buffer (0.1/0.2 M citrate/phosphate buffer, pH 4, and 22 mM $AgNO_3$) containing 1.5 mM 4-methylumbelliferyl-D-galactopyranoside (cat#38597-12-5, Sigma-Aldrich, St Louis, MO, USA) as fluorescent substrate. The reaction was terminated by adding 200 µL stop buffer (0.2 M glycine/NaOH, pH 10.6). The fluorescence of released 4-methylumbelliferone was measured on a DTX 880 Multimode Detector (Beckman Coulter, Brea, CA, USA) at excitation and emission wavelengths of 385 nm and 450 nm, respectively. Enzymatic activity was expressed as nmol/mg/h.

2.4. Psychosine Concentration

Psychosine levels were quantified using a previously described LC-MS method with slight modifications [13,14]. Psychosine (cat#1305) and N-acetyl-psychosine (cat#1325) (internal standard, IS) were obtained from Matreya Chemical Co. (Ann Arbor, MI, USA).

Fresh tissue samples were homogenized in ice cold methanol (20% *w*/*v*). An 80 µL aliquot of homogenate was mixed with 240 µL of a formic acid/ethanol/isopropanol/methanol solution (0.5:37.5:37.5:25, *v*/*v*/*v*/*v*) and 10 µL of 50 ng/mL IS. The mixture was centrifuged at 20,400× *g* for 5 min and 150 µL of the resulting supernatant was combined with 60 µL of 1,3-butanediol. The mixture was evaporated to dryness under a nitrogen stream at 60 °C, reconstituted in 300 µL of mobile phase B (0.1% formic acid in acetonitrile/methanol, 95:5, *v*/*v*), ultrasonicated for 5 min and centrifuged at 1220× *g* for 10 min. The supernatant was filtered through a 0.22 µm membrane filter and analyzed using an Agilent 1260 infinity LC system (Agilent Technologies, Sata Clara, CA, USA). Psychosine levels were expressed as pmole/mg weight.

2.5. Immunofluorescences

Immunofluorescence staining was performed on slide mounted mouse brain cryosections according to an optimized protocol. Briefly, tissues were fixed in 4% paraformaldehyde (PFA), sequentially cryoprotected in 30%, 40% and 60% sucrose solutions, embedded in optimal cutting temperature (OCT) compound, and snap-frozen in chilled isopentane.

The sections were allowed to equilibrate to room temperature and post-fixed in 4% PFA for 10 min. After three washes in PBS, antigen retrieval was performed using a DAKO antigen retrieval buffer (DAKO Omnis, Agilent Technologies, Sata Clara, CA, USA), according to the manufacturer's instructions. The slides were then allowed to cool at room temperature (RT) and washed in phosphate buffered saline (PBS). Permeabilization was performed with 0.3% Triton X-100 in PBS for 10 min, followed by blocking with 10% normal goat serum in 0.3% Triton X-100/PBS for 1 h at room temperature.

Sections were incubated overnight at 4 °C with primary antibodies targeting myelin proteolipid protein (PLP) (cat#ab28486, #ab9311, Abcam, Waltham, MA, USA, 1:200 dilution), glial fibrillary acid protein (GFAP) (cat#Z0334, DAKO Omnis, Agilent Technologies, Sata Clara, CA, USA, 1:300 dilution), ionized calcium binding adaptor molecule 1 (Iba1) (cat# 019-19741, FUJIFILM Wako Pure Chemical, Osaka, Japan, 1:200 dilution), CD68 (cat#ab53444, Abcam, Waltham, MA, USA, 1:200 dilution), sequestosome 1 (p62/SQSTM1) (cat#56416, Abcam, Waltham, MA, USA, 1:200 dilution) and ubiquitin (cat#7254, Abcam, Waltham, MA, 1:200 dilution). After washing, sections were incubated for 1 h at room temperature with appropriate Alexa Fluor 488 or Alexa Fluor 594 conjugated secondary antibodies (Thermal Fisher Scientific, Waltham, MA, USA, 1:500 dilution). The nuclei were counterstained with 4′,6-diamidino-2-phenylindole (DAPI). The stained sections were mounted with antifade medium and stored at 4 °C in the dark until imaging. Fluorescent images were captured using a Leica DM IL LED microscope (Leica, Wetzlar, Germany) for analysis.

2.6. X-Gal Histochemistry

The modified histochemical staining technique for the in situ localization of GALC activity was performed according to established protocols [15,16]. The cryosections were fixed in 4% paraformaldehyde for 15 min at RT and then equilibrated in citrate/phosphate buffer (C/P buffer, pH 4.2) for an additional 15 min at RT. The sections were then incubated with a solution containing taurodeoxycholic acid (TDCA, 5 mg/mL) and oleic acid (OA, 5 mg/mL) prepared in C/P buffer (pH 4.2). This was followed by staining with a X-Gal solution containing 2 mg/mL of X-Gal and 5 mM potassium ferricyanide/potassium ferrocyanide, which was supplemented with TDCA (5 mg/mL) and OA (5 mg/mL) in C/P buffer (pH 4.2) for 1.5 h at 37 °C. Finally, the sections were washed with phosphate buffered saline (PBS) and distilled water, followed by counterstaining with Nuclear Fast Red (Sigma-Aldrich, St Louis, MO, USA).

2.7. *In Situ Hybridization*

To confirm successful transfection of the AAV9-GALC vector into the brain, spinal cord, and sciatic nerve of Twi mice after gene therapy, GALC mRNA was detected using in situ hybridization with an Mm-GALC probe (cat#563541) designed via RNAscope™ technology (Advanced Cell Diagnostics, Newark, CA, USA).

The tissue cryosections mounted on slides were fixed in 4% paraformaldehyde pre-chilled in PBS at 4 °C for 15 min. After fixation, sections were sequentially dehydrated in graded ethanol solutions (50%, 70%, and 95%) for 5 min each at room temperature (RT) and air dried for 5 min. Using the RNAscope® 2.5 HD Detection Kit–BROWN (cat#322310), endogenous peroxidase activity was quenched with RNAscope® hydrogen peroxide. The slides were then treated with boiling 1X target recovery solution at 100 °C for 5 min, washed 3~5 times in distilled water at RT, rinsed in 100% ethanol, and air dried. Subsequently, the slides were treated with RNAscope Protease IV and incubated at 40 °C for 30 min. After washing with distilled water, hybridization was performed with the Mm-Galc probe at 40 °C for 2 h. The sections were washed with Wash Buffer for 2 min at RT and then sequentially hybridized with amplification reagents Amp1, Amp2, Amp3, Amp4, Amp5 and Amp6 for 30, 15, 30, 15, 30, and 15 min, respectively, with a 2-min Wash Buffer rinse at RT between each amplification step. The signal was detected using chromogenic DAB staining and the sections were counterstained with hematoxylin. Bright field microscopy was used to visualize the results.

2.8. *Quantification of Viral Genomes by ddPCR*

Quantification of AAV9 vector genome copies was performed using droplet digital PCR (ddPCR) on the Bio-Rad QX200 system (Bio-Rad Laboratories, Hercules, CA, USA) according to the manufacturer's instructions. Duplex ddPCR reactions were prepared using primer–probe sets targeting the chicken β-actin (CBA) intron sequence within the AAV vector genome and the mouse *Rpp30* gene as a single-copy endogenous reference for normalization. Droplets were generated with the Bio-Rad Automated Droplet Generator, thermocycled on a T1000 Thermal Cycler, and subsequently read on the QX200 Droplet Reader. Vector genome (VG) copies per diploid genome (VG/DG) were calculated as the ratio of AAV CBA intron copies to *Rpp30* copies divided by two, providing precise quantification of AAV9 biodistribution in mouse brain tissues following CNS delivery.

2.9. *Transmission Electron Microscopy*

Sciatic nerves were prepared for transmission electron microscopy following standard protocols. Briefly, the samples were immersed in a fixative solution containing 2.5% glutaraldehyde and 1% osmium tetroxide, then dehydrated through a series of graded ethanol. Subsequently, the dehydrated samples were infiltrated with Spurr resin at three separate intervals of two hours each, culminating in polymerization at 70 °C for 8 h. The semi-thin sections (1 μm) were then cut from the embedded samples and stained with 0.5% toluidine blue to facilitate the identification of areas of interest under a light microscope. From these identified regions, ultrathin sections (70~90 nm) were prepared and subjected to staining with 2% methanolic uranyl acetate, followed by Reynolds lead citrate. The stained ultrathin sections were subsequently examined using a transmission electron microscope (JEM-1200EXII, JEOL Co., Tokyo, Japan) to analyze the ultrastructural characteristics of the sciatic nerve tissue.

2.10. Survival and Phenotype

Untreated Twi mice and WT mice were designated as control groups for the assessment of lifespan and weight. Both AAV9-treated (AAV9-Twi) and untreated Twi mice were monitored under standard conditions and allowed to live freely until they reached the moribund stage. All mice were subjected to daily observations, with body weights recorded weekly.

Motor coordination, strength, and locomotor function were assessed using a battery of behavioral tests, including the wire maneuver, rod balance, pole test, negative geotaxis, and wire hang test, as previously described [17–19]. For the wire maneuver, mice were suspended by the tail and lowered onto a horizontal wire. Scoring was as follows: 0, mouse swung hind legs to grasp the wire; 1, grasped the wire with struggling; 2, unable to grasp with hind limbs; 3, fell within 3 s; 4, fell immediately. The balance rod was used to test balance and motor and locomotor function. The mouse was placed on top and in the middle of a wooden rod (90 cm in length and 1.2 cm in diameter) that was 24 cm above the surface. Scores were assigned as follows: 0, traversed rod to end; 1, fell before reaching end; 2, froze for 60 s; 3, fell within 3 s; 4, fell immediately. The pole test was performed with minor modifications from Matsuura et al. [19] and measured strength and motor and locomotor function. The mice were placed head-up on a vertical pole and given 60 s to descend. Scores were assigned as follows: 0, turned and climbed down within 10 s; 1, took longer than 10 s; 2, turned but slid down; 3, turned and fell; 4, held on for 60 s; 5, held on or more than 30 s; 6, held on or more than 5 s; 7, fell immediately. The wire hang test provided a simple measure of strength. Latency to fall from an inverted cage lid, 50 cm above the surface, onto the soft bedding was recorded (maximum time 60 s).

2.11. Statistical Analysis

All results were expressed as the mean ± SD, and analyzed using the GraphPad Prism 6.0 software package. The survival rates between groups were plotted and analyzed using the Kaplan-Meier method, with a p-value of less than 0.05 considered statistically significant.

3. Results

3.1. Survival and Weight and Behavioral Assessment

Untreated Twi mice exhibited a lifespan between 35 and 43 days, serving as a baseline for comparison with AAV9-treated twitcher (AAV9-Twi) mice. Administration of AAV9-delivered gene therapy resulted in a marked extension of survival, with treated Twi mice demonstrating a minimum lifespan of 207 days and a median survival of 530 days (Figure 1A). Notably, 90.5% of AAV9-treated twitcher mice survived beyond one year, and 66.7% attaining an advanced age exceeding 500 days. The two longest-lived AAV9-Twi mice survived for 726 and 814 days, respectively, before being euthanized to conclude the study.

The body weight of the experimental animals was recorded weekly and analyzed to assess the efficacy of the treatment (Figure 1B). The improvement benefit of AAV9-mediated gene therapy was also evident in terms of body weight. Untreated Twi mice exhibited a significant delay in weight gain beginning at P21, followed by progressive weight loss after P35, in contrast to WT and AAV9-Twi mice. Although AAV9-Twi mice displayed significantly lower body weight than WT mice after P42, their overall pattern of weight gain closely paralleled that of WT controls. In adulthood, AAV9-Twi mice reached a maximum body weight corresponding to approximately 75% of that observed in WT mice. Furthermore, AAV9-Twi mice preserved ambulatory function (Video S1) and feeding behavior throughout their lifespan, with only moderate weight loss observed prior to sudden death. Notably, two longest-lived AAV-Twi mice, surviving 726 and 814 days respectively, remained active and asymptomatic, displaying phenotypes comparable to age-matched WT mice. No age-related complications or late-onset adverse effects were observed in long-lived AAV9-Twi mice.

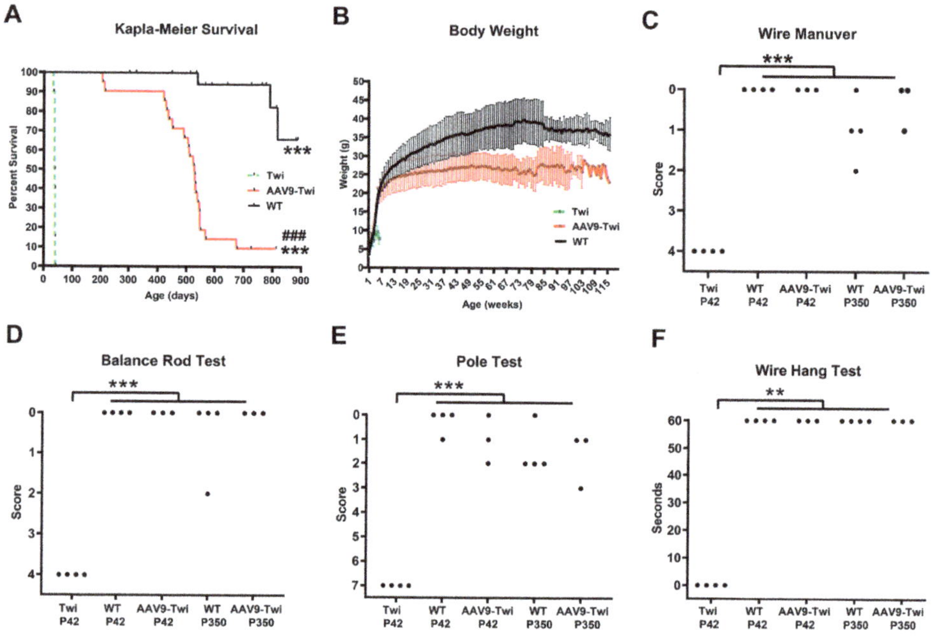

Figure 1. AAV9 Treatment Extends Survival and Improves Weight Gain and Behavioral Function in Twi Mice. (**A**) Kaplan-Meier survival curves demonstrate a significant extension of lifespan in AAV9-Twi mice ($n = 22$), reaching a maximum of 814 days, compared to untreated Twi mice ($n = 16$). WT mice ($n = 16$) are shown for reference. Vertical black ticks indicate censored animals, representing those removed at the study endpoint or for analysis. Statistical analysis was performed using Log-rank (Mantel-Cox) test. *** $p < 0.001$ compared to Twi, ### $p < 0.001$ compared to WT. (**B**) AAV9-Twi mice exhibit improved weight gain over time, in contrast to severely impaired growth in untreated Twi mice. Mean body weights for WT ($n = 16$), untreated Twi ($n = 16$), and AAV9-Twi mice are presented. (**C–F**) Behavioral assessments at P42 and P350 reveal that AAV9-Twi mice ($n = 3$) perform comparably to WT mice ($n = 4$) in wire maneuver, balance rod, pole, and wire hang tests. Untreated Twi mice display severe deficits at P42. Statistical analysis was performed using one-way ANOVA. ** $p < 0.01$; *** $p < 0.001$.

Functional assessments at P42 and P350 demonstrated preserved motor coordination, muscular strength, and locomotor activity in AAV9-Twi mice, as evaluated by wire maneuver (Figure 1C), rod balance (Figure 1D), pole test (Figure 1E), and wire hang assays (Figure 1F). Their performance was comparable to age-matched WT controls. In contrast, untreated Twi mice at P42 displayed profound impairments across these behavioral domains. Furthermore, AAV-Twi mice demonstrated normal spontaneous behaviors, including active locomotion (Video S1), drinking, and food consumption, indicative of sustained overall health and functional preservation following gene therapy. Normal behavior were supported by quantitative, validated behavioral assays and supplemental video documentation.

3.2. Supraphysiological Levels of GALC Activity

GALC activity was evaluated in multiple regions of the brain, including the cortex (Figure 2A), thalamus (Figure 2B), cerebellum (Figure 2C), brainstem (Figure 2D), spinal cord (Figure 2E), and sciatic nerve (Figure 2F) at P42 in near-moribund Twi mice and age-matched WT controls, as well as in aged (>P500) WT mice and aged (>P500) AAV9-Twi mice (Figure 2A–F). Tissue samples from Twi mice exhibited minimal detectable GALC activity compared to that of WT mice. GALC activity levels in aged WT mice were comparable to those observed at P42, indicating stable enzyme expression with age.

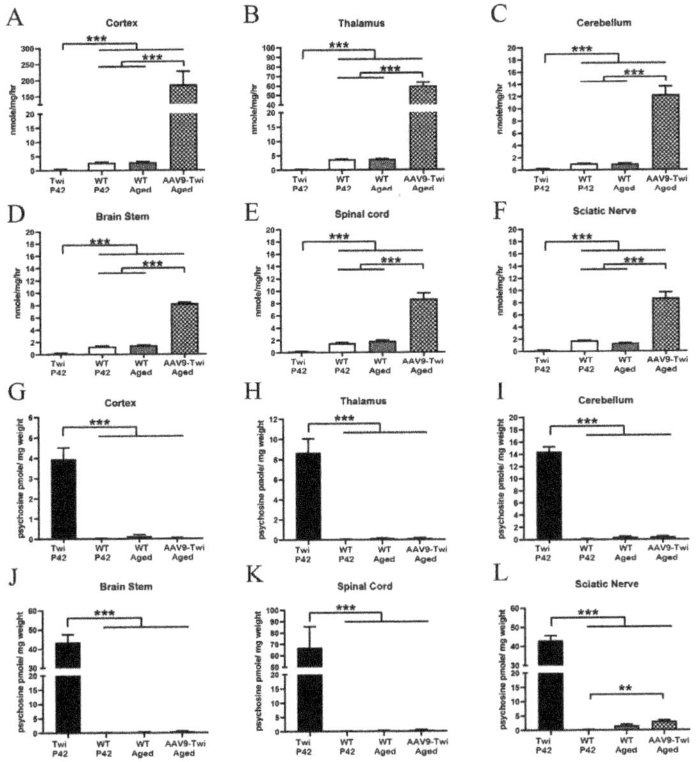

Figure 2. Supra-normal GALC Activity and Normalization of Psychosine Levels Following AAV9 Treatment in Twi Mice. (**A–F**) Aged (>P500) AAV9-Twi mice (n = 3) exhibit supra-normal GALC activity in the cortex, thalamus, cerebellum, brainstem, spinal cord, and sciatic nerve compared to P42 and aged (>P500) WT mice (n = 3 per group). In contrast, untreated Twi mice (n = 3, P42) display markedly reduced GALC enzymatic activity across these regions. (**G–L**) Psychosine concentrations in the cortex, thalamus, cerebellum, brainstem, spinal cord, and sciatic nerves are normalized in aged AAV9-Twi mice, reaching levels comparable to those observed in WT controls. Untreated Twi mice display substantial psychosine accumulation in all regions examined. The same tissue samples were used to measure GALC enzymatic activity in duplicate and psychosine levels in triplicate. Data are presented as mean ± SD. Statistical significance was determined by one-way ANOVA. ** $p < 0.01$; *** $p < 0.001$.

Remarkably, brain tissue from aged AAV9-Twi mice exhibited supranormal GALC activity relative to aged WT controls. In regions corresponding to AAV injection sites, GALC activity in the thalamus and cerebellum increased 20-fold and 12-fold, respectively, of the levels observed in aged wild-type mice, indicating efficient and robust transduction of AAV-GALC. Adjacent regions, including the cortex and brainstem, exhibited increases in GALC activity of up to 65-fold and 5.6-fold, respectively, relative to aged WT mice. Notably, distal regions such as the spinal cord and sciatic nerves also exhibited substantial increases, with GALC activity elevated approximately 4.5-fold and 6.5-fold, respectively, compared to aged WT mice. These findings of widespread, supranormal GALC activity in both CNS and PNS of AAV9-Twi mice indicate stable, robust, and sustained expression of AAV-delivered GALC throughout the life of the treated animals.

3.3. Normalization of Psychosine Concentration

In untreated Twi mice, lack of functional GALC leads to psychosine accumulation across tissues, notably in myelin-rich regions. To assess the relevance of GALC activity in catalyzing psychosine degradation, we measured psychosine levels in the same tissue samples used for the GALC activity assay. Mass spectrometry-based quantification revealed profound psychosine accumulation in the cortex (Figure 2G), thalamus (Figure 2H), cerebellum (Figure 2I), brainstem (Figure 2J), spinal cord (Figure 2K), and sciatic nerve (Figure 2L) of untreated Twi mice at P42, with the spinal cord exhibiting the highest levels. P42 and aged WT mice exhibited substantially lower psychosine levels across these regions (Figure 2G–L). Notably, aged AAV9-Twi mice (Figure 2G–L) showed psychosine concentrations in the cortex, thalamus, cerebellum, brainstem, and spinal cord comparable to those of P42 and aged WT mice. While psychosine levels in the sciatic nerve of AAV9-Twi mice were moderately elevated relative to P42 WT mice, they did not differ significantly from aged WT controls. These results demonstrate that AAV9-mediated GALC gene therapy effectively normalizes psychosine accumulation in both central and peripheral nervous system tissues, sustaining near-physiological levels throughout the lifespan of treated animals.

3.4. Intense and Broad GALC Expression

To visualize GALC enzymatic activity following AAV-mediated gene therapy, a modified X-Gal histochemical staining protocol was employed, as previously described, which selectively inhibits endogenous β-galactosidase activity to enable specific in situ detection of GALC activity in the CNS and PNS (Figure 3) [15].

In aged AAV9- Twi mice (Figure 3A–M), intense X-Gal staining was observed at the primary injection sites, including the thalamus and cerebellar white matter, corresponding to localized GALC enzymatic activity. Notably, robust and widespread staining extended from the olfactory bulb (Figure 3A), rostral infralimbic cortex (Figure 3B), cingulate cortex (Figure 3C), to caudal retrosplenial granular cortex and midbrain (Figure 3D). Intense staining was also detected in caudate putamen (Figure 3E), thalamus (Figure 3F) and hippocampus (Figure 3F). Within the hippocampus (Figure 3F), strong X-Gal signals were detected in neurons of the CA1, CA2, CA3 regions, and the dentate gyrus. Prominent staining was also observed in cerebellar white matter and Purkinje cells (Figure 3G).

In the brainstem and midbrain, variable staining intensities were detected in medulla (Figure 3H), with particularly strong signal in the substantia nigra (Figure 3I) and basilar pontine nuclei (Figure 3J). In sagittal spinal cord sections, abundant staining was evident along both the dorsal and ventral horns, and dorsal root of cervical (Figure 3K) and lumbar spinal cord (Figure 3L), and sciatic nerve (Figure 3M). Of note, X-Gal staining was also observed in the corpus callosum (Figure 3B–E), fornix (Figure 3E,F), and dorsal and lateral columns of the spinal cord (Figure 3K,L).

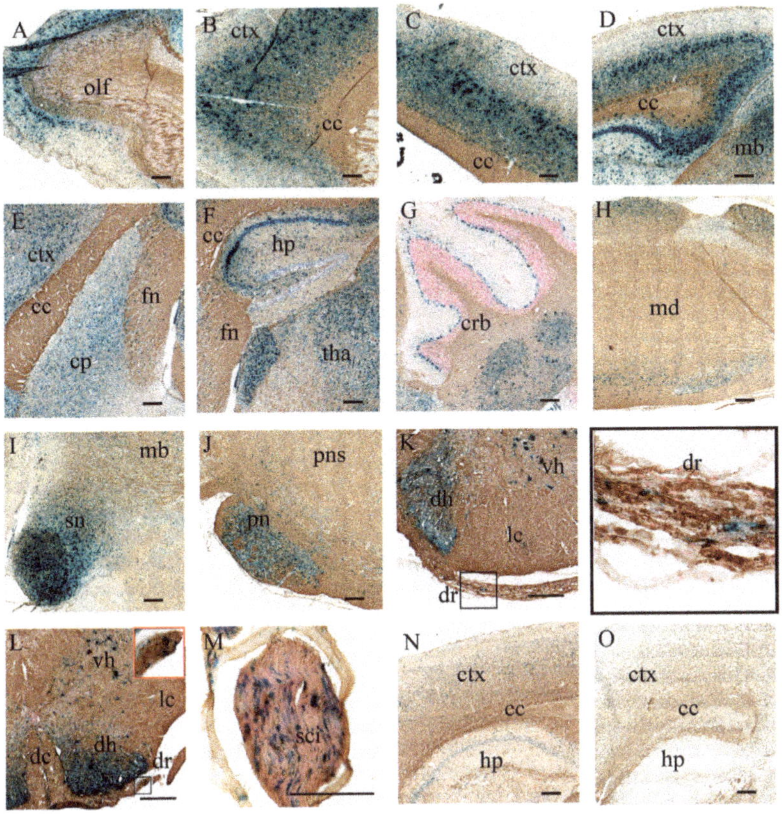

Figure 3. Widespread GALC Enzymatic Activity in the CNS and PNS of AAV9-Twi Mice. Modified X-Gal histochemical staining reveals robust GALC enzymatic activity throughout multiple regions of the central and peripheral nervous systems in aged (>P500) AAV9-Twi mice. Representative images show intense X-Gal staining in the (**A**) olfactory bulb; (**B**) prefrontal cortex; (**C**) cingulate cortex; (**D**) retrosplenial cortex; (**E**) caudate putamen; (**F**) hippocampus and thalamus; (**G**) cerebellar white matter and Purkinje cells; (**H**) medulla; (**I**) substantia nigra; (**J**) pons; dorsal and ventral horns of the (**K**) cervical and (**L**) lumbar spinal cord, including a higher magnification of intense GALC expression in the dorsal root (inset), respectively; and (**M**) longitudinal section of the sciatic nerve. Note the sparse X-Gal staining observed in the corpus callosum, fornix, and dorsal and lateral columns of the spinal cord. Cortical sections from (**N**) P42 WT and (**O**) P42 untreated Twi mice are shown as controls. Scale bars: 200 µm. Abbreviations: cc, corpus callosum; cp, caudate putamen; crb, cerebellum; ctx, cortex; dc, dorsal column; dh, dorsal horn; dr, dorsal root; fn, fornix; hp, hippocampus; lc, lateral column; mb, midbrain; md, medulla; olf, olfactory bulb; tha, thalamus; sn, substantia nigra; pns: pons; pn, pontine nucleus; vh, ventral horn; sci, sciatic nerve.

In WT mice at P42, X-Gal staining was evident in neurons of the cortex (Figure 3N), hippocampus (Figure 3N), and Purkinje cell layer. Moderate staining was also evident observed in the corpus callosum, mossy fibers of the caudate putamen, and cerebellar white matter. In contrast, no detectable X-Gal staining was present in the corresponding brain regions of untreated Twi mice (Figure 3O), consistent with the loss of GALC enzymatic function.

Collectively, our findings indicate widespread and sustained GALC expression following AAV9-mediated gene delivery, supporting effective transduction and enzymatic activity across multiple neuroanatomical regions of the CNS and PNS.

3.5. Sustained and Widespread Biodistribution

We used RNAscope technology, a highly sensitive and specific in situ hybridization method, to detect the delivery of therapeutic DNA molecules mediated by AAV9-based CNS-targeted gene therapy in the CNS and PNS of Twitcher mice. RNAscope utilizes target-specific probes to hybridize with RNA transcripts of interest, enabling single-molecule detection through a branched amplification system that generates intense and localized chromogenic signals.

Using an RNAscope probe designed for mGALC mRNA, the hybridized signals were amplified and visualized as chromatic puncta, representing individual RNA molecules (Figure 4).

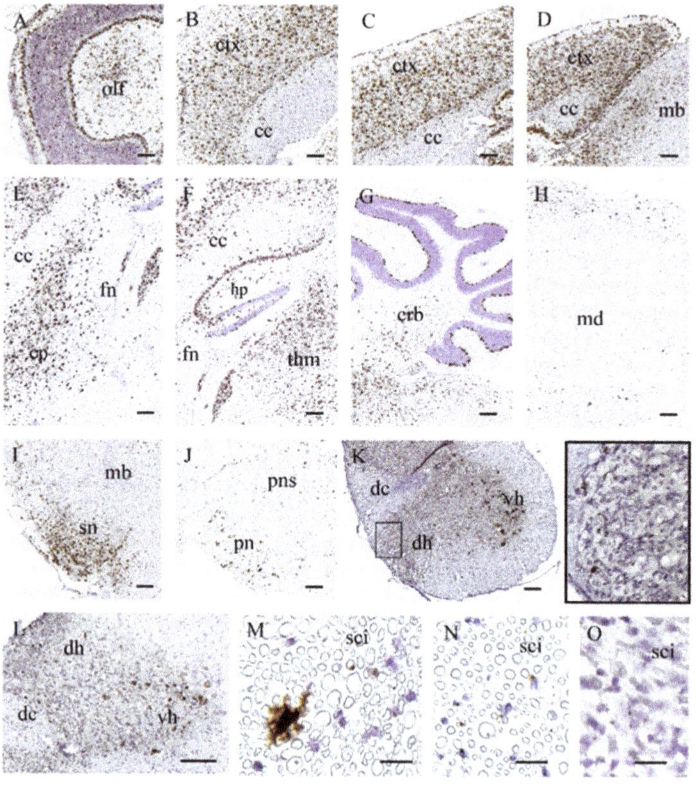

Figure 4. Sustained and Widespread Biodistribution in the CNS and PNS of AAV9-Twi Mice. Spatial biodistribution of AAV9-delivered GALC was assessed by in situ hybridization. Robust GALC mRNA expression is detected across multiple regions of the central and peripheral nervous systems in aged AAV9-Twi mice. Representative images show intense hybridization signals in the (**A**) olfactory bulb; (**B**) prefrontal cortex; (**C**) cingulate cortex; (**D**) retrosplenial cortex; (**E**) caudate putamen; (**F**) hippocampus and thalamus; (**G**) cerebellar white matter and Purkinje cells; (**H**) medulla; (**I**) midbrain; (**J**) pons; dorsal and ventral horns of the (**K**) cervical, including a higher magnification of the intense

hybridization signals in the dorsal horn (inset), and (**L**) lumbar spinal cord; and (**M**) cross section of the sciatic nerve. Sciatic nerve sections from (**N**) P42 WT and (**O**) P42 untreated Twi mice are included for comparison. Scale bars: 200 μm (**A–L**), 20 μm (**M–O**). Abbreviations: cc, corpus callosum; cp, caudate putamen; crb, cerebellum; ctx, cortex; dc, dorsal column; dh, dorsal horn; dr, dorsal root; fn, fornix; hp, hippocampus; lc, lateral column; mb, midbrain; md, medulla; olf, olfactory bulb; tha, thalamus; sn, substantia nigra; pns: pons; pn, pontine nucleus; vh, ventral horn; sci, sciatic nerve.

Robust and highly localized signals were detected in the brains of aged AAV9-Twi mice, demonstrating effective transgene delivery and expression (Figure 4A–M). Notably, the olfactory bubs (Figure 4A), rostral infralimbic cortex (Figure 4B), cingulate cortex (Figure 4C), caudal retrosplenial granular cortex (Figure 4D), midbrain (Figure 4D), caudate putamen (Figure 4E), and thalamus (Figure 4F) exhibited a broad and dense distribution of intense chromogenic signals. In the hippocampus (Figure 4F), strong signals were detected in the pyramidal cell layers of CA1, CA2, and CA3, as well as in the hilus and granular cell molecular layers of the dentate gyrus. Puncta signals were also observed in the corpus callosum (Figure 4B–F) and fornix (Figure 4E,F).

In the cerebellum (Figure 4G), prominent signals were localized to the Purkinje cell layer and white matter. Furthermore, intense signals were widely distributed in the medulla (Figure 4H), substantia nigra (Figure 4I), pons (Figure 4J), dorsal horn and ventral horn of cervical (Figure 4K) and lumbar (Figure 4L) spinal cord. Intriguingly, intense puncta signals were detected in the sciatic nerve (Figure 4M), underscoring the extensive reach of AAV9-mediated transgene delivery.

In comparison, clear and intense GALC transcript puncta were observed in sciatic nerve of WT mice (Figure 4N), while sciatic nerve from untreated Twitcher mice exhibited weakly stained GALC puncta and loss of myeline (Figure 4O).

These findings demonstrate that CNS-targeted AAV9 gene therapy facilitated widespread and robust delivery of therapeutic DNA, extending rostrally to the olfactory bulbs and caudally to the sciatic nerve. The use of the RNAscope provided critical insights into the spatial distribution and expression patterns of the transgene, underscoring its utility in evaluating the efficacy of gene therapy.

Consistent with the in situ hybridization results, quantitative analysis of viral vector biodistribution (Figure 5) revealed the highest AAV9 transduction in the cortex (3.15–8.17 GC/DG), followed by the thalamus (0.37–3.50 GC/DG), brainstem (0.17–0.60 GC/DG), spinal cord (0.07–0.70 GC/DG), cerebellum (0.03–0.47 GC/DG), and sciatic nerve (0.03–0.46 GC/DG). These findings corroborate the broad yet regionally graded distribution of AAV9-mediated GALC expression observed by in situ hybridization, reflecting efficient transduction in both central and peripheral nervous system compartments.

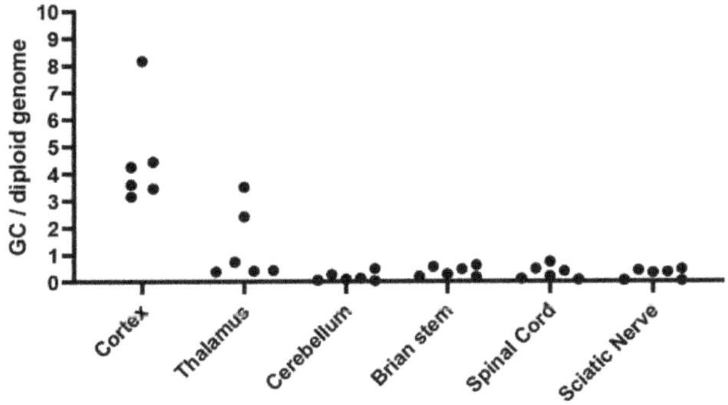

Figure 5. Quantification of Vector Genome Biodistribution in the Nervous System of Aged AAV9-Twi Mice. Vector genome copies were quantified by droplet digital PCR (ddPCR) in multiple regions of

the nervous system from aged AAV9-treated Twitcher (AAV9-Twi) mice ($p > 500$; $n = 3$; measured in duplicate). Quantitative data are normalized to the endogenous *Rpp30* reference gene and expressed as vector genome (VG) copies per diploid genome.

3.6. Reduction of Neuroinflammation

Neuroinflammation and demyelination constitute hallmark pathological features of GLD. In twitcher mice, progressive activation of microglia, astrocytosis, and macrophage infiltration occurs in a spatiotemporal pattern, initially affecting the cerebellar white matter and brainstem after P20, subsequently involving the cerebral white matter after P25, and extending to the cerebral gray matter by P30 [20,21]. To evaluate the efficacy of AAV9-GALC gene therapy in mitigating neuroinflammation, immunohistochemical analyses were performed targeting astrocytes and microglia using GFAP and Iba-1 markers, respectively.

At P42, Twi mice exhibited extensive reactive astrogliosis in the corpus callosum, cerebellar white matter, brainstem, and spinal cord, characterized by hypertrophic somas and thick stellate processes (Figure 6). In contrast, age-matched WT mice showed sparse, resting astrocytes with small somas and thin processes in the corpus callosum and cerebellar white matter (Figure 6). While aged WT mice displayed increased astrocytic density in these regions, most astrocytes remained in a resting state (Figure 6). Notably, aged AAV9-Twi mice demonstrated reduced astrocytic burden in the brain and spinal cord, predominantly composed of resting astrocytes, compared to aged WT mice (Figure 6).

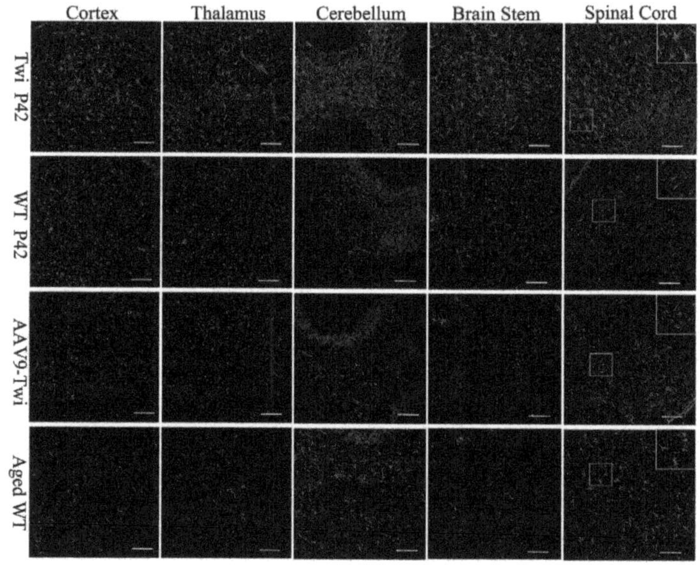

Figure 6. Reduction of astrocyte activation in the brain and spinal cord of aged AAV9-Twi mice. Immunohistochemical staining for glial fibrillary acidic protein (GFAP) with DAPI counterstaining reveals pronounced astrocyte activation in the cortex, thalamus, cerebellum, brain stem, and spinal cord of P42 Twi mice compared to P42 and aged (>P500) WT controls. Aged WT mice exhibit a moderate increase in astrocyte presence within the cerebellum and spinal cord relative to P42 WT mice. Aged (>P500) AAV9- Twi mice display a similar moderate level of astrocyte presence in these regions, comparable to aged WT controls, reflecting age-related gliosis and preserved glial homeostasis in the central nervous system following AAV9-mediated gene therapy. Insets in the spinal cord panels illustrate a higher magnification of hypertrophic reactive astrocytes in Twi, and resting astrocytes in WT, aged WT, and aged AAV9-Twi mice. Scale bars, 200 μm.

Microglial activation, a key contributor to multinucleated globoid cell formation and demyelination in GLD, was also evaluated [22,23]. At P42, the majority of microglia in the brains of WT mice remained in resting state, characterized by small somas and highly ramified processes (Figure 7). In contrast, untreated Twi mice exhibited a significant increase in reactive microglia, particularly in the brain stem and spinal cord (Figure 7). These microglia displayed hypertrophic somas with thickened bushy processes, along with amoeboid-like and multinucleated globoid cells, indicative of a highly activated phagocytic state (Figure 7). In both aged WT and AAV9-Twi mice, microglia largely retained resting morphologies, with only a moderate increase in hypertrophic microglia compared to younger WT mice, consistent with an ageing-associated low-grade pro-inflammatory state (Figure 7). The substantial reduction in neuroinflammation observed in AAV9-Twi mice at advanced age indicates the durable efficacy of gene therapy.

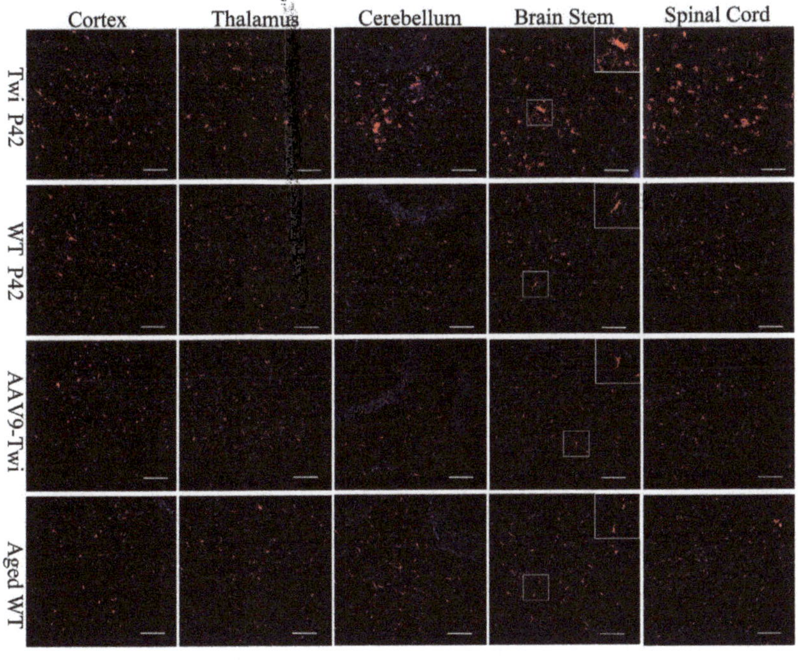

Figure 7. Inhibition of Globoid Cell Activation in the Brain and Spinal Cord of Aged AAV9-Twi Mice. Untreated Twi mice at P42 exhibit pronounced microglial activation and globoid cell formation in the cortex, thalamus, cerebellum, brainstem, and spinal cord, as demonstrated by immunohistochemical staining for ionized calcium-binding adapter molecule 1 (Iba1) with DAPI counterstaining, compared to P42 and aged (>P500) WT controls. Aged (>P500) AAV9-Twi mice exhibit markedly reduced Iba1+ staining and an absence of globoid cell clusters, comparable to levels observed in P42 and aged WT mice, indicating effective suppression of microglial activation and prevention of neuroinflammatory pathology following AAV9-mediated gene therapy. Insets in the brainstem panels illustrate a higher magnification of reactive microglia and globoid cells in Twi, and resting microglia in WT, aged WT, and aged AAV9-Twi mice. Scale bars, 200 μm.

Previous investigations have shown that CD68+ macrophage/microglia infiltration correlates with the progression of GLD disease and the severity of demyelination, occurring concomitantly with astrogliosis [24,25]. In the present study, immunohistochemical analyses revealed pronounced infiltration of CD68+ macrophages/microglia within the brain and spinal cord parenchyma of untreated Twi mice (Figure 8). In contrast, CD68 immunoreactivity was absent in both wild-type and aged AAV9-Twi mice, signifying effective suppression of neuroinflammatory responses through therapeutic GALC restoration (Figure 8). These findings underscore the pivotal role of CD68+ cells in GLD pathogenesis and highlight the capacity of AAV9-mediated gene therapy to modulate pathogenic immune activation.

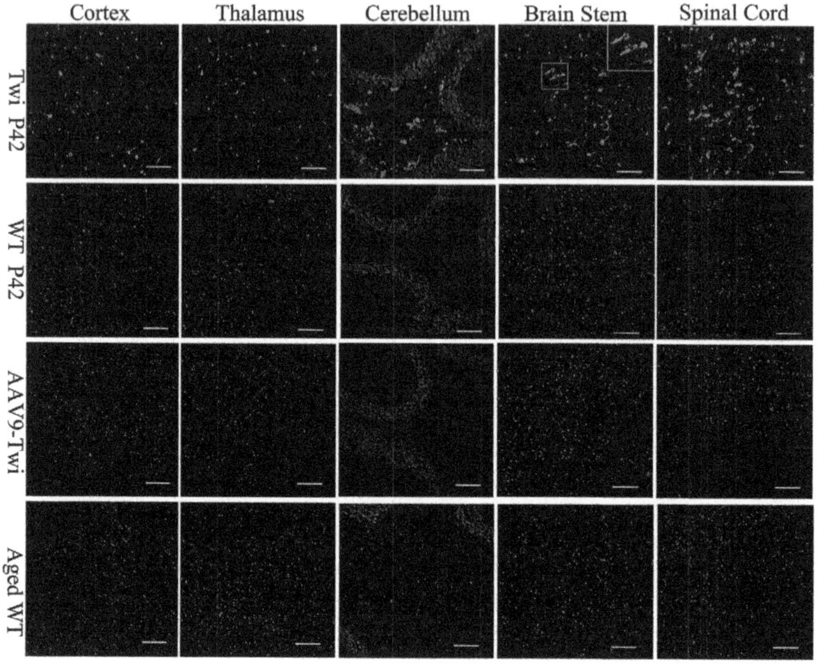

Figure 8. Ablation of Macrophage/Microglia Infiltration in the Brain and Spinal Cord of Aged Twi Mice Following AAV9 Treatment. Untreated Twi mice at P42 exhibit pronounced macrophage/microglia infiltration in the cortex, thalamus, cerebellum, brainstem, and spinal cord, as demonstrated by immunohistochemical staining for CD68 with DAPI counterstaining, compared to P42 and aged (>P500) WT controls. In contrast, aged (>P500) AAV9-Twi mice show an absence of macrophage/microglia infiltration in these regions, comparable to that observed in P42 and aged WT mice, indicating effective suppression of neuroinflammation and innate immune activation following AAV9-mediated gene therapy. Inset in the brainstem panel illustrates a higher magnification of morphology of CD68-positive macrophagy/microglia in Twi. Scale bars, 200 μm.

3.7. *Preservation of Proteostasis*

The efficacy of AAV9-mediated GALC gene therapy in restoring autophagy and UPS function in the CNS of twitcher mice was evaluated. Psychosine accumulation has been shown to impair autophagy and UPS in oligodendrocytes, leading to progressive aggregation of misfolded proteins in white matter regions [4,5,24]. In the current study, the therapeutic efficacy of AAV9-GALC was assessed by immunohistochemical detection of p62 (Figure 9) and ubiquitin-positive aggregates (Figure 10). At P42, untreated Twi mice exhibited a wide distribution of p62- (Figure 9) and ubiquitin-positive (Figure 10) aggregates throughout the brain, with the greatest burden localized to the spinal cord, brainstem, cerebellum, and corpus callosum. In contrast, P42 wild-type mice, aged WT controls, and aged AAV9-Twi mice exhibited no detectable aggregates in the brain or spinal cord (Figures 9 and 10). These results indicate that AAV9-GALC gene therapy effectively restores proteostasis and prevents aggregate accumulation, highlighting its therapeutic potential in mitigating autophagy and UPS dysfunction in GLD.

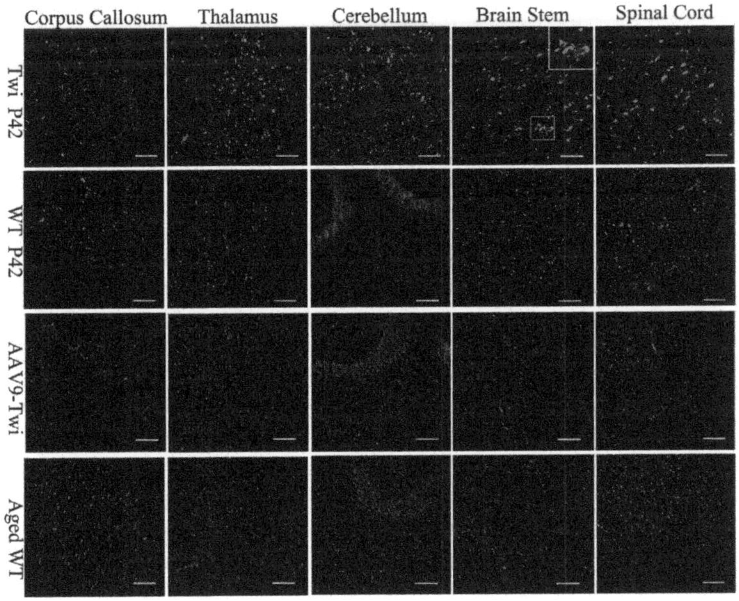

Figure 9. Preservation of Autophagy Function in Brain and Spinal Cord of Aged AAV9-Twi Mice. Immunohistochemical staining for p62 with DAPI counterstaining reveals extensive accumulation of p62-positive aggregates in the cortex, thalamus, cerebellum, brainstem, and spinal cord of untreated Twi mice at P42, compared to P42 and aged (>P500) WT controls. In contrast, aged (>P500) AAV9-treated Twi mice show an absence of p62 aggregates in these regions, comparable to that observed in P42 and aged WT mice, indicating preserved autophagy function and proteostasis in the nervous system following AAV9-mediated gene therapy. Inset in the brainstem panel illustrates a higher magnification of p62-aggregates in Twi. Scale bars, 200 μm.

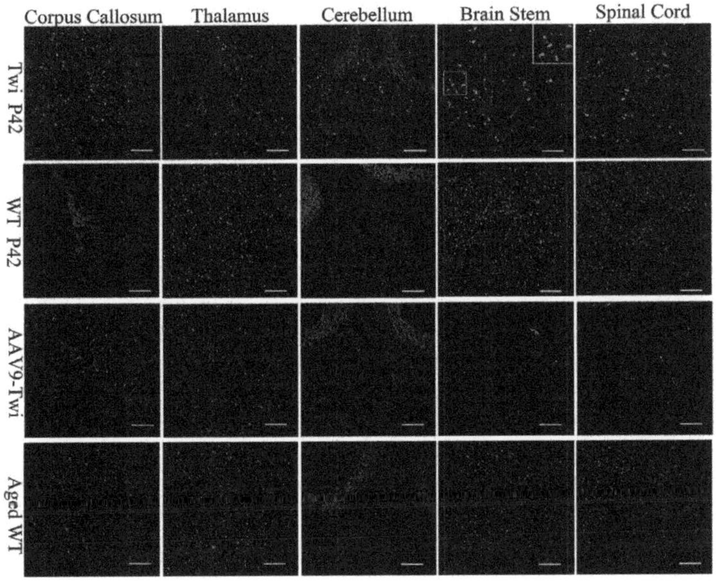

Figure 10. Preservation of the Ubiquitin-Proteasome System in the Brain and Spinal Cord of Aged AAV9-Twi Mice Following AAV9 Treatment. Immunohistochemical staining for ubiquitin with DAPI

counterstaining reveals extensive accumulation of ubiquitin-positive aggregates in the cortex, thalamus, cerebellum, brainstem, and spinal cord of untreated Twi mice at P42, compared to P42 and aged (>P500) WT controls. In contrast, aged (>P500) AAV9-Twi mice exhibit an absence of ubiquitin-positive aggregates in these regions, comparable to that observed in P42 and aged WT mice, indicating restoration of protein homeostasis and preservation of ubiquitin–proteasome system function in the nervous system following AAV9-mediated gene therapy. Inset in the brainstem panel illustrates a higher magnification of ubiquitin-positive aggregates in Twi. Scale bars, 200 μm.

3.8. Preservation of Myelination and Axon Integerity

Demyelination in twitcher mice typically begins between P15–P20, initially affecting the cerebellar white matter and brainstem, and progressively extending to the cerebral white matter, spinal cord, sciatic nerve, and eventually gray matter regions [20,26,27]. To assess the therapeutic efficacy of AAV9-mediated gene therapy in preserving CNS myelination, immunofluorescent histochemical analysis was performed. In untreated Twi mice at P42, extensive demyelination was evident across multiple CNS regions, including the subcortical white matter, corpus callosum, cerebellar white matter, brainstem, and spinal cord (Figure 11). In contrast, aged AAV9-Twi mice exhibited robust and widespread preservation of myelination throughout the brain and spinal cord, closely resembling the myelination patterns observed in age-matched WT controls (Figure 11).

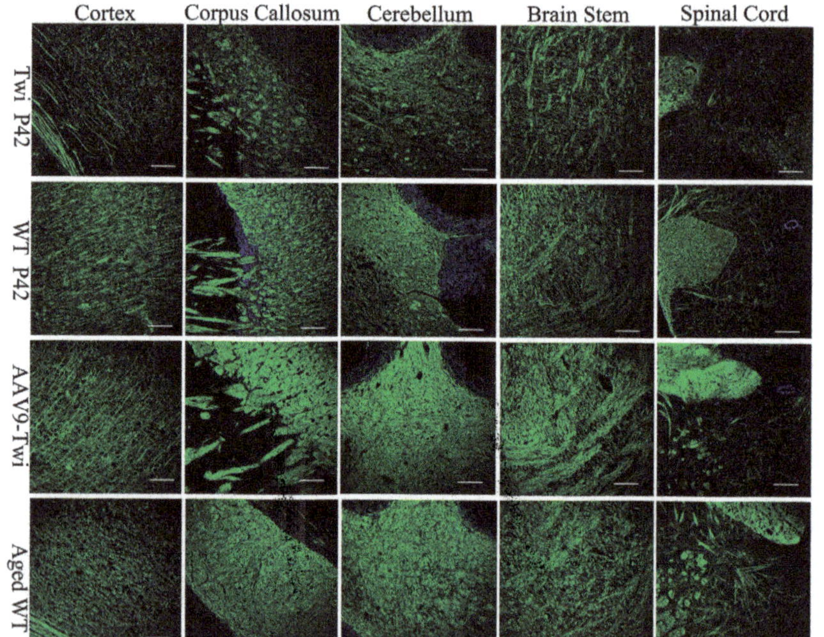

Figure 11. Preservation of Myelination in the Brain and Spinal Cord of Aged AAV9-Twi Mice. Myelination was evaluated by proteolipid protein (PLP) immunostaining with DAPI counterstaining in brain and spinal cord sections from P42 WT mice, P42 Twi mice, aged WT mice (>P500), and aged AAV9-Twi mice (>P500). AAV9 treatment results in sustained preservation of myelination throughout the brain and spinal cord of aged AAV9-Twi mice, comparable to that observed in aged WT controls. In contrast, untreated Twi mice at P42 exhibit severe demyelination in subcortical white matter, corpus callosum, cerebellar white matter, brainstem, and spinal cord. Scale bars, 200 μm.

To evaluate the impact of AAV9-GALC on PNS myelination, sciatic nerve ultrastructure was examined by transmission electron microscopy. Untreated Twi mice showed marked demyelination and axonal degeneration (Figure 12A), whereas aged AAV9-Twi mice exhibited well-preserved axonal morphology and compact myelin, resembling the WT phenotype (Figure 12B–D). Quantitative analysis of myelination based on g-ratio measurements revealed that P42 Twi mice had significantly reduced axon fiber diameter, axonal diameter, and myelin thickness, along with elevated g-ratio values, compared to WT, aged WT, and aged AAV9-Twi mice (Figure 12E–H). While aged WT mice showed increased g-ratios and axon dimensions relative to P42 WT mice, aged AAV9-Twi mice displayed normalized axon and fiber diameters, increased myelin thickness, and a g-ratio restored to levels observed in P42 WT mice.

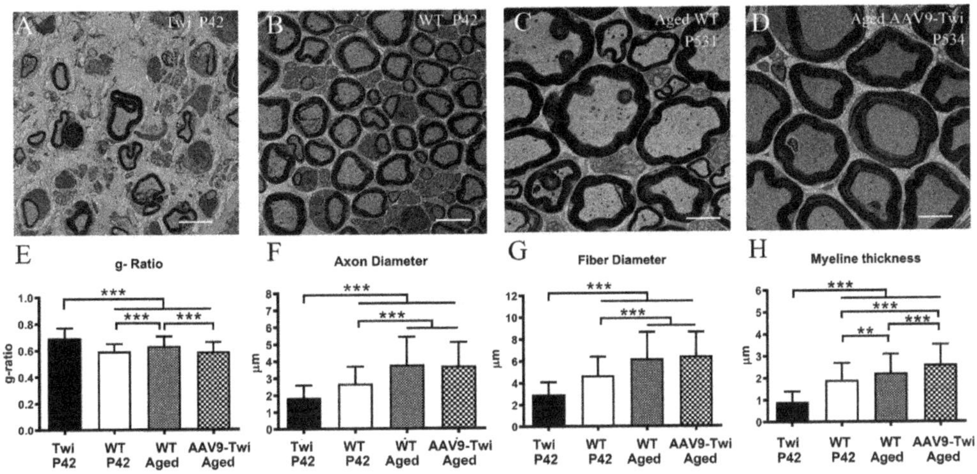

Figure 12. Correction of axonopathy in the sciatic nerves of aged AAV9-Twi mice. AAV9-mediated gene therapy preserves normal axonal morphology and myelination in the sciatic nerve (SN) of aged AAV9-treated Twi mice. Transmission electron microscopy (TEM) images reveal profound axonal degeneration and demyelination in the SN of (**A**) P42 Twi mice compared to (**B**) P42 WT and (**C**) aged WT mice. (**D**) Aged AAV9-treated Twi mice exhibit well-preserved myelination and intact axons comparable to those observed in aged WT controls. Quantitative analysis of myelinated axon structural integrity includes (**E**) g-ratio, (**F**) axon diameter, (**G**) fiber diameter, and (**H**) myelin thickness, measured from TEM images of SN samples from P42 Twi, P42 WT, aged WT (>P500), and aged (>P500) AAV9-treated Twi mice. AAV9 treatment results in sustained preservation of axonal integrity in the SN of aged Twi mice. Data represent mean ± SD derived from 150 axons per group, with $n = 3$ animals per group. Statistical significance was determined by one-way ANOVA. ** $p < 0.01$; *** $p < 0.001$. Scale bars, 5 µm.

Collectively, these findings demonstrate that AAV9-mediated GALC gene therapy effectively preserves CNS and PNS myelination and maintains axonal integrity in Twi mice.

4. Discussion

The multifaceted pathology of GLD, characterized by the involvement of both the CNS and the PNS, the accumulation of cytotoxic psychosine, profound neuroinflammation, impaired autophagy, and rapid progression of the disease, poses significant challenges for effective therapeutic intervention. Although gene therapy has demonstrated efficacy in alleviating symptoms and extending lifespan, prior monotherapy approaches utilizing viral vectors alone have been insufficient to comprehensively address all pathological aspects concurrently. The incorporation of HSCT, with or without SRT, has provided additional enzymatic and metabolic correction, further prolonging survival; nevertheless, achieving complete disease rescue remains an ongoing challenge. Importantly, the present study surpasses previous outcomes by demonstrating, for the first time, comprehensive pathological correction via minimal CNS-targeted monotherapy, providing compelling evidence of its potential as a standalone therapeutic strategy.

Early studies employing intracerebral administration of AAV1 (3×10^{10} vg) and AAV5 (2.4×10^{9} vg) in neonatal Twitcher mice modestly prolonged survival by 10 to 15 days, outperforming AAV2 and adenoviral vectors [17,28]. Previous work with AAV5 (2.4×10^{9} vg) delivery targeting the cortex, hippocampus and cerebellum extended the maximum survival to 66 days [29,30]. Subsequent refinements involving intrathecal (i.t.) and six-separate intracranial (i.c.) injections of AAV5 (3.6×10^{10} vg) further improved maximal survival to 78 days [31].

Advances in vector engineering introduced AAVrh10 and AAV9, characterized by a wider tropism of the CNS. Neonatal administration combining intracerebroventricular (i.c.v., 3×10^{9} vg), intracerebellar (1.5×10^{9} vg), and intravenous (i.v., 7.6×10^{9} vg) routes of AAVrh10 extended maximum survival to 240 days [32]. Further optimization with a tenfold increase in intravenous (i.v., 4×10^{14} vg/kg) dosing of AAVrh10 enhanced maximal survival 430 days [33]. Parallel studies employing AAV9 via 5-seperat i.c. (9×10^{9} vg), intrathecal (i.t., 8.25×10^{10} vg), and intravenous (i.v., 3.3×10^{11} vg) administration reported a maximum of 484 days [34].

To address limitations of single-modality gene therapy, combination strategies integrating AAV9-mediated CNS delivery (i.c. and i.t., 2.7×10^{10} vg) with HSCT and SRT achieved median survival of 404 days and maximum survival of 569 days [35]. Meanwhile, the combination of AAV9-mediated CNS administration (i.c. and i.t., 2.7×10^{10} vg) and HSCT alone extended median and maximal survival to 269 and 673 days, respectively [35]. Similarly, AAVrh10 i.v. (4×10^{13} vg/kg) administration in combination with HSCT further prolonged the median lifespan to 351 days, with a maximum survival of 750 days [33].

Previous studies (Table 1) employing multiple i.c. injections or combined CNS delivery routes (i.c., i.c.v., and i.t. administration) achieved only limited extension of lifespan in twitcher mice, primarily due to incomplete correction of both CNS and PNS pathology [17,28–31]. The addition of systemic i.v. AAV delivery (Table 1) improved outcomes by restoring GALC activity within the PNS and further prolonging survival [32–34]. Yet, the limited ability of i.v. AAV to cross blood-brain barrier constrained overall efficacy. When systemic i.v. high dose AAV delivery combined with HSCT, which mitigates neuroinflammation, this dual-therapy regimen (Table 1) achieved the longest survival previously reported [33]. Notably, our protocol with minimal and region-specific CNS delivery of high-titer AAV9 surpasses all prior benchmarks, achieving a median survival of 530 days and maximum survival of 814 days, approaching the WT lifespan. Critically, this streamlined approach circumvents the risks associated with HSCT, such as graft-versus-host disease, and avoids the complexities inherent to multi-route and multi-dose regimens, or high-dose systemic administration. The sustained therapeutic efficacy observed in long-lived animals further suggests that one-time treatment may confer a lifelong corrective effect. Those results underscore the transformative potential of our protocol as a standalone treatment for GLD, while remaining complementary to more complex multimodal strategies.

Table 1. Comparison of AAV-based therapeutic strategies in Twitcher mice.

Year	AAV Serotype	Postnatal Day	Route(s) & Sites	Vector Dose	Combination Therapy	Median Survival (Days)	Max. Survival (Days)	Reference
2005	AAV1	P0-1	i.c. (2 sites) i.c.v.	3×10^{10} vg 3×10^{10} vg	None None	55 55	66 66	[28]
2005	AAV2 AAV5	P3 P3	i.c. (6 sites) i.c. (6 sites)	4.4×10^{8} vg 2.4×10^{9} vg	None None	47 52	57 62	[17]
2007	AAV5	P3	i.c. (6 sites)	2.4×10^{9} vg	+BMT	111	151	[16]
2011	AAV5	P3	i.c. (6 sites) + i.t.	1.6×10^{10} vg, i.c. 2×10^{10} vg, i.t.	+BMT	71 123	78 282	[31]
2011	AAV5	P3	i.c. (6 site)	2.4×10^{9} vg	None	63	66	[29]
2012	AAVrh10	P2	i.c. (1 sites) + i.c.v. + i.v. (P2) + i.v. (P7)	1.5×10^{9} vg, i.c. 3.25×10^{9} vg, i.c.v. 7.6×10^{9} vg, i.v. (P2) 7.6×10^{9} vg, i.v. (P7)	None	104	240	[32]
2015	AAV5	P3	i.c. (6 sites)	2.4×10^{9} vg	None	60	63	[30]
2016	AAV9 AAVrh10	P10-11 P10-11	i.t. i.t.	2×10^{11} vg 2×10^{11} vg	+BMT +BMT	79 57	135 (Est.) 105 (Est.)	[36]
2018	AAV9	P0–P1	i.c. (5 sites) + i.t. + i.v.	9×10^{9} vg, i.c. 8.25×10^{10} vg, i.t. 3.3×10^{11} vg, i.v.	None +BMT	263 284	484 675	[34]
2021	AAVrh10	P8-9	i.v. i.v. i.v. i.v.	4×10^{13} vg/kg 4×10^{13} vg/kg 1.6×10^{14} vg/kg 4×10^{14} vg/kg	None +BMT None None	72 351 180 280	365 (Est.) 750 235 (Est.) 430	[33]
2021	AAV9	P0	i.c. (6 sites) + i.t.	1.2×10^{10} vg, i.c. 1.5×10^{10} vg, i.t.	None BMT BMT + SRT	66.5 269 404	83 675 569	[35]
2025	AAV9-	P3	i.c. (4 sites)	1.2×10^{12} vg, i.c.	None	530	814	Current study

BMT, bone marrow transplantation; Est, estimated; i.c., intracranial; i.c.v., intracerebroventricular; i.t., intrathecal; i.v., intravenous; SRT, substrate reduction therapy; vg, virus genome.

AAV9-mediated gene therapy has demonstrated substantial therapeutic efficacy in addressing pathological gene deficiencies, particularly in disorders affecting both the CNS and PNS. However, challenges persist in achieving spatially uniform transgene distribution, sustained expression, and adequate enzymatic activity in both the CNS and the PNS. In this study, intraparenchymal delivery of high-titer AAV9 targeting the thalamus and deep cerebellum resulted in widespread expression and activity of GALC throughout the brain, cerebellum, spinal cord, and sciatic nerves of aged AAV9-Twi mice. Histochemical staining confirmed robust transduction of neurons and glia across proximal CNS regions, as well as distal PNS tissues, facilitated by efficient anterograde and retrograde axonal transport mechanisms [37]. Notably, GALC activity reached supranormal levels in all neural compartments, with maximum activity observed in the cortex and thalamus, followed by elevated levels in the cerebellum, brainstem, spinal cord, and sciatic nerves. This extensive biodistribution underscores the ability of AAV9 to leverage axonal tranoport pathways for global transgene delivery without requiring multiple injection sites or invasive interventions. The advantages of AAV9 include its ability for efficient neuron/glia transduction, sustained enzymatic activity throughout the nervous system, and long-term therapeutic efficacy. Collectively, both thalamus and deep cerebellar nuclei are highly interconnected hubs with extensive projections to cortex, brainstem, and spinal cord [38,39], enabling robust anterograde and retrograde axonal transport of AAV9. Neonatal parenchyma also supports enhanced viral diffusion due to reduced myelination and greater extracellular permeability [40]. These mechanisms collectively explain how focal injections can yield widespread GALC expression without multi-route delivery. By obviating the need for systemic administration, multi-route, or multi-dose, current streamlined therapy offers a targeted yet comprehensive approach to addressing CNS/PNS pathologies while minimizing complexity of therapeutic approach and improving safety.

Psychosine accumulation due to GALC deficiency is central to the pathogenesis of GLD. The efficacy of AAV-mediated gene therapy in normalizing psychosine levels in Twitcher mice has been a critical focus in evaluating its therapeutic potential. Previous studies have reported incomplete normalization of psychosine within the nervous system following AAV-based interventions, whether administered alone or in combination with other therapies [31,34,35,41]. These limitations have been attributed to suboptimal AAV transduction efficiency, heterogeneous distribution of GALC activity in regions susceptible to residual psychosine accumulation, and insufficient GALC activity gradients to effectively counteract localized psychosine synthesis [34,35,41]. In the present study, for the first time in gene therapy of GLD, psychosine levels in both the CNS and the PNS of aged AAV9-treated Twitcher mice were comparable to those observed in aged wild-type mice. Importantly, these normalized levels remained stable throughout the lifespan of treated animals, indicating sustained transgene expression and adequate therapeutic GALC activity across all relevant CNS and PNS regions, thereby conferring durable therapeutic benefit. Furthermore, robust AAV9-GALC transduction was detected in CNS gray matter and spinal cord, consistent with AAV9 tropism, with additional transduction in white matter and sciatic nerves. These observations suggest that both local GALC production by transduced neurons and myelinating cells, as well as cross-correction from highly transduced neurons, contributed to supraphysiological GALC activity. Aligning with observations in previous long-term studies [33,34,41], the supraphysiological GALC levels were not associated histopathological or behavioral deficits throughout the extended longevity of AAV9-Twi mice. The sustained supranormal GALC activity ensures robust substrate clearance of psychosine and galactosylceramide in demyelination-prone regions over the life span, critical for preventing demyelination and neurodegeneration in Krabbe disease models. Collectively, these mechanisms underpin the global normalization of psychosine levels observed in AAV9-Twi mice.

In Twitcher mice, axonopathy precedes overt demyelination and neuronal loss, identifying axonal degeneration as an early and primary event in GLD pathogenesis [42]. Psychosine accumulation independently disrupts axonal transport and initiates dying-back neuropathy, effects that cannot be mitigated by glial GALC expression alone [36,42]. Additionally, Psychosine also directly impairs axonal transport and cytoskeletal organization, resulting in synaptic dysfunction and neurodegeneration [43,44]. These neurotoxic effects may persist despite remyelination if neuronal GALC activity is insufficient. Moreover, neuronal GALC deficiency leads to brainstem maldevelopment, axonal atrophy, neuroinflammation, and demyelination, underscoring its essential role in maintaining neuronal integrity [45]. In the present study, sustained psychosine clearance across CNS and PNS tissues of aged AAV9-treated Twi mice implicates effective correction in both neuronal and oligodendrocytic compartments. The preservation of compact myelin, normal axonal morphology in sciatic nerves, and restored motor behavior collectively confirm dual-compartment rescue. These results support that long-term therapeutic success in GLD requires robust GALC expression in both neurons and myelinating glia to prevent psychosine-mediated neurotoxicity and secondary demyelination.

Notably, the spatially heterogeneous distribution of transgenes in the CNS of Twitcher mice induces focal episomal vector depletion, allowing localized psychosine resurgence, demyelination, and neuroinflammatory cascades [41]. Psychosine exacerbates neurodegeneration by concurrently disrupting autophagy-lysosomal flux and UPS activity, driving cytoplasmic accumulation of cytotoxic aggregates of p62/ubiquitin and oligodendrocyte apoptosis [4]. In present study, aged AAV9-Twi mice lack detectable p62/ubiquitin aggregates in the brainstem and spinal cord and demonstrate resolution of neuroinflammation comparable to aged WT controls. This preservation of proteostasis and myelin integrity, evident in ultrastructurally normal sciatic nerves and sustained levels of myelin proteolipid protein, is correlated with the global GALC activity of the CNS/PNS and normalization of psychosine. Our data establishes that the current AAV9 delivery approach, by ensuring extensive enzyme distribution, preemptively abrogates psychosine's dual neuroinflammatory and proteotoxic effects, thereby interrupting the self-perpetuating cycle of oligodendrocyte loss and neuronal dysfunction. These findings underscore the critical necessity for spatially uniform, lifelong GALC expression to achieve durable therapeutic rescue in GLD.

This streamlined intracranial AAV9 gene therapy protocol achieves unprecedented efficacy in GLD without adjunctive therapy, HSCT, multi-dose and/or multi-route CNS administration, or high systemic dosing. Our protocol ensures efficient neuronal transduction and broad biodistribution in both CNS and PNS, leading to sustained global correction of metabolism, prevention of neuroinflammation, preservation of proteostasis and myelin integrity, restoration of motor function, and maximal survival to dat. Remarkably, treated mice achieve lifespans approaching those of WT controls, suggesting durable and potentially lifelong therapeutic benefit. Although late-stage behavioral assessments were based on a small cohort, the findings are reinforced by consistent motor performance, stable body weight, preserved histopathology, and video documentation of aged animals (Video S1), supporting the maintenance of meaningful neurological function. These outcomes redefine the therapeutic benchmark for GLD, demonstrating that a streamlined, intracranial region-specific AAV9 monotherapy can achieve comprehensive neurobiological correction with minimal invasiveness. Translationally, this protocol provides a clinically scalable framework for human intracerebral AAV delivery, enabling durable CNS correction while minimizing systemic exposure and procedural risk.

In present study, regional differences in vector genome copies likely reflect a combination of injection geometry, neuroanatomical connectivity, and tissue volume dilution. Proximal regions such as thalamus and cortex showed higher vector genome copies, consistent with direct parenchymal delivery and dense local transduction, whereas lower levels in spinal cord and sciatic nerve are in line with secondary distribution via long-range axonal transport and cross-correction rather than primary deposition. In human brains, amplified white matter compartmentalization and 1000-fold larger volumes will exacerbate gradients, necessitating dose escalation, trajectory planning, and possibly combination therapies for ensuring robust enzyme activity in distal and peripherally located target tissues.

From a safety perspective, AAV9-treated Twi mice maintained stable body weight, normal behavior, and intact locomotor activity without late-onset neurological decline throughout their >800-day lifespan, supporting a favorable long-term safety profile. Nonetheless, several considerations remain important for clinical translation. Off-target transduction and peripheral organ involvement were not fully assessed and will require comprehensive biodistribution and toxicology studies. Vector- and transgene-specific immune responses, such as pre-existing anti-AAV9 antibodies and potential T-cell activation, may limit redosing in humans and necessitate careful immune monitoring. Although supraphysiological GALC expression was well tolerated in mice, overexpression toxicity remains a theoretical risk, underscoring the need for dose optimization in clinical applications.

Overall, this streamlined protocol minimizes major known risks while achieving durable CNS and PNS correction, but also highlights key parameters to be addressed in future translational development.

5. Conclusions

While prior multi-route strategies and combination therapies have demonstrated meaningful benefits in GLD, particularly in enhancing PNS correction and mitigating neuroinflammation, they also involve substantial procedural complexity and systemic vector exposure. Our findings position the current single-dose, region-specific AAV9 approach as a complementary alternative that achieves comparable or superior long-term efficacy with reduced treatment burden. Recognizing the strengths of previous approaches underscores the significance of achieving such robust CNS–PNS correction through a streamlined and minimally invasive protocol.

Supplementary Materials: The following supporting information can be downloaded at: https://www.mdpi.com/article/10.3390/cells14241942/s1, Video S1: AAV9-Twi Motor Activity. Representative videos showing an untreated Twi mouse at P40, WT mice at P57 and P768, and AAV9-Twi mice at P63 and P789. AAV9-Twi mice exhibited body size and locomotor activity comparable to age-matched WT controls, whereas the untreated Twi mouse was markedly smaller and displayed severe motor impairment.

Author Contributions: D.-S.L. designed the study and conducted supervision, writing, editing, and reviewing of the manuscript. C.-S.H. coordinated studies with collaborators, analyzed immunofluorescence, histochemistry, and TEM. Y.-W.H. supervised and analyzed the GALC activity and psychosine concentration. T.-H.L. contributed to intraparenchymal injections in mice models, analysis of phenotype and collection of tissues. Z.-D.H. performed histochemistry, immunofluorescence and assay of enzyme activity. W.-C.C., T.-J.W. and S.-F.H. contributed to psychosine quantification. D.-S.L. conducted funding acquisition. All authors have read and agreed to the published version of the manuscript.

Funding: This work was supported by grants from the National Science and Technology Council (NSTC 112-2320-B-195 -001, 109-2314-B-195 -015 -MY3) and MacKay Memorial Hospital (MMH-E-114-02, MMH-E-111-02).

Institutional Review Board Statement: Approved by Institutional Animal Care and Use Committee (26 March 2020). The approval code is MMH-A-S-109-17.

Informed Consent Statement: Not applicable.

Data Availability Statement: All data generated or analyzed during this study are included in this published article.

Acknowledgments: We acknowledge the AAV Core Facility of Academia Sinica in generating recombinant AAV. We acknowledge the Academia Sinica Biological Electron Microscopy Core Facility for TEM technical support.

Conflicts of Interest: The authors declare no conflict of interest.

References

1. Suzuki, K. Globoid cell leukodystrophy (Krabbe's disease): Update. *J. Child Neurol.* **2003**, *18*, 595–603. [CrossRef]
2. Greco, M.R.; Lopez, M.A.; Beltran-Quintero, M.L.; Tuc Bengur, E.; Poe, M.D.; Escolar, M.L. Infantile Krabbe disease (0–12 months), progression, and recommended endpoints for clinical trials. *Ann. Clin. Transl. Neurol.* **2024**, *11*, 3064–3080. [CrossRef]
3. White, A.B.; Givogri, M.I.; Lopez-Rosas, A.; Cao, H.; van Breemen, R.; Thinakaran, G.; Bongarzone, E.R. Psychosine accumulates in membrane microdomains in the brain of krabbe patients, disrupting the raft architecture. *J. Neurosci.* **2009**, *29*, 6068–6077. [CrossRef]
4. Lin, D.S.; Ho, C.S.; Huang, Y.W.; Wu, T.Y.; Lee, T.H.; Huang, Z.D.; Wang, T.J.; Yang, S.J.; Chiang, M.F. Impairment of Proteasome and Autophagy Underlying the Pathogenesis of Leukodystrophy. *Cells* **2020**, *9*, 1124. [CrossRef]

5. Del Grosso, A.; Angella, L.; Tonazzini, I.; Moscardini, A.; Giordano, N.; Caleo, M.; Rocchiccioli, S.; Cecchini, M. Dysregulated autophagy as a new aspect of the molecular pathogenesis of Krabbe disease. *Neurobiol. Dis.* **2019**, *129*, 195–207. [CrossRef]
6. Suzuki, K.; Taniike, M. Murine model of genetic demyelinating disease: The twitcher mouse. *Microsc. Res. Tech.* **1995**, *32*, 204–214. [CrossRef]
7. Kondo, A.; Hoogerbrugge, P.M.; Suzuki, K.; Poorthuis, B.J.; Van Bekkum, D.W.; Suzuki, K. Pathology of the peripheral nerve in the twitcher mouse following bone marrow transplantation. *Brain Res.* **1988**, *460*, 178–183. [CrossRef]
8. Lee, W.C.; Courtenay, A.; Troendle, F.J.; Stallings-Mann, M.L.; Dickey, C.A.; DeLucia, M.W.; Dickson, D.W.; Eckman, C.B. Enzyme replacement therapy results in substantial improvements in early clinical phenotype in a mouse model of globoid cell leukodystrophy. *FASEB J.* **2005**, *19*, 1549–1551. [CrossRef] [PubMed]
9. Biswas, S.; Le Vine, S.M. Substrate-reduction therapy enhances the benefits of bone marrow transplantation in young mice with globoid cell leukodystrophy. *Pediatr. Res.* **2002**, *51*, 40–47. [CrossRef] [PubMed]
10. Heller, G.; Bradbury, A.M.; Sands, M.S.; Bongarzone, E.R. Preclinical studies in Krabbe disease: A model for the investigation of novel combination therapies for lysosomal storage diseases. *Mol. Ther.* **2023**, *31*, 7–23. [CrossRef] [PubMed]
11. Sakai, N.; Inui, K.; Tatsumi, N.; Fukushima, H.; Nishigaki, T.; Taniike, M.; Nishimoto, J.; Tsukamoto, H.; Yanagihara, I.; Ozono, K.; et al. Molecular cloning and expression of cDNA for murine galactocerebrosidase and mutation analysis of the twitcher mouse, a model of Krabbe's disease. *J. Neurochem.* **1996**, *66*, 1118–1124. [CrossRef]
12. Martino, S.; Tiribuzi, R.; Tortori, A.; Conti, D.; Visigalli, I.; Lattanzi, A.; Biffi, A.; Gritti, A.; Orlacchio, A. Specific determination of beta-galactocerebrosidase activity via competitive inhibition of beta-galactosidase. *Clin. Chem.* **2009**, *55*, 541–548. [CrossRef]
13. Matsumoto, S.I.; Sato, S.; Otake, K.; Kosugi, Y. Highly-sensitive simultaneous quantitation of glucosylsphingosine and galactosyl-sphingosine in human cerebrospinal fluid by liquid chromatography/tandem mass spectrometry. *J. Pharm. Biomed. Anal.* **2022**, *217*, 114852. [CrossRef]
14. Ribbens, J.J.; Moser, A.B.; Hubbard, W.C.; Bongarzone, E.R.; Maegawa, G.H. Characterization and application of a disease-cell model for a neurodegenerative lysosomal disease. *Mol. Genet. Metab.* **2014**, *111*, 172–183. [CrossRef]
15. Dolcetta, D.; Perani, L.; Givogri, M.I.; Galbiati, F.; Orlacchio, A.; Martino, S.; Roncarolo, M.G.; Bongarzone, E. Analysis of galactocerebrosidase activity in the mouse brain by a new histological staining method. *J. Neurosci. Res.* **2004**, *77*, 462–464. [CrossRef] [PubMed]
16. Lin, D.; Donsante, A.; Macauley, S.; Levy, B.; Vogler, C.; Sands, M.S. Central nervous system-directed AAV2/5-mediated gene therapy synergizes with bone marrow transplantation in the murine model of globoid-cell leukodystrophy. *Mol. Ther.* **2007**, *15*, 44–52. [CrossRef]
17. Lin, D.; Fantz, C.R.; Levy, B.; Rafi, M.A.; Vogler, C.; Wenger, D.A.; Sands, M.S. AAV2/5 vector expressing galactocerebrosidase ameliorates CNS disease in the murine model of globoid-cell leukodystrophy more efficiently than AAV2. *Mol. Ther.* **2005**, *12*, 422–430. [CrossRef] [PubMed]
18. Rogers, D.C.; Fisher, E.M.; Brown, S.D.; Peters, J.; Hunter, A.J.; Martin, J.E. Behavioral and functional analysis of mouse phenotype: SHIRPA, a proposed protocol for comprehensive phenotype assessment. *Mamm. Genome* **1997**, *8*, 711–713. [CrossRef]
19. Matsuura, K.; Kabuto, H.; Makino, H.; Ogawa, N. Pole test is a useful method for evaluating the mouse movement disorder caused by striatal dopamine depletion. *J. Neurosci. Methods* **1997**, *73*, 45–48. [CrossRef] [PubMed]
20. Taniike, M.; Suzuki, K. Spacio-temporal progression of demyelination in twitcher mouse: With clinico-pathological correlation. *Acta Neuropathol.* **1994**, *88*, 228–236. [CrossRef]
21. Ohno, M.; Komiyama, A.; Martin, P.M.; Suzuki, K. Proliferation of microglia/macrophages in the demyelinating CNS and PNS of twitcher mouse. *Brain Res.* **1993**, *602*, 268–274. [CrossRef]
22. Potter, G.B.; Petryniak, M.A. Neuroimmune mechanisms in Krabbe's disease. *J. Neurosci. Res.* **2016**, *94*, 1341–1348. [CrossRef]
23. Nicaise, A.M.; Bongarzone, E.R.; Crocker, S.J. A microglial hypothesis of globoid cell leukodystrophy pathology. *J. Neurosci. Res.* **2016**, *94*, 1049–1061. [CrossRef]
24. Cachon-Gonzalez, M.B.; Wang, S.; Cox, T.M. Expression of Ripk1 and DAM genes correlates with severity and progression of Krabbe disease. *Hum. Mol. Genet.* **2021**, *30*, 2082–2099. [CrossRef]
25. Wilson, I.; Vitelli, C.; Yu, G.K.; Pacheco, G.; Vincelette, J.; Bunting, S.; Siso, S. Quantitative Assessment of Neuroinflammation, Myelinogenesis, Demyelination, and Nerve Fiber Regeneration in Immunostained Sciatic Nerves from Twitcher Mice with a Tissue Image Analysis Platform. *Toxicol. Pathol.* **2021**, *49*, 950–962. [CrossRef] [PubMed]
26. Takahashi, H.; Suzuki, K. Demyelination in the spinal cord of murine globoid cell leukodystrophy (the twitcher mouse). *Acta Neuropathol.* **1984**, *62*, 298–308. [CrossRef] [PubMed]
27. Smith, B.; Galbiati, F.; Castelvetri, L.C.; Givogri, M.I.; Lopez-Rosas, A.; Bongarzone, E.R. Peripheral neuropathy in the Twitcher mouse involves the activation of axonal caspase 3. *ASN Neuro* **2011**, *3*, AN20110019. [CrossRef]
28. Rafi, M.A.; Rao, H.Z.; Passini, M.A.; Curtis, M.; Vanier, M.T.; Zaka, M.; Luzi, P.; Wolfe, J.H.; Wenger, D.A. AAV-mediated expression of galactocerebrosidase in brain results in attenuated symptoms and extended life span in murine models of globoid cell leukodystrophy. *Mol. Ther.* **2005**, *11*, 734–744. [CrossRef] [PubMed]

29. Lin, D.S.; Hsiao, C.D.; Liau, I.; Lin, S.P.; Chiang, M.F.; Chuang, C.K.; Wang, T.J.; Wu, T.Y.; Jian, Y.R.; Huang, S.F.; et al. CNS-targeted AAV5 gene transfer results in global dispersal of vector and prevention of morphological and function deterioration in CNS of globoid cell leukodystrophy mouse model. *Mol. Genet. Metab.* **2011**, *103*, 367–377. [CrossRef]
30. Lin, D.-S.; Hsiao, C.-D.; Lee, A.Y.-L.; Ho, C.-S.; Liu, H.-L.; Wang, T.-J.; Jian, Y.-R.; Hsu, J.-C.; Huang, Z.-D.; Lee, T.-H.; et al. Mitigation of cerebellar neuropathy in globoid cell leukodystrophy mice by AAV-mediated gene therapy. *Gene* **2015**, *571*, 81–90. [CrossRef]
31. Reddy, A.S.; Kim, J.H.; Hawkins-Salsbury, J.A.; Macauley, S.L.; Tracy, E.T.; Vogler, C.A.; Han, X.; Song, S.K.; Wozniak, D.F.; Fowler, S.C.; et al. Bone marrow transplantation augments the effect of brain- and spinal cord-directed adeno-associated virus 2/5 gene therapy by altering inflammation in the murine model of globoid-cell leukodystrophy. *J. Neurosci.* **2011**, *31*, 9945–9957. [CrossRef]
32. Rafi, M.A.; Rao, H.Z.; Luzi, P.; Curtis, M.T.; Wenger, D.A. Extended normal life after AAVrh10-mediated gene therapy in the mouse model of Krabbe disease. *Mol. Ther.* **2012**, *20*, 2031–2042. [CrossRef]
33. Rafi, M.A.; Luzi, P.; Wenger, D.A. Can early treatment of twitcher mice with high dose AAVrh10-GALC eliminate the need for BMT? *Bioimpacts* **2021**, *11*, 135–146. [CrossRef] [PubMed]
34. Marshall, M.S.; Issa, Y.; Jakubauskas, B.; Stoskute, M.; Elackattu, V.; Marshall, J.N.; Bogue, W.; Nguyen, D.; Hauck, Z.; Rue, E.; et al. Long-Term Improvement of Neurological Signs and Metabolic Dysfunction in a Mouse Model of Krabbe's Disease after Global Gene Therapy. *Mol. Ther.* **2018**, *26*, 874–889. [CrossRef]
35. Li, Y.; Miller, C.A.; Shea, L.K.; Jiang, X.; Guzman, M.A.; Chandler, R.J.; Ramakrishnan, S.M.; Smith, S.N.; Venditti, C.P.; Vogler, C.A.; et al. Enhanced Efficacy and Increased Long-Term Toxicity of CNS-Directed, AAV-Based Combination Therapy for Krabbe Disease. *Mol. Ther.* **2021**, *29*, 691–701. [CrossRef] [PubMed]
36. Karumuthil-Melethil, S.; Marshall, M.S.; Heindel, C.; Jakubauskas, B.; Bongarzone, E.R.; Gray, S.J. Intrathecal administration of AAV/GALC vectors in 10–11-day-old twitcher mice improves survival and is enhanced by bone marrow transplant. *J. Neurosci. Res.* **2016**, *94*, 1138–1151. [CrossRef] [PubMed]
37. Castle, M.J.; Gershenson, Z.T.; Giles, A.R.; Holzbaur, E.L.; Wolfe, J.H. Adeno-associated virus serotypes 1, 8, and 9 share conserved mechanisms for anterograde and retrograde axonal transport. *Hum. Gene Ther.* **2014**, *25*, 705–720. [CrossRef]
38. Allen, G.I.; Tsukahara, N. Cerebrocerebellar communication systems. *Physiol. Rev.* **1974**, *54*, 957–1006. [CrossRef]
39. Buckner, R.L.; Krienen, F.M.; Castellanos, A.; Diaz, J.C.; Yeo, B.T.T. The organization of the human cerebellum estimated by intrinsic functional connectivity. *J. Neurophysiol.* **2011**, *106*, 2322–2345. [CrossRef]
40. Saunders, N.R.; Habgood, M.D.; Dziegielewska, K.M. Barrier mechanisms in the brain, II. Immature brain. *Clin. Exp. Pharmacol. Physiol.* **1999**, *26*, 85–91. [CrossRef]
41. Heller, G.J.; Marshall, M.S.; Issa, Y.; Marshall, J.N.; Nguyen, D.; Rue, E.; Pathmasiri, K.C.; Domowicz, M.S.; van Breemen, R.B.; Tai, L.M.; et al. Waning efficacy in a long-term AAV-mediated gene therapy study in the murine model of Krabbe disease. *Mol. Ther.* **2021**, *29*, 1883–1902. [CrossRef] [PubMed]
42. Castelvetri, L.C.; Givogri, M.I.; Zhu, H.; Smith, B.; Lopez-Rosas, A.; Qiu, X.; van Breemen, R.; Bongarzone, E.R. Axonopathy is a compounding factor in the pathogenesis of Krabbe disease. *Acta Neuropathol.* **2011**, *122*, 35–48. [CrossRef] [PubMed]
43. Cantuti Castelvetri, L.; Givogri, M.I.; Hebert, A.; Smith, B.; Song, Y.; Kaminska, A.; Lopez-Rosas, A.; Morfini, G.; Pigino, G.; Sands, M.; et al. The sphingolipid psychosine inhibits fast axonal transport in Krabbe disease by activation of GSK3β and deregulation of molecular motors. *J. Neurosci.* **2013**, *33*, 10048–10056. [CrossRef] [PubMed]
44. Cantuti-Castelvetri, L.; Zhu, H.; Givogri, M.I.; Chidavaenzi, R.L.; Lopez-Rosas, A.; Bongarzone, E.R. Psychosine induces the dephosphorylation of neurofilaments by deregulation of PP1 and PP2A phosphatases. *Neurobiol. Dis.* **2012**, *46*, 325–335. [CrossRef]
45. Weinstock, N.I.; Kreher, C.; Favret, J.; Nguyen, D.; Bongarzone, E.R.; Wrabetz, L.; Feltri, M.L.; Shin, D. Brainstem development requires galactosylceramidase and is critical for pathogenesis in a model of Krabbe disease. *Nat. Commun.* **2020**, *11*, 5356. [CrossRef]

cells

MDPI

Review

Advancements in Hematopoietic Stem Cell Gene Therapy: A Journey of Progress for Viral Transduction

Aurora Giommetti [1,2] and Eleni Papanikolaou [1,3,*]

[1] Miltenyi Biotec B.V. & Co. KG, 51429 Bergisch Gladbach, Germany; aurorag@miltenyi.com
[2] Faculty of Biology, University of Freiburg, 79104 Freiburg, Germany
[3] Laboratory of Biology, School of Medicine, National and Kapodistrian University of Athens, 115 27 Athens, Greece
* Correspondence: elenip@miltenyi.com

Abstract: Hematopoietic stem cell (HSC) transduction has undergone remarkable advancements in recent years, revolutionizing the landscape of gene therapy specifically for inherited hematologic disorders. The evolution of viral vector-based transduction technologies, including retroviral and lentiviral vectors, has significantly enhanced the efficiency and specificity of gene delivery to HSCs. Additionally, the emergence of small molecules acting as transduction enhancers has addressed critical barriers in HSC transduction, unlocking new possibilities for therapeutic intervention. Furthermore, the advent of gene editing technologies, notably CRISPR-Cas9, has empowered precise genome modification in HSCs, paving the way for targeted gene correction. These striking progresses have led to the clinical approval of medicinal products based on engineered HSCs with impressive therapeutic benefits for patients. This review provides a comprehensive overview of the collective progress in HSC transduction via viral vectors for gene therapy with a specific focus on transduction enhancers, highlighting the latest key developments, challenges, and future directions towards personalized and curative treatments.

Keywords: gene therapy; hematopoietic stem cells (HSC); transduction; viral vectors; transduction enhancers; rare diseases

1. Introduction

In the area of medical science, the pursuit of innovative therapies has propelled researchers toward groundbreaking solutions to efficacious treatments that improve the patients' quality of life. One such frontier is the transformative field of hematopoietic stem cell (HSC) gene therapy. HSCs are defined by the remarkable ability of long-term self-renewal and differentiation into multiple blood cell lineages, which elucidates their profound clinical significance. Over the years, HSC transplantation (HSCT) has become a well-established and widely utilized procedure for the treatment of congenital metabolic diseases and blood-related disorders. The first successful applications of allogeneic HSCT were achieved in the treatment of X-linked severe combined immunodeficiency (X-SCID) [1] and Wiskott-Aldrich syndrome (WAS) [2], where patients received stem cells from a compatible donor. Discovering its curative potential, notable advancements have been made in allogeneic HSCT that include the identification of donors with compatible human leukocyte antigen (HLA), the expansion of donor registries, and the possibility of alternative donor sources, such as haplo-identical donors who are half-matched to the recipient. The refinements in conditioning regimens have also contributed to improved patient outcome, as well as a more effective control of graft-versus-host disease (GvHD) that involves the selective depletion of α/β T cells and naïve T cells [3]. Nevertheless, the successful applications of allogeneic HSCT can be constrained by the availability of suitable donors and a potential risk of morbidity and mortality due to the use of HLA-mismatched individuals.

Autologous HSCT represents a significant leap toward overcoming the risk of graft rejection and additional complications arising from alloreactivity [4]. Therefore, autologous HSCT, in combination with HSC gene therapy, has been explored as an alternative therapeutic strategy to treat not only hematological malignancies but also inherited diseases, including severe immunodeficiencies, hemoglobinopathies, and metabolic disorders. In this approach, the patient's own stem cells are harvested, cultured ex vivo, and genetically modified before being reinfused into the patient following an appropriate conditioning regimen to deplete progenitor and differentiated cells in the bone marrow and to favor engraftment. In this context, ex vivo HSC genetic engineering can be performed either through transduction with viral vectors delivering the therapeutic gene of interest or by targeted genome editing approaches that allow site-specific genome modifications depending on the disease. This review will analyze the current state of hematopoietic stem cell gene therapy, addressing the latest advancements, challenges, and therapeutic potentials of this application. After an overview of the gene transfer processes, we focus on the approaches that generated the first marketing authorizations, i.e., the vector-based modifications of HSCs and their optimization regarding efficacy and safety aspects, concluding with an overall discussion about the prospective developments in HSC gene therapy.

2. Gene Delivery in Hematopoietic Stem Cells

2.1. Overview

The most extensively investigated gene transfer techniques in hematopoietic stem cells to date are based on the ex vivo approach. This type of gene therapy involves the collection of HSCs from the patient that undergo genetic manipulation through viral transduction or gene editing to restore the correct phenotype. Following a preconditioning regimen, the engineered cells are infused back into the patient's body, where they self-renew and differentiate generating a long-term reservoir of modified HSCs giving rise to multiple blood lineages. This strategy allows the performance of the manipulation process in a controlled environment that enables the monitoring of cell characterization and functionality before transplantation, and it potentially represents a one-time curative treatment due to the engraftment capacity of gene-corrected HSCs. Therefore, it is a promising therapy to tackle hematological disorders and immune aberrations, as confirmed by the marketing authorization for advanced therapy medicinal products (ATMPs) based on engineered HSCs, namely Libmeldy™ for pediatric metachromatic leukodystrophy (MLD), Strimvelis® for the treatment of severe combined immunodeficiency due to adenosine deaminase deficiency (ADA-SCID), Zynteglo™ for β-thalassemia, Lyfgenia™ for the treatment of sickle cell disease (SCD), and Skysona® for early cerebral adrenoleukodystrophy (CALD) (Table 1).

Table 1. Hematopoietic stem cell gene therapy products approved on the market.

Product Name: Generic (Trade)	Applications	Manufacturer	Mechanism of Action	Approval Agency (Year)
STRIMVELIS®	ADA-SCID	Orchard Therapeutics	ADA gene addition via gamma retrovirus	EMA (2016)
Betibeglogene autotemcel (ZYNTEGLO™)	Transfusion-dependent B-thalassemia (TDT)	bluebird bio, Inc.	βA-T87Q-globin gene addition via lentivirus	EMA (2019) * FDA (2022)
Atidarsagene autotemcel (LIBMELDY®)	Metachromatic leukodystrophy (MLD)	Orchard Therapeutics	ARSA gene addition via lentivirus	EMA (2020) FDA (2024)
Lovotibeglogene autotemcel (LYFGENIA™)	Sickle cell disease (SCD)	bluebird bio, Inc.	βA-T87Q-globin gene addition via lentivirus	FDA (2023)

Table 1. *Cont.*

Product Name: Generic (Trade)	Applications	Manufacturer	Mechanism of Action	Approval Agency (Year)
Exagamglogene autotemcel (CASGEVY™)	TDT, SCD	Vertex Pharmaceuticals CRISPR Therapeutics	CRISPR/Cas9 technology	EMA (2023) FDA (2024)
Elivaldogene autotemcel (SKYSONA®)	CALD	bluebird bio, Inc.	ABCD1 gene addition via lentivirus	EMA (2021) * FDA (2022)

* Withdrawn at the request of the marketing-authorization holder.

2.2. Gammaretroviruses

Ex vivo gene manipulation employing viral vectors as delivery tools has been a major player in the field of HSC gene therapy. Following HSC collection and enrichment of $CD34^+$ cells, they undergo genetic modification through viral vector transduction before being reinfused to the host, where their engraftment will result in a sustained transgene function. This method exploits the innate ability of viruses to efficiently internalize their own genome into the target cells; however, the viral vectors are engineered from wildtype viruses by removal of most of the genes encoding for viral proteins from the viral genome to make them replication incompetent. A variety of viral vector classes have been used in several clinical trials conducted so far, including members of the *Retroviridiae* family (gammaretroviruses and lentiviruses), adenoviruses, and adeno-associated viruses, and each of these platforms has highlighted advantages and complications in preclinical and clinical testing phases. Gammaretroviruses (γRV) were initially the first vectors assessed for gene therapy and their popularity did not decrease overtime, mainly due to their low immunogenicity, ability to integrate the viral genome into the host cells, and the high efficiency of transduction in actively dividing target cells. Therefore, considering the quiescent state of HSCs, they are pre-stimulated to induce the cell cycle for an effective γRV transduction. However, gammaretroviral vectors preferentially integrate near transcriptional start sites and within CpG islands, and they have affinities toward proto-oncogenes, potentially leading to insertional oncogenesis and serious adverse events such as malignant transformation, as this was demonstrated in the original X-SCID study [5]. Thus, the constrained efficacy of γRV gene transfer into HSCs and the risk of genotoxicity prompted the advancement of self-inactivating lentiviral vectors (LVs) as a preferred delivery system due to their improved safety profile.

2.3. Lentiviruses

Lentiviruses employed in the clinic are usually based on human immunodeficiency virus (HIV) 1 devoid of its structural genes and commonly pseudotyped with a different envelope, and they have the ability to infect non-replicating cells due to the transfer of viral vector DNA in the nucleus via nuclear pores, allowing a faster transduction process and also harnessing their larger packaging capacity of up to 9 kb. Although the integration sites of LVs are generally unpredictable with hotspots in active transcription units, Biffi et al. [6] suggested that they cluster in the megabase-wide genomic regions without accumulation in specific genomic regions, in contrast to genotoxic integration sites that are not distributed along chromosomes but come across as isolated sharp peaks and always target a single gene, which is the culprit of oncogenesis. This study [6] provided a more comprehensive understanding of the preferential or common integration sites among retroviral vectors and showed that LVs present markedly consistent integration pattern. Alternative vector systems, such as adenoviruses and adeno-associated viruses, have encountered restricted success when applied to HSCs thus far due to the strong innate and adaptive immune responses induced by viral infection [7] and due to the challenging transient transgene expression even with high multiplicity of infection [8], respectively. Overall, genetic engineering of HSCs via viral vector transduction has been shown to be a crucial tool in the clinical setting for HSC gene therapy, achieving curative treatments in several clinical trials [9–11] and successful marketing approvals [12–14].

2.4. Genome Editing

Nonetheless, alongside HSC transduction, the identification and adaptation of programmable molecules such as nucleases, base editors, and prime editors have enabled targeted genetic modifications, allowing a major step forward in the field of genome editing. In this context, site-specific endonucleases are used to introduce a double-stranded break at a desired location at the DNA level that will recruit DNA repair proteins to correct the damage, establishing specific genetic changes which, depending on the type of indication, could result in gene disruption, gene correction or gene insertion. Although a detailed analysis of these methods falls beyond the scope of this review, it is worth mentioning the recent marketing approval of Casgevy™, the first CRISPR-Cas9 gene editing therapy, which aims to cure SCD and transfusion-dependent β-thalassemia (TDT) [14]. However, it should be noted that there is a growing body of literature evaluating the unexpected adverse events of the use of CRISPR technology for HSC modifications, including on-target genotoxicity such as deletions, translocations, micronuclei formation [15], and decreased long-term in vivo engraftment in terms of clonal dynamics [16]. A more detailed description of the CRISPR complications is presented in the paragraph "Safety Considerations" (see below).

3. Advancements in Transduction Technologies

3.1. Viral Vector Engineering

Even though HSC gene therapy utilizing viral vectors has evolved as a therapeutic alternative for various inherited diseases and gained success over the decades, achieving stable and clinically relevant gene transduction in hematopoietic stem cells still presents a significant challenge. Successful viral transduction of HSCs is hampered by several barriers, including low expression levels of viral receptors on the HSC surface, inefficient viral entry, as well as the presence of cellular restriction factors that inhibit viral replication. Due to their quiescent nature, differently from other target cell types such as T cells or natural killer cells, an efficient uptake of vectors by HSCs requires high multiplicity of infection (MOI) resulting in increased viral vector doses as the cell number grows, which can increase the risk of adverse events, and it is particularly relevant for large-scale production of engineered HSCs that will be translated into clinical applications.

A promising solution that is gaining steam is to enrich an HSC subpopulation by sorting the $CD34^{+}CD38^{-}$ cells which represent the small proportion of HSCs that actually contribute to long-term hematopoiesis, allowing a reduction in the amount of viral vector required for transduction without impacting the vector copy number (VCN) and the unification of the final characteristics of the infused HSCs that come from different sources such as bone marrow or mobilized peripheral blood. However, although there was initial enthusiasm, this approach was not widely applied in the clinic due to the difficulties in enriching pure $CD34^{+}CD38^{-}$ populations, which made the overall process laborious, lengthy, and eventually inefficient.

In a different approach, the quest for more effective gene transfer techniques employed evaluation of several vector designs with a particular focus on the investigation of different envelope proteins for vector pseudotyping to check which one can increase their binding and uptake by target HSCs. Initial attempts to pseudotype the vectors with retroviral envelopes deriving from amphotropic murine retrovirus or the Gibbon Ape Leukemia Virus (GALV) led to low viral titers which triggered their replacement with more efficient envelopes, most notably the glycoprotein derived from the vesicular stomatitis virus (VSV-G) that confers broad tropism and enhances the vector stability, enabling effective HSC transduction. Moreover, VSV-G pseudotyping not only allows the concentration of vectors at high titers, but its robust fusion activity also facilitates entry into HSCs, exploiting the abundantly expressed low-density lipoprotein receptor (LDL-R) [17,18], bypassing the need for specific uncommon cellular receptors, which may be a limiting factor for successful transduction. Furthermore, a novel envelope protein derived from baboon endogenous retrovirus (BaEV) holds significant promise for HSC gene therapy applications. The BaEV envelope offers several benefits, including its unique natural tropism for human CD34^{+} hematopoietic stem cells and a favorable safety profile [19]. This inherent specificity reduces the risk of off-target effects and enhances the efficiency of gene delivery, making it particularly well suited for HSC transduction. In addition, although the titer measurements were lower than the VSV-G pseudotyped viral vectors, BaEV is more effective at an equivalent amount [19] and displays minimal immunogenicity, reducing the likelihood of immune responses that could compromise the success of the therapy.

On the basis of the promising envelope improvement, researchers are poised to develop safer and more targeted approaches to manipulate vectors for HSC transduction. A cutting-edge strategy to improve the vector design consists of the integration of recombinant membrane proteins leading to the expression of cytokines (e.g., stem cell factor and thrombopoietin) on the surface of the vector virions that will specifically recognize and bind to the respective receptors present on the HSC surface (c-kit, c-mpl). This system was designed to promote high levels of transduction of the most immature CD34^{+} cells, crucial for clinical application, with a selective and minimal HSC stimulation that is already supplied by the cytokines expressed on the virions, avoiding the addition of hematopoietic growth factors in the medium [20]. In agreement, a greater preference of lentiviruses in clinical trials has been currently observed, compared to their retroviral or adenoviral counterparts due to the ability of LVs to integrate into non-dividing cells that do not require prolonged cytokine stimulation to activate the HSC cell cycle before transduction, thereby circumventing intense cell proliferation in culture that could progressively compromise engraftment potential [21]. The current focus mainly lies in achieving an efficient lentiviral vector transduction of long-term repopulating quiescent HSCs, which are resistant to genetic manipulation but an ideal gene therapy target, with minimal in vitro culturing to avoid extensive cell stimulation and cell cycle commitment [22].

3.2. Transduction Enhancers

Along with the progress in engineered vector design and ex vivo culture conditions, several reagents have been tested aiming to achieve a clinically translatable transduction efficiency without interfering with HSC self-renewal and differentiation. A promising strategy is the addition of transduction enhancers which encompass a diverse array of small molecules that modulate cellular pathways involved in viral entry, intracellular trafficking, endosomal escape, or nuclear import (Figure 1). An overview of several transduction enhancers is shown in Table 2.

Prostaglandin E2 (PGE2) has been investigated as an adjuvant that promotes lentiviral transduction of CD34^{+} cells and modestly increases the VCN both in vitro and in an NOD/SCID xenotransplantation mouse model without evidence of in vivo toxicity [23] while reducing the duration of ex vivo culture. The mechanism leading to higher transduction levels is still under investigation, but PGE2 may act by improving the reverse transcription and hence replication of the vector inside the cell prior to nuclear entry and integration, since an increase in late RT copies was detected within 6 h after transduction [24,25]. Additional studies reported the beneficial effect of PGE2 on HSC transduction and VCN [26,27]; however, Poletti et al. [28] showed that it causes a significant reduction in stem cell clonogenic capacity once transplanted in humanized immunodeficient mice in a competitive repopulation assay despite its safety in increasing cord blood engraftment being demonstrated in the clinic [29] and its favorable effects being corroborated in a clinical trial for the treatment of Hurler disease [30]. It should be noted, however, that there was a lot of skepticism in the gene therapy field about the use of PGE2 because of the preexisting evidence of potential reduction in stemness [28]. The leukemic events in the bluebird clinical trial for SCD [31] also corroborated the initial doubts on the grounds of generation of leukemic phenotypes as a result of graft failure since the percentage of leukemic events in the gene therapy setting was equivalent to the leukemias presented in the allogeneic setting for SCD when graft failure is observed [32]. In particular, the working hypothesis was that after rejection or gene therapy, the stress from switching from homeostatic to regenerative hematopoiesis by autologous cells drives clonal expansion and leukemogenic transformation of preexisting premalignant clones, eventually resulting in hematological dysplasias. Nevertheless, in December 2023, the U.S. Food and Drug Administration (FDA) granted marketing authorization for Lyfgenia™, the lentiviral approach to treat SCD by bluebird bio, but with a warning for blood cancer [33]. It should be also noted that, in view of the exact same manufacturing process utilized both in the case of Lyfgenia™ and Zynteglo™, more gravity is given to the specific pathophysiology of SCD and not to the vector-cell interactions. Finally, the efficiency of lentiviral vector transduction on HSCs exhibited a significant increase also when PGE2 was tested in combination with polybrene [26], a surfactant polycation widely and successfully utilized as a transduction enhancer from the early days due to its interaction with the negatively charged cellular membrane which leads to charge shielding between the vector and the cell surface [34], but this approach did not reach clinical applicability.

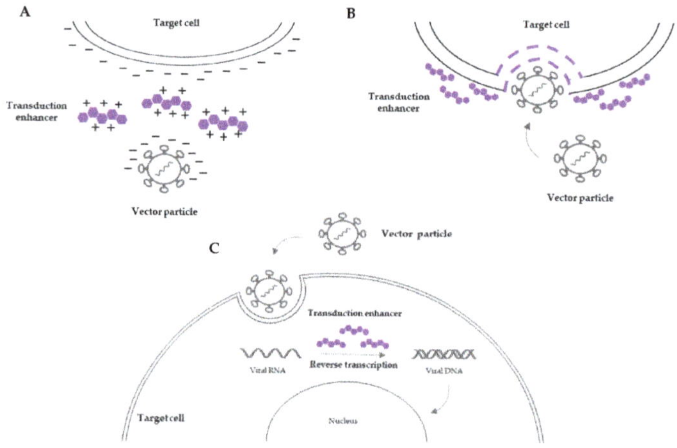

Figure 1. Overview of the mechanisms of action of transduction enhancers. (**A**) Lower charge repulsion between the vector particle and the target cell surface facilitates cell-to-vector interaction. (**B**) Increased permeability of the target cell surface membrane facilitates viral entry. (**C**) Influence on intracellular processes (e.g., reverse transcription of viral RNA) prior to vector integration into the host genome facilitates vector trafficking and integration.

Traditionally, protamine sulfate represents another cationic additive which produces an optimal transduction-enhancing effect by neutralization of the cell membrane charge, and its approval for human use by the FDA, together with its low toxicity on a range of cell types [35], highlights its versatility and potential for clinical translation. A similar mechanism of action is observed when poloxamer is supplemented in culture, where it influences the physiochemical properties of the cell membrane, also promoting transmembrane transport. Different sizes of poloxamers have been investigated, among which P118, P338 [36], and P407 which result in a similar increment in both the percentage of transduced cells and the number of vector copies per cell without significant toxicity [25]. Lately, the most commonly used transduction enhancer is termed LentiBOOST™ and consists of a combination of poloxamer 338 and Pluronic F108 and is considered an entry enhancer because it seems to increase the permeability of the target cell surface facilitating the entry of viral particles [37]. LentiBOOST™ outperformed the aforementioned enhancers, leading to a strong effect in terms of vector expression at low MOI with an acceptable VCN increase [38] and maintaining HSC differentiation potential, and was also demonstrated in xenotransplantion experiments [27] with no signs of toxicity in vivo [39]. Another peptide that enables high levels of gene transfer with various retroviral and lentiviral pseudotypes into CD34⁺ HSCs is Vectofusin-1® [40]. Through the formation of alpha-helical nanofibrils, it fosters the adhesion of viral vectors to targeted receptors on the cell surface and facilitates endocytosis, ultimately leading to increased transduction efficiency of HSCs in vitro that is preserved also in their progeny (T and B cells) after engraftment in an NSG mouse model [41]. Specifically, Vectofusin-1® does not alter the cell viability and functionality and the safety of the transduction process, and since it is soluble in water, it allows avoidance of the pre-coating step required for similar compounds such as Retronectin, making it an ideal candidate for clinical settings and scalable gene therapy protocols [42]. Furthermore, a deeper understanding of the mechanisms regulating HSC proliferation, self-renewal, and quiescence has led to the detection of rapamycin, a macrolide compound with immunosuppressive properties, as a potential transduction enhancer for HSC engineering. While the precise underlying mechanisms still need to be elucidated, rapamycin promotes efficient viral transduction of both human and murine HSCs via the inhibition of the mTOR signaling pathway, significantly boosting the frequency of long-term engrafting cells in mice [43] and ex vivo long-term hematopoietic reconstitution [44]. Moreover, rapamycin's well-established safety profile and clinical use in other therapeutic contexts, including prevention of allograft rejection and cancer treatment, underscores its potential application in clinically relevant viral transduction protocols.

Table 2. List of reagents employed to enhance transduction efficiency.

Reagent	Mechanism of Action	Side Effects	Side Effects in Gene Therapy	Clinical Applications in Gene Therapy	References
Prostaglandin E2	Improvement of reverse transcription (under investigation)	Nausea, vomiting, diarrhea, abdominal pain	Reduction of HSC clonogenic potential	Hurler syndrome, β-thalassemia	[21,26,28]
Protamine sulfate	Lower charge repulsion between the vector and the cell surface	Low blood pressure, allergic reactions, vomiting	Cell toxicity (concentrations higher than 10 μg/mL)	N/A	[33]
Poloxamers	Membrane fluidization and reduction in electrostatic barriers	Dehydration, abdominal discomfort	N/A	N/A	[34]
LentiBOOST™	Increased permeability of the target cell surface	N/A	N/A	X-SCID, Artemis-SCID	[35]
Vectofusin-1®	Enhanced adhesion and fusion of viral particles to the cell membrane	N/A	N/A	N/A	[38]
Rapamycin	Inhibition of mTOR signaling pathway (immunosuppression)	Anemia, increased blood pressure, muscle pain	N/A	N/A	[41,43]
Cyclosporin A Cyclosporin H	Inhibition of cyclophilin A (immunosuppression) Inhibition of IFITM3	Blurred vision, back pain, dizziness, decreased appetite	N/A	N/A	[43,45,46]

N/A: not applicable.

Along with rapamycin, cyclosporin A (CsA) is an additional immunosuppressive compound that acts by inhibiting the activity of the cellular protein cyclophilin A, which is known to interact with the viral capsid protein of retroviruses and lentiviruses during transduction. By blocking this interaction, CsA enhances the efficiency of viral vector entry into target cells, including HSCs, thereby improving transduction efficiency without adversely affecting their colony-forming capacity. Importantly, increased transduction efficiencies were maintained long term in vivo and no negative effects on HSC engraftment were observed [45]. Additionally, CsA has been shown to mitigate the inhibitory effects of cellular antiretroviral restriction factors, such as TRIM5α, on viral transduction, further enhancing gene delivery to HSCs. These results were corroborated by Evans and colleagues [46], who reported that TRIM5α transcript levels in human CD34$^+$ cells correlate with donor variability in transduction efficiency with lentiviral vectors. From the same class of compounds of CsA, cyclosporin H (CsH) has gained attention as a transduction enhancer operating with a mechanism of action similar to CsA. In detail, CsH reduced the innate resistance mechanism against LV infection performed through the interferon-induced transmembrane protein 3 (IFITM3) constitutive inhibition of viral entry by degrading IFITM3, leading to significant improvement in gene transfer levels both in murine and human HSCs [47,48].

Overall, the use of transduction enhancers will have profound implications for clinical practice. By enhancing the efficiency of gene delivery, these strategies allow minimization of the multiplicity of infection of the viral vector, improving safety and therapeutic efficacy of HSC gene therapy applications. Furthermore, the development of targeted and customizable transduction enhancers may enable tailored approaches for specific patient populations and disease contexts, minimizing off-target effects and optimizing treatment outcomes.

4. Safety Considerations

Ensuring the safety of HSC transduction is paramount to sustained clinical efficacy and the long-term success of therapeutic interventions and it represents a crucial step not only for manufacturing of engineered HSCs but also from a pure clinical perspective. Indeed, the HSC gene therapy field was initially severely hampered because of safety concerns deriving from insertional mutagenesis (creating the risk of leukemia) and/or immunogenicity.

4.1. Genotoxicity and Leukemias

In the notorious French X-SCID clinical trial, four out of seven patients treated initially with gamma RV developed lymphocytic leukemia [49,50], which was associated with vector integration in the vicinity of the LMO2 gene, leading to its upregulation, which was suggested to be the determining event for the onset of the blood cancer. The development of malignancies resulted in a temporary interruption of this trial, which eventually resumed, but only for patients who had failed standard transplant therapy. In the early 2000s, other gene therapy trials were also delayed due to the potential risk of leukemia but eventually continued as the risk-versus-benefit ratio was deemed to be in favor of the patients [51].

Unfortunately, leukemias were not observed only during the French X-SCID trial. Cancer transformation was observed in other clinical trials employing γRV vectors for X-linked Chronic Granulomatous Disease (X-CGD) [52] and WAS [53] as a result of vector integration close to and activation of proto-oncogenes. At that time, the field's response focused on the following actions: a) further advancement of viral vector engineering within the context of self-inactivating (SIN) LVs as a vehicle for gene delivery and b) a deeper understanding of the vectors' integration sites, alongside rigorous preclinical safety testing to predict potential adverse effects and mitigate the long-term risk of genotoxicity. Additionally, in vitro assays such as the In Vitro Immortalization assay (IVIM) and the Surrogate Assay for Genotoxicity Assessment (SAGA) have been developed in an effort to predict or quantify the pre-clinical genotoxicity of integrating vectors. Notably, IVIM quantifies the mutagenic potential of retroviruses based on the acquisition of a proliferation advantage under limiting dilution conditions of murine hematopoietic stem and progenitor cells transduced with mutagenic vectors [54]. Although this approach is relatively specific for the detection of mutants with insertions near the *Mecom* locus (also known as *Evi1*) or its close relative *Prdm16*, both of which were shown to be clinically relevant as inducers of clonal imbalance in clinical trials for X-CGD [55,56], X-SCID [49], WAS [57], it has been accepted by regulatory authorities as part of the pre-clinical safety assessment. A more accurate prediction is performed with the SAGA approach that classifies integrating retroviral vectors using machine learning algorithms to detect the activation of gene expression pathways connected to oncogenesis during the course of in vitro cell immortalization [58]. However, due to the specific culture conditions, both assays present an intrinsic myeloid bias, and thus Bastone and colleagues [54] have introduced the SAGA-XL assay that follows a similar bioinformatic strategy enabling the identification of lymphoid genotoxicity predictors. Notably, it should be mentioned that the onset of leukemias in HSC gene therapy is not always the result of vector-induced insertional mutagenesis, since in the case of the Lyfgenia™ trials, the leukemic blasts were devoid of vector genetic material, clearly suggesting that the dysplasias were independent of the vector in this specific setting and were rather associated with the overall stemness/fitness of the graft and/or the specific pathophysiology of SCD.

Genotoxicity poses concerns also in the CRISPR field, as a growing line of evidence shows the occurrence of unexpected on-target genotoxic events, including deletions and chromosomal translocations, which may compromise the genomic integrity and functionality of edited HSCs. Studies, such as Kosicki et al. [59], have highlighted that DNA breaks introduced by CRISPR-Cas9 editing can resolve into onsite large deletions as well as crossover events and lesions distal to the cut site, which may constitute a first carcinogenic 'hit' in stem cells and progenitors that have a long replicative lifespan [59]. Further works reported that p53-mediated DNA damage response activated by double-strand breaks induced by CRISPR could lead to selection for cells with mutations in the p53 pathway [60], potentially contributing to oncogenesis. Recently, Lee et al. [16] have shown that the homology-directed repair (HDR) pathway induced by CRISPR-Cas9 led to decreased short- and long-term multilineage HSC engraftment and graft clonality in a competitive rhesus macaque autologous transplantation model. In the same study, the authors demonstrated that CRISPR/HDR-edited cells showed lower viability, cell proliferation, and markedly decreased long-term engraftment compared to lentiviral transduced cells, suggesting increased toxicity of editing [16]. It should be noted, however, that the latest results by Zeng and colleagues [15] indicated a better outcome for the CRISPR-engineered HSCs in terms of genotoxicity and micronucleation in the context of short-term ex vivo culture in the absence of cytokines (hence avoiding the induction of cell cycle through ex vivo cytokine stimulation).

4.2. Immunogenicity

Other possible obstacles that could trigger undesirable events can arise from innate and adaptive immunity against reagents used during manufacturing and immune reactions against neoantigens introduced into HSCs by genetic engineering [61] due to gene disruptions and/or translocations. To address these concerns, which are triggered by the transduction/gene editing per se and are expressed long term after the administration of the genetically corrected graft, will require further in-depth investigations, always taking into consideration the vector system, the engineering approach, and the transplantation settings applied in each different scenario.

5. Conclusions and Future Perspectives

Over the last twenty years, the HSC gene therapy field has witnessed notable clinical achievements that resulted in remarkable marketing approval of ATMPs by the EMA and/or FDA. A scientific breakthrough came in 2016 when Strimvelis® received marketing authorization for the treatment of ADA-SCID, employing a gamma retroviral vector carrying the sequence of the therapeutic gene. The approval of Zynteglo™ offers patients with TDT the possibility to be treated with a lentiviral vector encoding a β-globin transgene, which is mutated or absent in these patients. Lastly, there was also the marketing authorization of Skysona® for early CALD. In terms of the availability and prices of these drugs, it is necessary to underline that ex vivo HSC gene therapy is an extremely personalized treatment that presents significant technical challenges. A prime example is Strimvelis®, which, owing to its fresh formulation, can be administered only at the approved manufacturing facility in Europe in Milan, where the patient and their family have to stay for around 4 months. In this specific case, the overall costs are covered by the national insurance of the patient's country, but this does not apply for Zynteglo™ and Skysona®, which, despite obtaining marketing authorization, have been withdrawn from Europe due to their extremely high price tag per patient [62]. Besides gene therapy products, another milestone treatment for TDT and for severe SCD patients is Casgevy™, the first gene editing technology based on CRISPR-Cas9 system on the market, although discussion around the reimbursement from public health budgets or insurance companies is still ongoing. Thus, although there are significant number of marketing authorizations in spite of the aforementioned limitations, several HSC gene therapy trials involving viral vectors and genome editing are still ongoing (Table 3). These ongoing trials are diversified compared to the initial ones, either by differential patient stratification (e.g., in the KL003 trial, TDT patients are stratified based on the levels of serum ferritin) or by addition of transduction enhancers (NCT03538899, NCT01306019), indicating that while significant progress has been made in improving HSC transduction efficiency and long-term safety, challenges remain, limiting the clinical applicability of gene therapy. To this end, one might argue that the development of transduction enhancers holds promise for overcoming barriers to HSC transduction towards improving therapeutic outcomes because of the long-standing clinical experience with lentiviral vectors in the field.

Table 3. Currently ongoing hematopoietic stem cell gene therapy clinical trials.

Clinical Trial Registry Number	Disease	Intervention	Sponsor	Phase
NCT04797260	RAG1-SCID	Autologous $CD34^+$ cells transduced with the pCCL.MND.coRAG1.wpre LV	Leiden University Medical Center	I/II
NCT05071222	Artemis-SCID	Autologous $CD34^+$ cells transduced with the G2ARTE LV expressing the DCLRE1C cDNA	Assistance Publique—Hôpitaux de Paris/Genethon	I/II
NCT02559830	MLD, ALD	Autologous $CD34^+$ cells transduced with a LV encoding the human ARSA(for MLD)/ ABCD1(for ALD) cDNA	Shenzhen Second People's Hospital	I/II
NCT05860595	TDT	Autologous $CD34^+$ cells transduced with the βA-T87Q-globin gene LV (KL003)	Institute of Hematology and Blood Diseases Hospital, China/Kanglin Biotech	N/A
NCT05762510	TDT	Autologous $CD34^+$ cells transduced with the GMCN-508B (LentiRed) LV	First Affiliated Hospital of Guangxi Medical University	Early I
NCT05432310	ADA-SCID	Autologous $CD34^+$ cells transduced with the EFS-ADA LV encoding the ADA enzyme	University of California, Los Angeles	I/II
NCT06149403	Hurler syndrome	Autologous $CD34^+$ cells transduced with LV encoding the human IDUA gene	Orchard Therapeutics	III
NCT05265767	Hemophilia A	Autologous $CD34^+$ cells transduced with LV encoding a novel coagulation factor VIII transgene	Christian Medical College, Vellore, India	I
NCT03818763	Hemophilia A	Autologous $CD34^+$ cells transduced with LV encoding the *ITGA2B* gene promoter for ectopic expression of human B-domain-deleted factor VIII	Medical College of Wisconsin	I
NCT06155500	SCD	Observational: long-term follow-up of patients treated with CRISPR/Cas9-edited HSPCs from NCT04443907	Novartis Pharmaceuticals	I
NCT01306019	X-SCID	Autologous $CD34^+$ HSC with VSV-G pseudotyped LV CL20- 4i-EF1alpha-hgammac-OPT	National Institute of Allergy and Infectious Diseases (NIAID)	I/II
NCT03538899	Artemis-SCID	Autologous $CD34^+$ cells transduced with LV (AProArt) encoding the corrected DCLRE1C gene	University of California, San Francisco	I/II
NCT05757245	TDT	Autologous $CD34^+$ cells transduced with GMCN-508A LV	First Affiliated Hospital of Guangxi Medical University	I
2014-000274-20	WAS	Observational: long-term follow-up of patients treated with w1.6_hWASP_WPRE (VSVg) LV transduced autologous HSCs	Genethon	II
2019-004266-18	TDT	Observational: long-term follow-up of patients treated with βA-T87Q LV (LentiGlobin BB305) transduced autologous HSCs	bluebird bio, Inc.	III

Table 3. *Cont.*

Clinical Trial Registry Number	Disease	Intervention	Sponsor	Phase
2020-000517-33	Leukocyte adhesion deficiency I	Autologous CD34$^+$ cells transduced with LV encoding the ITGB2 gene	Rocket Pharmaceuticals, Inc.	I/II
2017-001366-14	TDT	Observational: long-term follow-up of patients treated with GSK2696277	GlaxoSmithKline Research and Development	II
2017-002430-23	Hurler syndrome	Autologous CD34$^+$ cells transduced with IDUA LV encoding the human α-L-iduronidase gene	Ospedale San Raffaele	I/II
2018-001404-11	Glioblastoma multiforme	Autologous CD34$^+$ cells transduced with LV encoding the interferon-α2 gene	Genenta Science S.r.l	I/IIa
2013-002245-11	Hemoglobinopathies	Observational: long-term follow-up of patients treated with LentiGlobin BB305 Drug Product	bluebird bio, Inc.	III

Data taken from www.clinicaltrials.gov and www.clinicaltrialsregister.eu (both accessed on 31 May 2024) for recruiting and ongoing (respectively) clinical trials based on the search term 'haematopoietic stem cell' AND 'gene therapy'. N/A: not applicable.

Finally, advancements in gene editing technologies with designer nucleases offer exciting opportunities for precise genome modification in HSCs, paving the way for personalized and curative treatments. In this context, over the last decade, nanoparticles (NPs) have also emerged as an attractive tool to deliver therapeutic agents with sharp specificity and versatility. In particular, NPs could be equipped with targeting motifs specific to HSCs and can potentially overcome the need for ex vivo manipulation of patient HSCs [63]. Among the investigated NPs, lipid nanoparticles (LNPs) have been reported to substantially decrease electroporation-induced toxicity and to generate higher yields of edited cells [64].

By leveraging emerging technologies and prioritizing early development of options that will feasibly become drugs which are economically competitive compared to the standard of care treatment, further products will reach the market to bridge the gap between bench research and patient's bedside.

To conclude, toward safer gene therapies for blood disorders, the focus should be on the potency of the graft both in terms of engraftment and clonogenic capacity and in terms of functionality, i.e., its ability to produce a functional amount of the therapeutic protein.

As for any other cell product manufactured by genetic modification, the HSC-engineered grafts have to comply with the overall EMA/FDA regulation for ATMPs, which pose a long-term follow-up for at least 15 years to monitor the persistence of their clonal dynamics to detect potential adverse events (e.g., leukemogenesis or blood dysplasias) as a result of clonal hematopoiesis due to the engineering approach (vector transduction or genome editing). Therefore, risk mitigation performed both by applying safety strategies during manufacturing (SIN vectors, optimized ex vivo protocols, use of transduction enhancers, etc.) and by implementing careful monitoring and long-term pharmacovigilance measures has the potential to promote gene therapy approaches to medical routine.

Finally, further attempts should aim toward alleviating the "financial" toxicities which currently severely limit the wider applicability of these therapies.

Author Contributions: Conceptualization, E.P. and A.G.; writing—original draft preparation A.G.; writing—review and editing, E.P. All authors have read and agreed to the published version of the manuscript.

Funding: Funded by the European Union. Views and opinions expressed are, however, those of the author(s) only and do not necessarily reflect those of the European Union or the European Research Executive Agency. Neither the European Union nor the European Research Executive Agency can be held responsible for them.

Conflicts of Interest: E.P. and A.G. are employees of Miltenyi Biotec.

Abbreviations

ADA-SCID, severe combined immunodeficiency due to adenosine deaminase deficiency; ALD, adrenoleukodystrophy; CALD, cerebral adrenoleukodystrophy; ATMPs, advanced therapy medicinal products; BaEV, baboon endogenous retrovirus; Cas9, CRISPR-associated protein 9; CRISPR, clustered regularly interspaced short palindromic repeats; CsA, cyclosporin A; CsH, cyclosporin H; EMA, European Medicines Agency; FDA, U.S. Food and Drug Administration; γRV, gammaretrovirus; GvHD, graft-versus-host disease; HDR, homology-directed repair; HIV, human immunodeficiency virus; HLA, human leukocyte antigen; HSCs, hematopoietic stem cells; HSCT, hematopoietic stem cell transplantation; IVIM, in vitro immortalization; LV, lentivirus; MLD, metachromatic leukodystrophy; MOI, multiplicity of infection; NOD/SCID, nonobese diabetic/severe combined immunodeficiency; NSG, NOD SCID gamma; PGE2, prostaglandin E2; SAGA, surrogate assay for genotoxicity assessment; SCD, sickle cell disease; TDT, transfusion-dependent β-thalassemia; VCN, vector copy number; VSV-G, vesicular stomatitis virus G; WAS, Wiskott-Aldrich syndrome; X-CGD, X-linked chronic granulomatous disease; X-SCID, X-linked severe combined immunodeficiency.

References

1. Gatti, R.A.; Meuwissen, H.J.; Allen, H.D.; Hong, R.; Good, R.A. Immunological reconstitution of sex-linked lymphopenic immunological deficiency. *Lancet* **1968**, *292*, 1366–1369. [CrossRef] [PubMed]
2. Bach, F.H.; Albertini, R.J.; Joo, P.; Anderson, J.L.; Bortin, M.M. Bone-marrow transplantation in a patient with the Wiskott-Aldrich syndrome. *Lancet* **1968**, *2*, 1364–1366. [CrossRef] [PubMed]
3. Prezioso, L.; Manfra, I.; Bonomini, S.; Schifano, C.; Segreto, R.; Monti, A.; Sammarelli, G.; Todaro, G.; Sassi, M.; Bertaggia, I.; et al. Haploidentical hematopoietic stem cell transplantation in adults using the αβTCR/CD19-based depletion of G-CSF-mobilized peripheral blood progenitor cells. *Bone Marrow Transplant.* **2019**, *54*, 698–702. [CrossRef] [PubMed]
4. Hatzimichael, E.; Tuthill, M. Hematopoietic stem cell transplantation. *Stem Cells Cloning* **2010**, *3*, 105–117. [CrossRef] [PubMed]
5. Hacein-Bey-Abina, S.; Garrigue, A.; Wang, G.P.; Soulier, J.; Lim, A.; Morillon, E.; Clappier, E.; Caccavelli, L.; Delabesse, E.; Beldjord, K.; et al. Insertional oncogenesis in 4 patients after retrovirus-mediated gene therapy of SCID-X1. *J. Clin. Investig.* **2008**, *118*, 3132–3142. [CrossRef] [PubMed]
6. Biffi, A.; Bartolomae, C.C.; Cesana, D.; Cartier, N.; Aubourg, P.; Ranzani, M.; Cesani, M.; Benedicenti, F.; Plati, T.; Rubagotti, E.; et al. Lentiviral vector common integration sites in preclinical models and a clinical trial reflect a benign integration bias and not oncogenic selection. *Blood* **2011**, *117*, 5332–5339. [CrossRef] [PubMed]
7. Shirley, J.L.; de Jong, Y.P.; Terhorst, C.; Herzog, R.W. Immune Responses to Viral Gene Therapy Vectors. *Mol. Ther.* **2020**, *28*, 709–722. [CrossRef] [PubMed]
8. Nathwani, A.C.; Hanawa, H.; Vandergriff, J.; Kelly, P.; Vanin, E.F.; Nienhuis, A.W. Efficient gene transfer into human cord blood CD34+ cells and the $CD34^{+}CD38^{-}$ subset using highly purified recombinant adeno-associated viral vector preparations that are free of helper virus and wild-type AAV. *Gene Ther.* **2000**, *7*, 183–195. [CrossRef]
9. Cicalese, M.P.; Ferrua, F.; Castagnaro, L.; Pajno, R.; Barzaghi, F.; Giannelli, S.; Dionisio, F.; Brigida, I.; Bonopane, M.; Casiraghi, M.; et al. Update on the safety and efficacy of retroviral gene therapy for immunodeficiency due to adenosine deaminase deficiency. *Blood* **2016**, *128*, 45–54. [CrossRef]
10. Locatelli, F.; Thompson, A.A.; Kwiatkowski, J.L.; Porter, J.B.; Thrasher, A.J.; Hongeng, S.; Sauer, M.G.; Thuret, I.; Lal, A.; Algeri, M.; et al. Betibeglogene Autotemcel Gene Therapy for Non-β^0/β^0 Genotype β-Thalassemia. *N. Engl. J. Med.* **2022**, *386*, 415–427. [CrossRef]
11. Magrin, E.; Semeraro, M.; Hebert, N.; Joseph, L.; Magnani, A.; Chalumeau, A.; Gabrion, A.; Roudaut, C.; Marouene, J.; Lefrere, F.; et al. Long-term outcomes of lentiviral gene therapy for the β-hemoglobinopathies: The HGB-205 trial. *Nat. Med.* **2022**, *28*, 81–88. [CrossRef] [PubMed]
12. Schimmer, J.; Breazzano, S. Investor Outlook: Rising from the Ashes; GSK's European Approval of Strimvelis for ADA-SCID. *Hum. Gene Ther. Clin. Dev.* **2016**, *27*, 57–61. [CrossRef]
13. Horgan, C.; Watts, K.; Ram, D.; Rust, S.; Hutton, R.; Jones, S.; Wynn, R. A retrospective cohort study of Libmeldy (atidarsagene autotemcel) for MLD: What we have accomplished and what opportunities lie ahead. *JIMD Rep.* **2023**, *64*, 346–352. [CrossRef]
14. Schuessler-Lenz, M.; Enzmann, H.; Vamvakas, S. Regulators' Advice Can Make a Difference: European Medicines Agency Approval of Zynteglo for Beta Thalassemia. *Clin. Pharmacol. Ther.* **2020**, *107*, 492–494. [CrossRef]
15. Zeng, J.; Nguyen, M.A.; Liu, P.; Ferreira da Silva, L.; Lin, L.Y.; Justus, D.G.; Petri, K.; Clement, K.; Porter, S.N.; Verma, A.; et al. Gene editing without ex vivo culture evades genotoxicity in human hematopoietic stem cells. *bioRxiv* **2023**. [CrossRef]
16. Lee, B.C.; Gin, A.; Wu, C.; Singh, K.; Grice, M.; Mortlock, R.; Abraham, D.; Fan, X.; Zhou, Y.; AlJanahi, A.; et al. Impact of CRISPR/HDR editing versus lentiviral transduction on long-term engraftment and clonal dynamics of HSPCs in rhesus macaques. *Cell Stem Cell* **2024**, *31*, 455–466. [CrossRef]

17. Finkelshtein, D.; Werman, A.; Novick, D.; Barak, S.; Rubinstein, M. LDL receptor and its family members serve as the cellular receptors for vesicular stomatitis virus. *Proc. Natl. Acad. Sci. USA* **2013**, *110*, 7306–7311. [CrossRef] [PubMed]
18. Amirache, F.; Lévy, C.; Costa, C.; Mangeot, P.E.; Torbett, B.E.; Wang, C.X.; Nègre, D.; Cosset, F.L.; Verhoeyen, E. Mystery solved: VSV-G-LVs do not allow efficient gene transfer into unstimulated T cells, B cells, and HSCs because they lack the LDL receptor. *Blood* **2014**, *123*, 1422–1424. [CrossRef] [PubMed]
19. Girard-Gagnepain, A.; Amirache, F.; Costa, C.; Lévy, C.; Frecha, C.; Fusil, F.; Nègre, D.; Lavillette, D.; Cosset, F.L.; Verhoeyen, E. Baboon envelope pseudotyped LVs outperform VSV-G-LVs for gene transfer into early-cytokine-stimulated and resting HSCs. *Blood* **2014**, *124*, 1221–1231. [CrossRef]
20. Verhoeyen, E.; Wiznerowicz, M.; Olivier, D.; Izac, B.; Trono, D.; Dubart-Kupperschmitt, A.; Cosset, F.L. Novel lentiviral vectors displaying "early-acting cytokines" selectively promote survival and transduction of NOD/SCID repopulating human hematopoietic stem cells. *Blood* **2005**, *106*, 3386–3395. [CrossRef]
21. Glimm, H.; Oh, I.H.; Eaves, C.J. Human hematopoietic stem cells stimulated to proliferate in vitro lose engraftment potential during their $S/G_2/M$ transit and do not reenter G_0. *Blood* **2000**, *96*, 4185–4193. [CrossRef]
22. Valeri, E.; Unali, G.; Piras, F.; Abou-Alezz, M.; Pais, G.; Benedicenti, F.; Lidonnici, M.R.; Cuccovillo, I.; Castiglioni, I.; Arévalo, S.; et al. Removal of innate immune barriers allows efficient transduction of quiescent human hematopoietic stem cells. *Mol. Ther.* **2024**, *32*, 124–139. [CrossRef] [PubMed]
23. Heffner, G.C.; Bonner, M.; Christiansen, L.; Pierciey, F.J.; Campbell, D.; Smurnyy, Y.; Zhang, W.; Hamel, A.; Shaw, S.; Lewis, G.; et al. Prostaglandin E_2 Increases Lentiviral Vector Transduction Efficiency of Adult Human Hematopoietic Stem and Progenitor Cells. *Mol. Ther.* **2018**, *26*, 320–328. [CrossRef] [PubMed]
24. Zonari, E.; Desantis, G.; Petrillo, C.; Boccalatte, F.E.; Lidonnici, M.R.; Kajaste-Rudnitski, A.; Aiuti, A.; Ferrari, G.; Naldini, L.; Gentner, B. Efficient Ex Vivo Engineering and Expansion of Highly Purified Human Hematopoietic Stem and Progenitor Cell Populations for Gene Therapy. *Stem Cell Rep.* **2017**, *8*, 977–990. [CrossRef] [PubMed]
25. Uchida, N.; Nassehi, T.; Drysdale, C.M.; Gamer, J.; Yapundich, M.; Demirci, S.; Haro-Mora, J.J.; Leonard, A.; Hsieh, M.M.; Tisdale, J.F. High-Efficiency Lentiviral Transduction of Human $CD34^+$ Cells in High-Density Culture with Poloxamer and Prostaglandin E2. *Mol. Ther. Methods Clin. Dev.* **2019**, *13*, 187–196. [CrossRef]
26. Masiuk, K.E.; Zhang, R.; Osborne, K.; Hollis, R.P.; Campo-Fernandez, B.; Kohn, D.B. PGE2 and Poloxamer Synperonic F108 Enhance Transduction of Human HSPCs with a β-Globin Lentiviral Vector. *Mol. Ther. Methods Clin. Dev.* **2019**, *13*, 390–398. [CrossRef]
27. Jang, Y.; Kim, Y.S.; Wielgosz, M.M.; Ferrara, F.; Ma, Z.; Condori, J.; Palmer, L.E.; Zhao, X.; Kang, G.; Rawlings, D.J.; et al. Optimizing lentiviral vector transduction of hematopoietic stem cells for gene therapy. *Gene Ther.* **2020**, *27*, 545–556. [CrossRef]
28. Poletti, V.; Montepeloso, A.; Pellin, D.; Biffi, A. Prostaglandin E2 as transduction enhancer affects competitive engraftment of human hematopoietic stem and progenitor cells. *Mol. Ther. Methods Clin. Dev.* **2023**, *31*, 101131. [CrossRef]
29. Cutler, C.; Multani, P.; Robbins, D.; Kim, H.T.; Le, T.; Hoggatt, J.; Pelus, L.M.; Desponts, C.; Chen, Y.B.; Rezner, B.; et al. Prostaglandin-modulated umbilical cord blood hematopoietic stem cell transplantation. *Blood* **2013**, *122*, 3074–3081. [CrossRef]
30. Gentner, B.; Tucci, F.; Galimberti, S.; Fumagalli, F.; De Pellegrin, M.; Silvani, P.; Camesasca, C.; Pontesilli, S.; Darin, S.; Ciotti, F.; et al. Hematopoietic Stem- and Progenitor-Cell Gene Therapy for Hurler Syndrome. *N. Engl. J. Med.* **2021**, *385*, 1929–1940. [CrossRef]
31. Leonard, A.; Tisdale, J.F. A pause in gene therapy: Reflecting on the unique challenges of sickle cell disease. *Mol. Ther.* **2021**, *29*, 1355–1356. [CrossRef] [PubMed]
32. Jones, R.J.; DeBaun, M.R. Leukemia after gene therapy for sickle cell disease: Insertional mutagenesis, busulfan, both, or neither. *Blood* **2021**, *138*, 942–947. [CrossRef] [PubMed]
33. Parums, D.V. Editorial: First Regulatory Approvals for CRISPR-Cas9 Therapeutic Gene Editing for Sickle Cell Disease and Transfusion-Dependent β-Thalassemia. *Med. Sci. Monit.* **2024**, *30*, e944204. [CrossRef] [PubMed]
34. Davis, H.E.; Rosinski, M.; Morgan, J.R.; Yarmush, M.L. Charged polymers modulate retrovirus transduction via membrane charge neutralization and virus aggregation. *Biophys. J.* **2004**, *86*, 1234–1242. [CrossRef] [PubMed]
35. Cornetta, K.; Anderson, W.F. Protamine sulfate as an effective alternative to polybrene in retroviral-mediated gene-transfer: Implications for human gene therapy. *J. Virol. Methods* **1989**, *23*, 187–194. [CrossRef] [PubMed]
36. Anastasov, N.; Höfig, I.; Mall, S.; Krackhardt, A.M.; Thirion, C. Optimized Lentiviral Transduction Protocols by Use of a Poloxamer Enhancer, Spinoculation, and scFv-Antibody Fusions to VSV-G. *Methods Mol. Biol.* **2016**, *1448*, 49–61. [CrossRef] [PubMed]
37. Hauber, I.; Beschorner, N.; Schrödel, S.; Chemnitz, J.; Kröger, N.; Hauber, J.; Thirion, C. Improving Lentiviral Transduction of $CD34^+$ Hematopoietic Stem and Progenitor Cells. *Hum. Gene Ther. Methods* **2018**, *29*, 104–113. [CrossRef] [PubMed]
38. Schott, J.W.; León-Rico, D.; Ferreira, C.B.; Buckland, K.F.; Santilli, G.; Armant, M.A.; Schambach, A.; Cavazza, A.; Thrasher, A.J. Enhancing Lentiviral and Alpharetroviral Transduction of Human Hematopoietic Stem Cells for Clinical Application. *Mol. Ther. Methods Clin. Dev.* **2019**, *14*, 134–147. [CrossRef]
39. Delville, M.; Soheili, T.; Bellier, F.; Durand, A.; Denis, A.; Lagresle-Peyrou, C.; Cavazzana, M.; Andre-Schmutz, I.; Six, E. A Nontoxic Transduction Enhancer Enables Highly Efficient Lentiviral Transduction of Primary Murine T Cells and Hematopoietic Stem Cells. *Mol. Ther. Methods Clin. Dev.* **2018**, *10*, 341–347. [CrossRef]

40. Fenard, D.; Ingrao, D.; Seye, A.; Buisset, J.; Genries, S.; Martin, S.; Kichler, A.; Galy, A. Vectofusin-1, a new viral entry enhancer, strongly promotes lentiviral transduction of human hematopoietic stem cells. *Mol. Ther. Nucleic Acids* **2013**, *2*, e90. [CrossRef]
41. Radek, C.; Bernadin, O.; Drechsel, K.; Cordes, N.; Pfeifer, R.; Sträßer, P.; Mormin, M.; Gutierrez-Guerrero, A.; Cosset, F.L.; Kaiser, A.D.; et al. Vectofusin-1 Improves Transduction of Primary Human Cells with Diverse Retroviral and Lentiviral Pseudotypes, Enabling Robust, Automated Closed-System Manufacturing. *Hum. Gene Ther.* **2019**, *30*, 1477–1493. [CrossRef]
42. Piovan, C.; Marin, V.; Scavullo, C.; Corna, S.; Giuliani, E.; Bossi, S.; Galy, A.; Fenard, D.; Bordignon, C.; Rizzardi, G.P.; et al. Vectofusin-1 Promotes RD114-TR-Pseudotyped Lentiviral Vector Transduction of Human HSPCs and T Lymphocytes. *Mol. Ther. Methods Clin. Dev.* **2017**, *5*, 22–30. [CrossRef] [PubMed]
43. Wang, C.X.; Sather, B.D.; Wang, X.; Adair, J.; Khan, I.; Singh, S.; Lang, S.; Adams, A.; Curinga, G.; Kiem, H.P.; et al. Rapamycin relieves lentiviral vector transduction resistance in human and mouse hematopoietic stem cells. *Blood* **2014**, *124*, 913–923. [CrossRef]
44. Luo, Y.; Li, L.; Zou, P.; Wang, J.; Shao, L.; Zhou, D.; Liu, L. Rapamycin enhances long-term hematopoietic reconstitution of ex vivo expanded mouse hematopoietic stem cells by inhibiting senescence. *Transplantation* **2014**, *97*, 20–29. [CrossRef] [PubMed]
45. Petrillo, C.; Cesana, D.; Piras, F.; Bartolaccini, S.; Naldini, L.; Montini, E.; Kajaste-Rudnitski, A. Cyclosporin a and rapamycin relieve distinct lentiviral restriction blocks in hematopoietic stem and progenitor cells. *Mol. Ther.* **2015**, *23*, 352–362. [CrossRef] [PubMed]
46. Evans, M.E.; Kumkhaek, C.; Hsieh, M.M.; Donahue, R.E.; Tisdale, J.F.; Uchida, N. TRIM5α variations influence transduction efficiency with lentiviral vectors in both human and rhesus $CD34^+$ cells in vitro and in vivo. *Mol. Ther.* **2014**, *22*, 348–358. [CrossRef]
47. Petrillo, C.; Thorne, L.G.; Unali, G.; Schiroli, G.; Giordano, A.M.S.; Piras, F.; Cuccovillo, I.; Petit, S.J.; Ahsan, F.; Noursadeghi, M.; et al. Cyclosporine H Overcomes Innate Immune Restrictions to Improve Lentiviral Transduction and Gene Editing in Human Hematopoietic Stem Cells. *Cell Stem Cell* **2018**, *23*, 820–832.e9. [CrossRef] [PubMed]
48. Olender, L.; Bujanover, N.; Sharabi, O.; Goldstein, O.; Gazit, R. Cyclosporine H Improves the Multi-Vector Lentiviral Transduction of Murine Haematopoietic Progenitors and Stem Cells. *Sci. Rep.* **2020**, *10*, 1812. [CrossRef]
49. Hacein-Bey-Abina, S.; von Kalle, C.; Schmidt, M.; Le Deist, F.; Wulffraat, N.; McIntyre, E.; Radford, I.; Villeval, J.L.; Fraser, C.C.; Cavazzana-Calvo, M.; et al. A serious adverse event after successful gene therapy for X-linked severe combined immunodeficiency. *N. Engl. J. Med.* **2003**, *348*, 255–256. [CrossRef]
50. Marshall, E. Gene therapy. Second child in French trial is found to have leukemia. *Science* **2003**, *299*, 320. [CrossRef]
51. Kaiser, J. RAC Hears a Plea for Resuming Trials, Despite Cancer Risk. *Science* **2003**, *299*, 991. Available online: https://www.jstor.org/stable/i371100 (accessed on 15 April 2024). [CrossRef] [PubMed]
52. Ott, M.G.; Schmidt, M.; Schwarzwaelder, K.; Stein, S.; Siler, U.; Koehl, U.; Glimm, H.; Kühlcke, K.; Schilz, A.; Kunkel, H.; et al. Correction of X-linked chronic granulomatous disease by gene therapy, augmented by insertional activation of MDS1-EVI1, PRDM16 or SETBP1. *Nat. Med.* **2006**, *12*, 401–409. [CrossRef] [PubMed]
53. Boztug, K.; Schmidt, M.; Schwarzer, A.; Banerjee, P.P.; Díez, I.A.; Dewey, R.A.; Böhm, M.; Nowrouzi, A.; Ball, C.R.; Glimm, H.; et al. Stem-cell gene therapy for the Wiskott-Aldrich syndrome. *N. Engl. J. Med.* **2010**, *363*, 1918–1927. [CrossRef] [PubMed]
54. Bastone, A.L.; Dziadek, V.; John-Neek, P.; Mansel, F.; Fleischauer, J.; Agyeman-Duah, E.; Schaudien, D.; Dittrich-Breiholz, O.; Schwarzer, A.; Schambach, A.; et al. Development of an in vitro genotoxicity assay to detect retroviral vector-induced lymphoid insertional mutants. *Mol. Ther. Methods Clin. Dev.* **2023**, *30*, 515–533. [CrossRef] [PubMed]
55. Modlich, U.; Bohne, J.; Schmidt, M.; von Kalle, C.; Knöss, S.; Schambach, A.; Baum, C. Cell-culture assays reveal the importance of retroviral vector design for insertional genotoxicity. *Blood* **2006**, *108*, 2545–2553. [CrossRef] [PubMed]
56. Modlich, U.; Navarro, S.; Zychlinski, D.; Maetzig, T.; Knoess, S.; Brugman, M.H.; Schambach, A.; Charrier, S.; Galy, A.; Thrasher, A.J.; et al. Insertional transformation of hematopoietic cells by self-inactivating lentiviral and gammaretroviral vectors. *Mol. Ther.* **2009**, *17*, 1919–1928. [CrossRef] [PubMed]
57. Braun, C.J.; Boztug, K.; Paruzynski, A.; Witzel, M.; Schwarzer, A.; Rothe, M.; Modlich, U.; Beier, R.; Göhring, G.; Steinemann, D.; et al. Gene therapy for Wiskott-Aldrich syndrome—Long-term efficacy and genotoxicity. *Sci. Transl. Med.* **2014**, *6*, 227. [CrossRef] [PubMed]
58. Schwarzer, A.; Talbot, S.R.; Selich, A.; Morgan, M.; Schott, J.W.; Dittrich-Breiholz, O.; Bastone, A.L.; Weigel, B.; Ha, T.C.; Dziadek, V.; et al. Predicting genotoxicity of viral vectors for stem cell gene therapy using gene expression-based machine learning. *Mol. Ther.* **2021**, *29*, 3383–3397. [CrossRef] [PubMed]
59. Kosicki, M.; Tomberg, K.; Bradley, A. Repair of double-strand breaks induced by CRISPR-Cas9 leads to large deletions and complex rearrangements. *Nat. Biotechnol.* **2018**, *36*, 765–771. [CrossRef]
60. Haapaniemi, E.; Botla, S.; Persson, J.; Schmierer, B.; Taipale, J. CRISPR-Cas9 genome editing induces a p53-mediated DNA damage response. *Nat. Med.* **2018**, *24*, 927–930. [CrossRef]
61. Charlesworth, C.T.; Hsu, I.; Wilkinson, A.C.; Nakauchi, H. Immunological barriers to haematopoietic stem cell gene therapy. *Nat. Rev. Immunol.* **2022**, *22*, 719–733. [CrossRef] [PubMed]
62. Dunleavy, K. With the Pricing Situation 'Untenable' in Europe, Bluebird Will Wind down Its Operations in the 'Broken' Market. Fierce Pharma. 2021. Available online: https://www.fiercepharma.com/pharma/situation-untenable-bluebird-will-wind-down-its-operations-broken-europe (accessed on 16 April 2024).

63. Cruz, L.J.; Rezaei, S.; Grosveld, F.; Philipsen, S.; Eich, C. Nanoparticles targeting hematopoietic stem and progenitor cells: Multimodal carriers for the treatment of hematological diseases. *Front. Genome Ed.* **2022**, *4*, 1030285. [CrossRef] [PubMed]
64. Vavassori, V.; Ferrari, S.; Beretta, S.; Asperti, C.; Albano, L.; Annoni, A.; Gaddoni, C.; Varesi, A.; Soldi, M.; Cuomo, A.; et al. Lipid nanoparticles allow efficient and harmless ex vivo gene editing of human hematopoietic cells. *Blood* **2023**, *142*, 812–826. [CrossRef] [PubMed]

cells

MDPI

Review

Engineering Liver-Specific Promoters: A Comprehensive Review of Design, Mechanisms, and Clinical Applications in Gene Therapy

Valentin Artemyev [1,*], Anastasiia Iu. Paremskaia [2], Amina A. Dzhioeva [1], Daria Mishina [2], Viktor Bogdanov [2], Julia Krupinova [2,3], Ali Mazloum [2], Sofya G. Feoktistova [2], Olga N. Mityaeva [2] and Pavel Yu. Volchkov [2,3]

1 Moscow Center for Advanced Studies, Kulakova Str. 20, Moscow 123592, Russia; dzhioeva.aa@genlab.llc
2 Federal Research Center for Innovator and Emerging Biomedical and Pharmaceutical Technologies, Moscow 125315, Russia; ne.w.ay1357@gmail.com (A.I.P.); mishina.dm@genlab.llc (D.M.); bogdanov@genlab.llc (V.B.); krupinova_ua@academpharm.ru (J.K.); mazlum.a@genlab.llc (A.M.); feoktistova_sg@academpharm.ru (S.G.F.); mityaeva_on@academpharm.ru (O.N.M.); volchkov_py@academpharm.ru (P.Y.V.)
3 Moscow Clinical Scientific Center N.A. A.S. Loginov, Moscow 111123, Russia
* Correspondence: artemev.vv@genlab.llc

Abstract

The liver is a primary metabolic hub and a pivotal target for gene therapy, owing to its capacity for protein secretion, role in metabolic homeostasis and immune tolerance. Liver-directed gene therapies are used to treat numerous inherited metabolic disorders and coagulation factor deficiencies including hemophilia (A and B), Crigler–Najjar syndrome, mucopolysaccharidoses, phenylketonuria, Fabry, Gaucher, Wilson and Pompe diseases. The efficacy and safety of liver-directed gene therapy rely on the use of strong tissue-specific promoters. To date, there are many different liver-specific promoters used in preclinical and clinical studies, including novel completely synthetic promoters. This review provides a comprehensive analysis of the design, engineering and application of liver-specific promoters. Furthermore, we discuss fundamental principles of gene expression regulation in the liver and the physiological and immunological characteristics that make it a suitable target organ for gene therapy delivery.

Keywords: liver-specific promoters; synthetic promoters; promoters design; gene regulation; gene therapy; viral vectors; immune privilege

1. Introduction

The liver is a primary metabolic center in the body, regulating vital processes including metabolism (amino acids, carbohydrates, lipids, hormones, vitamins, etc.), detoxification (including xenobiotics) and the synthesis of bile, and other vital compounds. It is also responsible for the synthesis of a large number of proteins, many of which are secreted into the blood, where they mediate processes such as blood coagulation, lipid transport, and others. Thus, proteins synthesized in the liver can serve either purely intrahepatic functions (e.g., alpha1-antitrypsin, fumaroylacetoacetate hydrolase) or extrahepatic functions (e.g., factors VIII and IX) [1].

Hereditary metabolic disorders constitute a heterogeneous group of genetic diseases that primarily manifest in childhood, with a prevalence of approximately 1 in 784 newborns [2]. Many hereditary diseases, including hemophilia, hereditary angioedema, and alpha-1 antitrypsin deficiency (AATD), are associated with the secretion of defective proteins in hepatocytes, making the liver a pivotal target for treatment [3]. Metabolic diseases that require liver-targeted therapy are usually split into two main groups. The first group includes disorders associated with protein secretion in hepatocytes that indirectly affect the whole body, such as glycogen storage disease Ia, acute intermittent porphyria, Crigler–Najjar syndrome, ornithine transcarbamylase deficiency, Wilson's disease, and others. The second group includes diseases with dysfunction of protein synthesis affected by multiple organs at once. Restoring expression of these proteins in the liver is expected to significantly attenuate disease severity. Such diseases include Fabry disease, type II glycogen storage disease, mucopolysaccharidosis types I, II, IIIA and VI [3,4].

Numerous treatment approaches have been developed for liver diseases, including cell-based, pharmaceutical, and gene therapies [3,5,6]. This review discusses in detail the design and engineering of liver-specific promoters for the development of liver-directed gene therapies for hereditary diseases, emphasizing the liver's characteristics as a target organ, the mechanisms of gene expression regulation, and the origin and application of the main promoter systems used in liver-directed gene therapy.

2. Liver as a Target for Gene Therapy

Hepatocytes are the parenchymal cells of the liver representing over 80% of the liver's volume [7]. Non-parenchymal cells represent approximately 6% of the liver's volume and comprise hepatic macrophages (2.1%, Kupffer cells), sinusoidal endothelial cells (2.8%), hepatic stellate cells (1.5%, Ito cells), natural killers (NK) cells, cells residing in the Disse space and others [8]. Most gene therapies targeting the liver are specifically directed at hepatocytes, as they are responsible for the synthesis of the majority of proteins secreted by the liver. Even when a protein is expressed by other cell types, targeting hepatocytes is preferable due to their quantitative advantage. For example, coagulation factor VIII (FVIII) is normally expressed by liver sinusoidal endothelial cells, which are less abundant and challenging to target with vectors; therefore, gene therapy for hemophilia A primarily targets hepatocytes [9]. Thus, we refer to targeting hepatocytes in the further discussion of liver-directed gene therapies. In the following sections, we will examine the characteristics of the liver and hepatocytes as targets for gene therapies.

2.1. Vector Size Limitations

As a highly vascularized organ, the liver receives a large volume of blood from throughout the body, enabling viral particles introduced into the bloodstream to rapidly reach hepatic cells [10]. This fact allows the use of various vector administration methods including systemic delivery (via the tail vein), portal vein injection, and direct intraparenchymal administration, etc. [11–13]. However, the liver's capillaries are lined with sinusoidal endothelial cells that contain small pores called fenestrae, which act as a filter permitting only particles smaller than the fenestrae to access parenchymal cells [14]. Consequently, there are limitations on the size of delivery vectors, especially large ones such as adenoviruses (approximately 93 nm with 30 nm fibers), when delivered to the liver via the bloodstream, depending on the model organism and liver region. The fenestrae size of human and mouse sinusoidal cells is about 100 nm (50–300 nm), and that of rabbits is 60 nm [15,16].

2.2. Low Hepatocyte Proliferation Rate—An Advantage and an Obstacle for Vectors

Early studies established that resection of two-thirds of hepatocytes leads to a maximal regeneration rate, which then gradually declines as regeneration proceeds [17]. Despite the well-known regenerative capacity of the liver, differentiated hepatocytes in a healthy liver are long-lived cells with low basal DNA synthesis, where approximately only 1 in

https://doi.org/10.3390/cells15010014

20,000 cells are undergoing mitosis at any given time [18]. Consequently, liver cells are typically in a quiescent state in vivo, which hinders nuclear entry and integration of certain retroviruses, such as Moloney's murine leukemia virus (MoMuLV), that cannot pass through nuclear pores [19]. In some cases, liver regeneration can be induced by partial hepatectomy, but this is rarely used in gene therapy. In contrast, adeno-associated viral (AAV) vectors efficiently transduce both dividing and non-dividing cells. Therefore, the low rate of hepatocyte proliferation makes the liver a suitable target for long-term AAV-mediated gene therapy, supporting persistent and stable transgene expression [4]. An exception is found in rapidly proliferating hepatocytes during liver development. In the developing mouse liver, transgene expression declines rapidly, with stable residual expression after 2 weeks in only 4–8% of hepatocytes [20]. Thus, the low proliferation rate of hepatocytes simultaneously poses a barrier for viral vectors unable to integrate into the nucleus of non-dividing cells, yet this represents a key advantage for vectors capable of efficiently transducing quiescent cells.

2.3. Immune System Barriers to Liver-Directed Gene Therapy

As previously mentioned, the liver contains a large number of resident immune cells, and 90% of all resident macrophages in the body are widely represented in the liver Kupffer cells [21]. The presence of Kupffer cells in the liver poses an additional obstacle to gene therapy due to the substantial uptake of delivery vectors by these macrophages. Several studies report methods to overcome this barrier by the use of liposomes to deplete Kupffer cells [22]. Another significant problem with liver-targeted gene therapy, as with other organs, is the development of an immune response to both the expressed transgene and the delivery vector. Pre-existing immunity to delivery vectors, particularly to viral capsids, is frequently encountered Transgene expression can be significantly reduced as a result of capsid neutralization by antibodies and activation of CD8+ cytotoxic T-cells that attack transduced cells [23,24]. To overcome the neutralization of viral vectors, modifications of the capsid protein amino acid sequence and chemical modification are currently being developed [25,26]. Some recent studies have employed a novel approach to reduce the neutralization of viral particles by antibodies, which involves encapsulating AAV in a lipid nanoparticle that imitates natural enveloped viruses, helping to evade the immune response [27,28]. To mitigate the immune response mediated by activation of CD8+ cytotoxic T cells against viral capsids, it is essential to maximize transgene expression while minimizing the vector dose. Various approaches are currently used to improve the efficiency of vector-mediated transgene expression, including codon optimization of the transgene, creation of strong promoters, use of self-complementary adeno-associated viral vectors (scAAV), and transgene variants with greater activity [29–31]. As a general principle, the immune response to transgenes occurs more frequently in genotypes with protein reading frame alteration due to deletions, nonsense mutations, gene inversions, etc., whereas genotypes with missense mutations usually do not trigger an immune response and are characterized by more stable expression [32,33]. The use of a more efficient promoter leads to increased expression of the functional product, allowing for a reduction in the vector dose. This strategy lowers the viral capsid load delivered to hepatocytes while maintaining high transgene expression levels [34]. The balance between CD8+ T cell-mediated clearance of AAV-transduced hepatocytes and immune tolerance is dose-dependent. Specifically, a higher viral antigen load can lead to more significant immune clearance. This is associated with a downregulation of T-cell negative checkpoint markers, e.g., the programmed death 1 receptor, and upregulated expression of relevant cytokines [35]. More detailed mechanisms of transgene expression suppression related to the immune response against the viral capsid are discussed in a dedicated review [36].

https://doi.org/10.3390/cells15010014

2.4. Liver Immune Privilege in Gene Therapy

Immune privilege refers to the status of an organ in which the presence of antigens does not trigger an inflammatory immune response. Typically, immune-privileged organs include the brain and central nervous system, eyes, and the pregnant uterus. In addition to limited regenerative capacity, features of immune privilege include restricted antigen drainage to lymph nodes, low expression of MHC I, the elimination of inflammatory cells that enter the organ (e.g., through the FasL pathway), and immune deviation (such as ACAID and BRAID) [37]. Recently, other organs and tissues, including the testes, liver, hair follicles, and even the intestinal mucosa, as well as tumors, are also considered immune-privileged [38]. The liver has a special form of immune privilege called hepatic tolerance, where liver transplantation leads to donor-specific T-cell tolerance [39]. This phenomenon was demonstrated in an early study on allogeneic liver transplantation in pigs without prior immunosuppression [40].

The liver serves a barrier function and produces numerous neoantigens; therefore, the risk of immune response activation in the liver is very high. To moderate this, the liver employs several key mechanisms of immune tolerance:

1. Non-parenchymal liver cells (including stellate cells and plasmacytoid dendritic cells) produce a large number of immunosuppressive anti-inflammatory cytokines, such as IL-10, TGF-beta [41];
2. Liver natural killer cells express a negative T-lymphocyte costimulatory–the programmed cell death ligand (PD-L1) [42];
3. Hepatocytes themselves also contribute to immune tolerance by producing PD-L1 [43].

Another aspect of the special liver immune privilege is induced immune tolerance to antigens introduced either directly into hepatocytes or systemically through the portal vein [39]. It is suggested that regulatory CD4+CD25+ Tregs and the tolerogenic properties of the liver, including the extensive expression of anti-inflammatory cytokines such as IL-10 and TGF-beta, play a central role in this mechanism. These cytokines suppress both antibody and T-cell immune responses to the endogenously expressed transgene [44,45]. Vector delivery to the liver results in lower titers of neutralizing antibodies compared to the previously common muscle-directed delivery. This is supported by studies in animal models of hemophilia. Blood coagulation factor IX (FIX) gene transfer to mice and dogs via intramuscular injection of AAV vectors led to rapid inhibitor formation and was effective only in combination with immunosuppression [46]. In another study, AAV-FIX delivery via the mesenteric and portal veins provided a lower level of neutralizing IgG production or its absence in some cases [47].

It was also demonstrated that AAV-mediated liver-directed gene therapy was successfully delivered to mice pre-immunized with FIX protein and thus possessing antibodies against it. This approach resulted in long-term disease correction and a dramatic reduction in antibody titers, even with the repeated protein presentation [9,48]. This strategy was also applied to other metabolic disorders and lysosomal storage diseases [49,50]. Currently, the immune tolerance induction to a transgene delivered to the liver is just beginning to be studied in clinical trials.

Successful induction of immune tolerance in animal models has enabled the first human trials to be conducted [48,51]. BMN-270 Phase I/II clinical trial by BioMarin Pharmaceutical was started in 2020 and subsequently showed promising preliminary results in the treatment hemophilia A patients with pre-existing immunity to coagulation factor VIII (FVIII) [52]. Two patients with a history of FVIII inhibitors and two patients with active inhibitors received AAV-mediated liver-directed gene therapy for endogenous FVIII expression. In patients with a history of FVIII inhibitors, no antibody reappearance occurred, and the treatment proved effective. In one patient with active FVIII inhibitors

https://doi.org/10.3390/cells15010014

a decrease in inhibitor titers was observed, along with an increase in FVIII concentration and activity in the blood by week 28 [53]. The therapy is currently known as Valoctocogene roxaparvovec (Roctavian). Thus, targeting transgene expression to the liver not only provides therapy but also induces specific immune tolerance to the transgene, which is particularly important for patients with pre-existing immunity. To sum up, liver-directed delivery is a promising approach to reduce the immune response to the transgene due to the special immune privilege of the liver as an organ.

3. Liver Gene Expression Regulation

3.1. Genomic Regulatory Elements

Gene expression is controlled by regulatory elements, including promoters, enhancers, insulators, and silencers, which can interact with transcription factors (TFs) and co-regulators [54,55]. In gene therapy, a promoter usually refers to a promoter construct–a combination of regulatory elements sufficient to drive transgene expression. Promoter activity is typically quantified as the level of gene expression in terms of mRNA and protein, with the assumption that higher mRNA/protein levels correspond to greater promoter activity. Increasing promoter activity is key to the effectiveness of liver gene therapies, so it is important to consider the mechanisms that regulate them. Increased transcription levels can be achieved by targeting regulatory elements that recruit transcription factors or cofactors and promote pre-initiator complex stabilization, accelerating transcription initiation [56].

In molecular biology, promoters are DNA regions located near transcription start sites (TSS), where transcription initiation occurs (Figure 1). In a natural eukaryotic promoter, two main regions are distinguished: the core promoter (minimal promoter) and the proximal promoter [54,57].

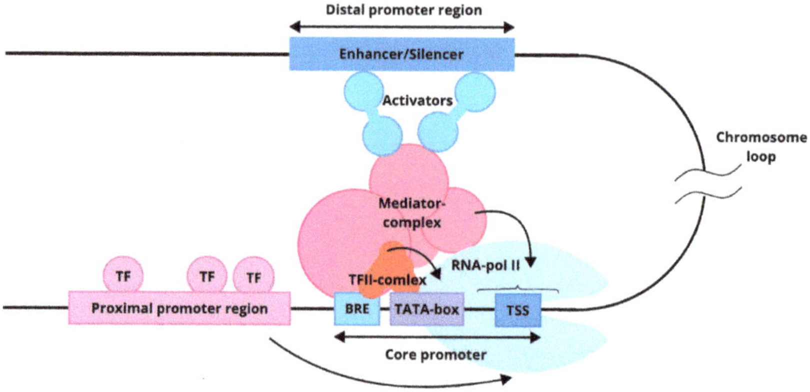

Figure 1. Functional model of a eukaryotic promoter. The core promoter recruits a set of general transcription factors (TFII complex) and RNA polymerase II. The proximal promoter region, composed of distinct regulatory elements, binds additional transcription factors that either enhance or suppress transcription. The distal region, represented by an enhancer or silencer, can be located several thousand nucleotides away from the core promoter. Arrows indicate the influence of the proximal and distal regions on the core promoter.

The core promoter is a set of sequences sufficient for assembling the pre-initiation complex, which includes RNA polymerase II and associated general transcription factors (GTFs) [57]. In vitro localization of the core promoter is sufficient to determine the TSS. The structure of a minimal promoter and the set of sequences it contains are diverse. Common and well-known regulatory sequences present in a core promoter include the Inr (initiator

https://doi.org/10.3390/cells15010014

sequence), TATA box, BRE (TFIIB recognition element), DPE, and others [58]. The binding of GTFs (TFIIA, TFIIB, TFIID, TFIIE, TFIIF, and TFIIH) enables the core promoter to initiate transcription, although it typically exhibits low basal activity [59]. Located upstream of the core promoter, the proximal promoter spans several hundred nucleotides and contains regulatory elements that bind transcription factors (TFs). These TFs can either repress or enhance transcription, thereby modulating core promoter activity [55].

The combination of regulatory elements and the availability of cell-type-specific TFs determine promoter specificity. Promoters are thus broadly categorized as tissue-specific or ubiquitous. Tissue-specific promoters, such as hAAT (human alpha-1 antitrypsin) (liver) or MCK (muscle), facilitate targeted expression in specific tissues or cells, and their function directly depends on the presence of TFs in the respective tissue. Ubiquitous promoters, such as cytomegalovirus (CMV) immediate early promoter, chicken β actin promoter, and promoter of ubiquitin C, drive widespread expression across all tissue [60]. Additionally, promoters are traditionally classified as either constitutive or inducible. Constitutive promoters maintain a constant level of gene transcription, independent of specific external signals. This type of regulation is typical for genes encoding proteins involved in fundamental cellular processes. The expression of genes under an inducible promoter depends on some external chemical or physical stimuli [61].

An enhancer is a distal regulatory region that modulates transcription levels from a promoter, irrespective of distance relative orientation, as their interaction is facilitated by the three-dimensional architecture of chromatin [62,63]. Enhancer sequences can be identified by chromatin modifications, including histone modifications [64] as well as by bidirectional transcription of enhancer RNA from these regions [54,65,66]. Enhancers increase gene expression through two complementary mechanisms involving chromatin architecture and specific motif organization: At the chromatin level, enhancers overcome the energetic barrier of nucleosomes through cooperative binding of multiple transcription factors with proper spacing, orientation, and positioning (the "enhancer grammar"), which recruit coactivator complexes (p300/CBP, SAGA, Mediator); these coactivators perform acetylation and methylation of histones (H3K27ac, H3K4me1) and recruit chromatin-remodeling factors that physically displace nucleosomes, opening DNA access for RNA polymerase II. The number of transcription factor motifs positively correlates with nucleosome eviction intensity and gene expression output, following either rigid "enhanceosome" architecture (with fixed spacing and orientation) or flexible "billboard" models (with variable motif arrangement), with most enhancers occupying a spectrum between these extremes. At the three-dimensional level, enhancers establish spatial contacts (loops) with target gene promoters within topological associated domains (TADs), directing recruited coactivators and transcription factors to transcription initiation sites and enhancing RNA polymerase II recruitment and transcriptional burst frequency. Furthermore, multiple enhancer copies ("shadow enhancers") can function additively, synergistically, or competitively, providing regulatory robustness and ensuring precise, reproducible gene expression levels across developmental and evolutionary contexts [67].

Furthermore, more than 1000 liver genes are expressed under the control of super-enhancers, including TFs (*HNF4A*, *C/EBPa*, and *HNF1B*), cytochromes (*CYP2E1* and *CYP8B1*), albumin (*ALB*), blood coagulation factors, and others [68]. A super-enhancer is a cluster of enhancers characterized by a high density of TFBS, active chromatin marks (H3K4me1, H3K27ac, and P300), and association with the transcription activators Mediator Complex Subunit 1 (MED1) and Bromodomain-containing Protein 4 (BRD4). These elements are located at a considerable distance from each other but within a range not exceeding 12.5 kb [69,70]. The Mediator complex performs multiple functions during transcription, with its primary role being the transmission of regulatory signals from

https://doi.org/10.3390/cells15010014

enhancer regions to the transcriptional machinery through activators [71]. In addition, the complex regulates chromosomal spatial architecture by supporting three-dimensional contacts between the core promoter and distant regulatory regions [72]. MED1 and BRD4 participate in the formation of biomolecular condensates and promote the assembly of dynamic transcriptional complexes through liquid–liquid phase separation (LLPS) [73]. LLPS is a physicochemical process in which molecules organize into dense and dilute phases. The phase transition occurs abruptly, providing a key functional property of super-enhancers—the ability to rapidly and stably activate gene expression in response to minor fluctuations in transcription factor concentration [70,74]. It should be noted that groups of enhancers regulating the expression of the same gene are characterized by functional redundancy [75,76], meaning they perform similar or overlapping functions. This mechanism ensures the robustness of transcriptional regulation, preventing major changes in expression upon the loss or disruption of activity of one of the elements. Moreover, super-enhancers have been shown to play an important role in the regulation of microRNA transcription as well as in their maturation, including the processing of primary microRNAs [77].

Transcription factors play a key role in the development and maintenance of cellular identity, normal cell function, under stress conditions, and during the development of pathological processes. A group of pioneer TFs is distinguished by their ability to bind to compacted ("closed") chromatin, promote the remodeling of adjacent regions, and increase their accessibility to other TFs. The structure of TFs varies, but in most cases, they include several functional domains. Among them are a domain responsible for the specific recognition of DNA sequences and an effector domain that modulates the expression of the target gene. Typically, transcription factor-binding sites (TFBS) are enriched in promoter and enhancer regions [78]. The direction of gene expression regulation (activation or repression) is determined by the availability of co-activators or co-repressors with which a TF can interact. For example, the interaction between SIRT6 and FOXA2 inhibits the activity of the Zeb2 promoter in hepatocellular carcinoma cells [79]. In contrast, a study using a mouse model showed that the PGC-1β–FOXA2 complex enhances the transcription of genes encoding mitochondrial β-oxidation enzymes (CptI, Mcad, and Vlcad) [80]. Pioneer TFs are often less selective and can form paired interactions with a large number of TFs available in the cell.

Some TF pairs do not interact directly, but are capable of cooperatively binding to neighboring DNA sites and ensuring joint gene regulation. In most cases, such interaction is typical for pioneer TFs. For example, analysis of the ALB enhancer has shown that the cooperative binding of FOXA1 and GATA4 increases chromatin accessibility by repositioning nucleosome N1 at the NS-A1 and eF sites [81]. According to the affinity theory proposed by Zhao et al., TFs exhibit varying degrees of pioneer activity, determined by their ability to bind DNA [82]. In their study, the authors demonstrated that FOXA1 and HNF4A can regulate the expression of liver-specific genes both independently and cooperatively. Notably, the density of TFBS is higher in closed chromatin regions, which likely reflects the need to overcome greater energetic barriers to initiate transcription.

The spatial organization of regulatory elements and their accessibility for binding TFs play a critical role in the regulation of gene expression. In this regard, the identification of natural regulatory sequences and the design of synthetic ones represent important areas of research in molecular and synthetic biology. To develop gene therapies targeting the liver, key components of the liver's transcriptional regulatory network are being studied. The main transcription factors regulating gene expression in the liver, as well as their molecular interactions, will be discussed below.

https://doi.org/10.3390/cells15010014

3.2. Liver-Enriched Transcription Factors

Hepatocyte nuclear factor 4 alpha (HNF4A) is a member of the nuclear receptor superfamily, covering cis-regulatory regions of at least 42% of actively transcribed genes in hepatocytes and binding to DNA as a homodimer [83]. This TF is involved in the regulation of many liver functions, participating in the metabolism of xenobiotics, bile acid synthesis, lipid homeostasis, gluconeogenesis, cell proliferation, apoptosis, etc. HNF4A is involved in establishing and maintaining active chromatin marked by H3K4me1 and H3K4me3 signatures [84]. The *HNF4A* gene is encoded on chromosome 20 and regulated by two promoters, P1 and P2. Alternative splicing from P1 generates six isoforms (HNF4A1-6), which are characteristic of normal adult liver and kidney tissues, while six isoforms from the P2 promoter (HNF4A7-12) are expressed normally in fetal liver, pancreas, and in adult liver in response to starvation, as well as in various pathological conditions. The role of HNF4A in the cellular identity of hepatocytes is interesting: P1-HNF4A acts as a tumor suppressor, whereas increased expression of P2-HNF4A is associated with the progression of hepatocellular carcinoma [85]. Both isoform groups are also expressed in the colon epithelium and pancreatic ducts, where an increase in P2 isoforms is associated with potential cancer progression [86]. HNF4A consists of five structural domains: the N-terminal A/B domain, the C domain, which includes the functional DNA-binding domain, the D and E domains that form the ligand-binding domain, and the C-terminal F-domain. The N-terminal and C-terminal domains have a disordered structure and do not participate in DNA motif binding; however, they contribute to the differences between isoforms [87,88]. Interacting with co-regulators, HNF4A can exhibit both repressive and activation activities. For example, the interaction with PGC1α [21], SRC-1, and GRIP1 [89], as well as PPARα, significantly enhances transcription of the target gene. In contrast, IRF2BP2 [90] and DAX-1 [91] act as co-repressors. HNF4A predominantly exerts its activity during the night, directly regulating the circadian expression of downstream genes by acting as a repressor of the CLOCK:BMAL1 heterodimer [92,93].

Hepatocyte nuclear factor 1 alpha (HNF1A) is a transcription factor from the Pit–Oct–Unc (POU) family, encoded on chromosome 12. The protein consists of several domains: a N-terminal dimerization domain, an acidic amino acid region, a POU homeodomain that includes two subdomains, POUs and POUh, and a C-terminal transactivation domain [94]. HNF1A regulates the expression of *HNF4A* through an auto-regulatory positive feedback loop [95]. Somatic mutations in the POU domain lead to reduced transcriptional activity of the P1-HNF4A and P2-HNF4A isoforms, which is associated with decreased promoter affinity [96]. HNF1A is involved in lipid metabolism, maintaining cholesterol homeostasis through the regulation of PCSK9 and miR-122-dependent activation of SREBP-2 [97]. It inhibits the expression of SREBP-1c and the activity of the STAT3 signaling pathway [98], thereby preventing lipid accumulation in the liver. Moreover, HNF1A is associated with MODY3, as it normally enhances IRS-1 and AKT phosphorylation and activates the insulin signaling pathway [96]. HNF1A regulates the expression of complement pathway components such as C5, C8A, and factor D [99,100]. A decrease in hepatic HNF1A expression during fibrosis leads to the activation of inflammatory signaling pathways, including NF-κB and JAK/STAT, which in turn form a positive feedback loop that further suppresses HNF1A [101,102]. Thus, HNF1A represents one of the key factors involved in lipid and carbohydrate metabolism as well as in modulating the hepatic immune response

Prospero-related homeobox 1 (Prox1) is a transcription factor that plays a critical role in liver development and regeneration, promoting the recovery of the hepatocyte population [103–105]. It is required for the migration and differentiation of hepatocytes and cholangiocytes during organogenesis. This factor can directly repress the expression of genes unrelated to liver differentiation, thereby helping to maintain the hepatocyte

https://doi.org/10.3390/cells15010014

phenotype [106]. The regulatory role of Prox1 as a tumor suppressor has been discussed in detail in a recent review by Lee and Ma [107]. Genes under its regulatory influence also include other hepatic transcription factors, as illustrated by HNF4A [108]. In addition to direct Prox1 binding, an interaction has been demonstrated between the N-terminal LXXLL motif of Prox1 and the activation function 2 domain of HNF4A [109].

Members of the Forkhead box (FOX) transcription factor subfamily, FOXA, also known as HNF3, constitute a group of pioneer transcription factors and include three main proteins: FOXA1, FOXA2, and FOXA3. FOXA is expressed in various tissues, including the liver, pancreas, intestine, lungs, and prostate. FOXA3 regulates bile acid transport by inhibiting the NTCP and OATP1 transporters, preventing excessive lipid accumulation in the liver, and suppressing inflammatory responses through inhibition of TNF-α, interleukin-1β, Icam1, Nfκb1, Cd68, and Jnk1. FOXA3 and HNF4A co-regulate the cell cycle via activation of the TP53 gene promoter [110–112], accompanied by increased histone H3K9 acetylation at the promoter region [110]. FOXA1 and FOXA2 have partially overlapping functions and play key roles in the development and maintenance of hepatocyte identity [113–115]. Literature data indicate that their knockdown reduces the expression of major hepatic transcription factors and redirects the cellular phenotype toward alternative differentiation trajectories, showing increased neuroectodermal and pluripotency markers [116].

The ONECUT (OC) transcription factor family, also known as HNF6, exists in three isoforms: OC-1 (HNF6A) and OC-2 (HNF6B), which are predominantly expressed in the liver, and OC-3, which is less studied but primarily expressed in neural tissue. HNF6 has been extensively reviewed in a recent publication [117]. In this work, we focus only on the main aspects of its regulatory activity. HNF6 is known to interact differently with the promoters of *TTR* and *FOXA2*. In the *FOXA2* promoter, only the cut-domain binds DNA, while its LSDLL motif and F48M50 dyad of the homeodomain are available to interact with the co-activator CREB-binding protein (CBP). In contrast, in the *TTR* promoter, binding involves both domains of HNF6A: the cut domain and the homeodomain, recruiting the co-activator P300/CBP-associated factor. Notably, changing only two nucleotides in the DNA motif leads to such a substantial alteration in the activation mechanism [118].

The transcription factor Activator Protein-1 (AP-1) binds DNA as a dimer. The components of this dimer can include proteins from various families, such as JUN (c-Jun, JunB, and JunD), FOS (c-Fos, FosB, Fra-1, and Fra-2), MAF (musculoaponeurotic fibrosarcoma) protein family, and activating transcription factor (ATF) [119]. The composition and functional outcome of AP-1 TFs depend on upstream signaling pathway activity (e.g., MAPK and JNK) and cytokine availability. For example, interleukin-6 (IL-6) induction enhances the transcription of insulin-like growth factor-binding protein 1 through the HNF-1 site via AP-1 (c-Fos/c-Jun) and STAT3 [120]. A similar mechanism, AP-1/HNF-1, increases the expression of C-reactive protein [121]. Thus, AP-1 TFs are involved in cellular responses to oxidative stress [122], inflammation [123], and high-fat intake [124]; AP-1 TFs can also induce apoptosis [125] and regulate proliferation [126].

The C/EBP family of transcription factors can form homo- or heterodimers and bind DNA via a leucine zipper domain. Members of the C/EBP family are expressed in many tissues; however, in the liver, the primary regulatory roles are carried out by two transcription factors: C/EBPa and C/EBPb. C/EBPa exists in several isoforms: the full-length p42, the truncated p30, and an extended isoform, expressed at lower levels, which initiates translation from an alternative start codon (CUG in vertebrates, GUG in humans) [127].

C/EBPb, also known as interleukin-6-dependent DNA-binding protein or nuclear factor interleukin-6 (NF-IL6) [128], is also represented by three isoforms: LAP1 and LAP2, which generally act as transcriptional activators, and the shorter LIP isoform, which more

https://doi.org/10.3390/cells15010014

often functions as a transcriptional inhibitor. C/EBPa exhibits anti-proliferative effects [129] and is involved in the regulation of lipogenesis [130] and gluconeogenesis through the co-activator CBP [131]. Overexpression of C/EBPa in hepatic stellate cells increases the expression of ATG5 and Beclin1, thereby promoting autophagy [132].

C/EBPb is induced by IL-6 and plays a key role in the regulation of immune responses [133]. In addition, this transcription factor acts as an effector of endocrine stimuli, particularly thyroid-stimulating hormone (TSH). TSH increases the expression of miR-374b, which subsequently suppresses C/EBPb transcription and directly targets the 3′-untranslated region of C/EBP, creating a negative feedback loop [134].

It should be emphasized that the regulatory activity of TFs is determined by several interconnected factors. First, it is constrained by the ~150 bp span of nucleosomes and exhibits a periodicity of approximately 10.5 bp, which reflects the helical turns of DNA. TFBS affinity also strongly influences specificity: cooperative effects of low-affinity sites can promote tissue specificity, whereas high-affinity sites may reduce it. DNA-mediated cooperativity is frequently observed; in this case, adjacent sites and defined spacing between them stabilize simultaneous binding. This property is used in the design of synthetic promoters (as described in the corresponding section). The mechanisms of cooperative interactions are discussed in greater detail in the review [135].

In addition, the composition and arrangement of TFs on enhancers and promoters can be flexible, reflecting diverse mechanisms of cooperativity and enabling TFs to function across different regulatory contexts. Thus, understanding the regulatory landscape is essential for the development of applied solutions in gene therapy.

4. Safety Advantages of Liver-Specific over Ubiquitous Promoters

Most gene therapy studies and commercial genetic platforms utilize natural promoters within their delivery constructs. These include ubiquitous and tissue-specific eukaryotic promoters, as well as viral promoters [58]. Numerous gene therapy studies, in particular, employ viral promoters such as the early CMV promoter [136], the simian virus 40 (SV40) promoter [137], the adenoviral major late promoter (MLP) [138], and various retroviral long terminal repeat (LTR) promoters [139]. In their native context, strong viral promoters are essential for efficient viral replication and therefore often drive substantially higher transcription levels than native eukaryotic promoters. Moreover, they are generally more compact than eukaryotic counterparts, making them easier to manipulate and utilize in gene therapy vectors [29]. In addition to natural viral promoters, synthetic ubiquitous promoters such as CBA and CAG are widely used; these contain both viral and eukaryotic regulatory elements. A comprehensive review on the clinical landscape of AAV-based gene therapies reported that between 2015 and 2019, 45% of clinical trials employing constructs with disclosed promoters used CMV, CAG, or CBA promoters [140].

The primary limitation of using viral promoter use in gene therapy is their propensity to trigger immune activation, resulting in rapid decline in transgene expression. Silencing of viral promoters is partly associated with the induction of pro-inflammatory cytokines, such as TNF-α (tumor necrosis factor-alpha) and IFN-γ (interferon-gamma), during the activation of the innate immune response [29,141]. Several studies suggest that inhibition occurs at the level of transgenic mRNA, and cytokines do not cause degradation of vector DNA or suppress general cellular protein synthesis. The analysis indicates that IFN-γ and TNF-α lead to promoter-specific, post-transcriptional destruction of viral mRNA. It was demonstrated that viral promoters are more sensitive to the inhibitory action of inflammatory cytokines compared to eukaryotic promoters [142]. An early study reported that expression of factor IX under the elongation factor 1α (EF1α) promoter in mouse liver

https://doi.org/10.3390/cells15010014

persisted for at least six months, whereas expression driven by the CMV promoter was eliminated by week 5 [143].

Another issue associated with the use of viral regulatory elements, particularly those of retroviral origin, is their targeted methylation in eukaryotic cells, which inhibits transgene expression [144]. Viral promoters (e.g., CMV) are rich in CpG dinucleotides, which makes them targets for DNA methyltransferases and leads to epigenetic silencing. In contrast, many endogenous liver-specific promoters have low CpG content or protected chromatin contexts, which provides them with resistance to hypermethylation and more stable expression [145]. This mechanism is discussed in detail in the another review [146]. It was also reported that the epigenetic status of the transduced cell with AAV vectors effects transgene expression [147].

An alternative solution was to use eukaryotic ubiquitous promoters such as EF1α and PGK. However, ubiquitous eukaryotic promoters also present significant challenges, including the difficulty in achieving physiological expression levels and the risk of ectopic expression in non-target cells and tissues. Ectopic transgene expression often leads to adverse outcomes such as inflammation and other immune responses associated with de novo expression in non-privileged tissues and antigen-presenting cells [148].

In contrast, such as ApoE/HCR1-hAAT (human alpha1-antitrypsin promoter coupled with hepatic control region 1 of apolipoprotein gene cluster), and TBG (thyroxine-binding globulin promoter), provide higher transgene expression levels compared to ubiquitous eukaryotic and viral promoters, and do not lead to the transgene-specific antibodies formation [149,150].

Thus, a crucial requirement in the development of effective and safe gene therapies is not only the level of expression but also its localization within target tissues. The selection of a suitable promoter for targeted expression becomes particularly important, and tissue-specific promoters are increasingly being used in gene therapy development, as supported by recent data [140]. Therefore, the following section will focus on the most commonly used liver-specific promoters in gene therapies and the key stages in the creation of their improved synthetic variants (Table 1, Supplementary Material S1). More information about liver-directed gene therapies are presented in Supplementary Table S1.

Table 1. Liver-specific promoters utilized in gene therapy.

Promoter	Gene of Origin	Length, bp	Features	Disease	Gene Therapy	Ref.
hAAT	Human *AAT*	305	−264/+41 hAAT promoter	Wilson disease	VTX801	[151–153]
EalbAAT	Human *AAT*, murine *Alb*	673	−264/+20 hAAT promoter with 376 bp mAlb enhancer	AIP (acute intermittent porphyria)	rAAV2/5-PBGD	[154–156]
ApoE/HCR1-hAAT	Human *ApoE/HCR1* and *AAT*	727	−355/+42 hAAT promoter with 321 bp ApoE/HCR1 enhancer	Hemophilia B	Beqvez (fidanacogene elaparvovec, SPK-9001)	[157,158]
ApoE/HCR1-hAAT	Human *ApoE/HCR1* and *AAT*	732	−353/+50 hAAT promoter with 321 bp ApoE/HCR1 enhancer	Phenylketonuria (PKU)	BMN 307	[159,160]
ApoE/HCR1-hAAT	Human *ApoE/HCR1* and *AAT*	725	−355/+43 hAAT promoter with 321 bp ApoE/HCR1 enhancer	Phenylketonuria (PKU)	NGGT002	[161]
ApoE/HCR1-hAAT	Human *ApoE/HCR1* and *AAT*	723	−355/+38 hAAT promoter with 321 bp ApoE/HCR1 enhancer	Fabry Disease	ST-920	[162–164]

https://doi.org/10.3390/cells15010014

Table 1. *Cont.*

Promoter	Gene of Origin	Length, bp	Features	Disease	Gene Therapy	Ref.
ApoE/HCR1-hAAT	Human *ApoE/HCR1* and *AAT*	727	−355/+42 hAAT promoter with 321 bp ApoE/HCR1 enhancer	Crigler-Najjar syndrome	GNT0003	[165–167]
ApoE/HCR1-hAAT	Human *ApoE/HCR1* and *AAT*	732	−355/+42 hAAT promoter with 326 bp ApoE/HCR1 enhancer	Pompe disease	SPK-3006 (vanglusagene ensiparvovec)	[168,169]
FRE76	Human *ApoE/HCR1* and *AAT*	728	−347/+43 hAAT promoter with 321 bp ApoE/HCR1 enhancer Same as ApoE/HCR1-hAAT	Gaucher disease type 1	FLT201	[170]
LP1	Human *ApoE/HCR1* and *AAT*	448	−212/+43 hAAT promoter with 192 bp ApoE/HCR1 enhancer	Hemophilia B	Hemgenix (AMT-061, CSL222, etranacogene dezaparvovec)	[171,172]
LP1	Human *ApoE/HCR1* and *AAT*	448	−212/+43 hAAT promoter with 192 bp ApoE/HCR1 enhancer	Hemophilia B	scAAV2/8-LP1-hFIXco	[173–176]
LP1	Human *ApoE/HCR1* and *AAT*	448	−212/+43 hAAT promoter with 192 bp ApoE/HCR1 enhancer	Phenylketonuria (PKU)	HMI-102	[177,178]
LP1	Human *ApoE/HCR1* and *AAT*	448	−212/+43 hAAT promoter with 192 bp ApoE/HCR1 enhancer	Phenylketonuria (PKU)	HMI-103	[160,179]
HLP	Human *ApoE/HCR1* and *AAT*	252	−247/−216 and −143/+43 hAAT promoter with 34 bp ApoE/HCR1	Hemophilia A	AAV-HLP-hFVIII-V3 (GO-8)	[180,181]
HLP	Human *ApoE/HCR1* and *AAT*	252	−247/−216 and −143/+43 hAAT promoter with 34 bp ApoE/HCR1	Hemophilia A	Roctavian (valoctocogene roxaparvovec, BMN 270)	[182–184]
FRE1 (HLP2)	Human *ApoE/HCR1* and *AAT*	335	−247/−216 and −143/+43 hAAT promoter with 117 bp ApoE/HCR1 enhancer	Hemophilia B	FLT180a (verbrinacogene setparvovec)	[185–187]
FRE1 (HLP2)	Human *ApoE/HCR1* and AAT	335	−247/−216 and −143/+43 hAAT promoter with 117 bp ApoE/HCR1 enhancer	Fabry disease	FLT190	[187,188]
Em-hAATsh	Human *AAT*, synthetic enhancer	139	Shorten −133/+51 hAAT promoter divided on 4 parts with synthetic enhancer composed of hepatocyte TF binding sites	Hemophilia A	ZS802	[189,190]
mTTR mut	Murine *TTR*	223	−138/−135 gact>tgtg mutant mTTR promoter	Hemophilia A	NGGT003	[191]
mTTR mut	Murine *TTR*	223	−138/−135 gact>tgtg mutant mTTR promoter	Hemophilia A	SPK-8011 (dirloctocogene samoparvovec)	[192–194]
mTTR enhancer/promoter	Murine *TTR*	330	−204/+5 mTTR promoter with 100 bp mTTR enhancer in antisense orientation	Hemophilia B	AskBio009 (BAX 335)	[195,196]
mTTR enhancer/promoter	Murine *TTR*	330	−204/+5 mTTR promoter with 100 bp mTTR enhancer in antisense orientation	Hemophilia A	TAK-754 (BAX 888)	[197–199]
mTTR enhancer/promoter	Murine *TTR*	372	−202/+27 mTTR promoter with modified mTTR enhancer in antisense orientation Shorten ET promoter	Hemophilia B	ANB-002	[200,201]
E03.TTR	Murine and human TTR	296	−189/+1 hTTR promoter with 100 bp mTTR enhancer	Hemophilia A	DTX201 (BAY2599023)	[202–204]

https://doi.org/10.3390/cells15010014

Table 1. *Cont.*

Promoter	Gene of Origin	Length, bp	Features	Disease	Gene Therapy	Ref.
E03.TTR	Murine and human *TTR*	290	−189/+1 hTTR promoter with 100 bp mTTR enhancer	Wilson disease	UX701	[205,206]
AlMB2-mTTR482	Murine *TTR*, human *AMBP*	671	−203/+21 mTTR modified promoter with modified 92 bp mTTR enhancer and 2 copies of modified 162 bp hAMBP enhancer	Phenylketonuria (PKU)	SAR444836	[207,208]
CRMSBS2-mTTR	Murine *TTR*, human *AAT*	307	−202/+21 mTTR promoter with modified −122/−51 hAAT in antisense orientation	Hemophilia A	PF-07055480, formerly SB-525 (giroctocogene fitelparvovec)	[209–211]
3xCRM8-enTTR-mTTR	Murine *TTR*, human *AAT*	548	−204/+5 mTTR promoter with 100 bp mTTR enhancer and 3 copies of CRM8 (−122/−51 hAAT in antisense orientation)	Hemophilia B	TAK-748 (SHP648)	[196,212]
3xCRM8-enTTR-mTTR	Murine *TTR*, human *AAT*	520	−202/+1 mTTR core promoter with 100 bp mTTR enhancer and 3 copies of CRM8 (−122/−51 hAAT in antisense orientation)	Hemophilia B	VGB-R04	[213]
HCB	*Xenopus laevis Alb*, human *AMBP*	146	−67/−26 xAlb promoter (SynO region) with AbpShort (region of human AMBP shortened to 56 bp), and predicted conservative TSS	Hemophilia A	ASC618	[214–216]
GT001 (vector title)	*Xenopus laevis Alb*, canine *AAT*, human *ApoE/HCR1*	266	16 bpHCR1 enhancer with modified canine AAT and −66/+38 Xenopus laevis Alb	Hemophilia A	GS1191-0445	[217,218]
LSP	Human *TBG* and *AMBP*	698	−474/+3 TBG with 2 copies of 101 bp AMBP enhancer	MPS VI (Mucopolysaccharidosis Type VI)	AAV2/8.TBG.h-ARSB	[219,220]
LSP	Human *TBG* and *AMBP*	747	−475/+4 TBG promoter with 2 copies of modified 98 bp AMBP enhancer with 3 point mutations	Pompe disease	ACTUS-101	[221,222]
LSP	Human *TBG* and *AMBP*	734	−475/+4 TBG promoter with 2 copies of 98 bp AMBP enhancer with 3 point mutations	Hemophilia B	DTX101	[158,223]
LSP *	Human *TBG* and *AMBP*	698	−474/+3 TBG promoter with 2 copies of 101 bp AMBP enhancer	Ornithine Transcarbamylase Deficiency	DTX301	[224–227]
LXP2.1	Completely synthetic	188	Consists of hepatocyte TF binding sites	Hemophilia A	BBM-H803 (BBM 002)	[228]
LXP2.1	Completely synthetic	188	Consists of hepatocyte TF binding sites	Hemophilia B	BBM-H901	[229]
G6PC1	Human *G6PC1*	2864	−2786/+78 hG6PC1 native promoter	Glycogen storage disease type I (GSDIa)	DTX401	[230,231]
C7	NS	NS	NS	Fabry Disease	AMT-191	[232]

* The DTX301 clinical trial claims to use the TBG promoter despite the presence of two copies of the AMBP enhancer.

5. Promoters Based on Human *SERPINA1* Gene

Alpha 1-antitrypsin (AAT) is the main inhibitor of serine proteases, encoded by the *SERPINA1* (Serpin Family A Member 1) gene approximately 12.2 kb in length. The healthy plasma AAT level is between 0.9 and 2 g/L, but it can be increased fivefold relative to normal during an acute-phase response (APR) [233]. AAT is expressed in the various isoforms and is controlled by two promoters for different tissues. The promoter at the 5′-end of the gene with TSSs in the untranslated exon 1A drives expression in monocytes, macrophages, and lungs, while the liver-specific promoter is located closer to the coding

https://doi.org/10.3390/cells15010014

region with a TSS mapped in the untranslated exon 1C [234,235]. Since in this paper we discuss the use of promoters for liver-specific expression, we will focus on describing the *SERPINA1* promoter with TSS in exon 1C, and will use base numbering relative to the corresponding TSS.

In the 1980s, sequencing of the coding and non-coding regions of the *SERPINA1* gene (formerly termed *AAT*) enabled a detailed studies of its promoter regulation [236,237]. Subsequent studies identified multiple transcript variants of this gene and two principal promoters located approximately ~2 kb apart. The upstream promoter is responsible for expression in macrophages, while the downstream promoter provides expression in hepatocytes (Figure 2A) [238]. The inducibility of the promoter is determined by the action of cytokine IL-6, mediated through transcription factor NF-IL6, whose binding site is located in the 3′ enhancer. The 3′ enhancer region also contains AP1 sites for binding to Fox/Jun and an Oct-1 binding site. IL-6 induction has minor effect on the 5′ promoter separately from the 3′ enhancer, suggesting complex spatial regulation of inducible promoter activation [238–240].

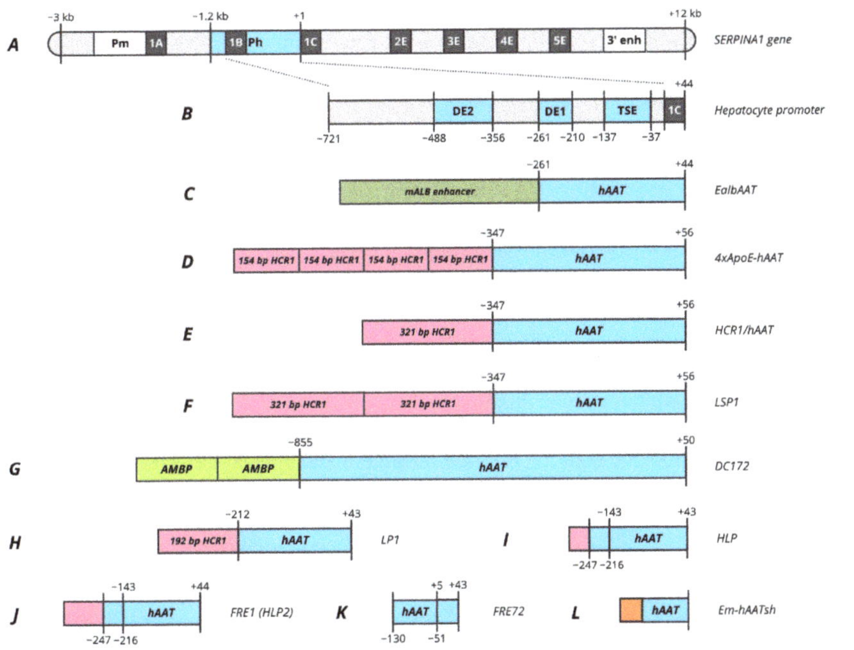

Figure 2. Promoters based on human SERPINA1 (AAT) gene. (**A**)—human SERPINA1 gene. (**B**)——721/+44 hAAT promoter. (**C**)—EalbAAT promoter composed of hAAT promoter with murine Alb enhancer. (**D**)—4xApoE-hAAT promoter composed of hAAT promoter with four copies of 154 bp HCR1 enhancer. (**E**)—HCR1-hAAT promoter composed of hAAT promoter with 321 bp HCR1 enhancer. (**F**)—LSP1 promoter composed of hAAT promoter with 2 copies of 321 bp HCR1 enhancer. (**G**)—DC172 promoter composed of hAAT promoter with 2 copies of 160 bp AMBP enhancer. (**H**)—LP1 promoter composed of hAAT promoter with 192 bp HCR1 enhancer. (**I**)—HLP promoter composed of two hAAT promoter parts with 34 bp HCR1 enhancer. (**J**)—FRE1 (HLP2) promoter composed of two hAAT promoter parts with 117 bp HCR1 enhancer. (**K**)—FRE72 promoter composed of two hAAT promoter parts. (**L**)—Em-hAATsh promoter composed of four hAAT promoter parts with synthetic Em enhancer. Pm, monocytes and macrophages promoter; Ph, hepatocytes promoter; 1A, 1B and 1C, untranslated exons; 2E, 3E, 4E and 5E, translated exons; 3′ enh, enhancer located downstream from coding sequence; DE2, distal element 2; DE1, distal element 2; TSE, tissue-specific element.

https://doi.org/10.3390/cells15010014

However, the 5′ promoter of hepatocytes exhibits high basal activity even in the absence of IL-6 induction and activation by the 3′ enhancer. In a pilot study by Ciliberto et al., a −1200/+44 bp region of hAAT promoter was reported to provide a high level of specific expression in hepatocytes [237]. Within this promoter various regions with different enhancer activity and specificity were determined based on the analysis of truncated variants (Figure 2B). The most proximal region, located at −137/−37, titled as a tissue-specific element (TSE), is capable of specifically enhancing the heterologous SV40 promoter 25-fold in hepatocytes. HNF4 and HNF1 binding sites have been identified in this region, and a mutation in one of them leads to a complete elimination of promoter activity [141,238]. Within the TSE, CRM8 (cis-acting regulatory modules) was later characterized, which has a high homology among many species and is essential for promoter function [241]. C/EBP, HNF1, HNF4, HNF6 and two HNF3 sites with different binding affinities were identified within −202/−70 region [242,243]. The intermediate region −261/−210 (DE1) is capable of enhancing the heterologous SV40 promoter in 40–50 times, which contrasts markedly with the fact that shortening the native promoter to −210 reduces activity only in 4–5 times. A core enhancer site was identified in this region, which is also found in many liver genes as well as viral enhancers. The site is presented in the 3′ enhancer and flanked by the AP1 and C/EBP sites [141,238,240]. The third major regulatory region DE2 of the promoter is located at −488/−356 and exhibits strong but non-specific enhancer activity [238].

Most further studies focused on the −721/+44 truncated promoter variant, since this promoter has an analogous activity to −1200/+44 promoter (Figure 2B) [237]. In one of the earliest studies focused on the comparison of promoter activity in primary hepatocytes the −732/+44 hAAT promoter demonstrated relatively weak activity compared to the 500 bp hAlb (human albumin) promoter and viral promoters [244]. However, in retroviral vectors the truncated −347/+56 hAAT promoter was superior to 820 bp mAlb both in vitro on differentiated hepatocytes and in vivo during hAAT expression in mice [245]. A similar result was obtained by comparison of the truncated −261/+44 hAAT promoter with −180/+16 mAlb where the hAAT promoter without enhancer showed higher activity, representing 40% of the CMV promoter activity. In the same study, the addition of 376 bp mAlb enhancer to the hAAT promoter increased the expression level of AAT from 20 to 70 ug/mL in mouse plasma (Figure 2C, Table 1) [141,246]. Therefore, both promoters found their clinical applications in which −261/+44 hAAT was used in AAV3B-mediated gene therapy VTX801 for Wilson's disease, while EalbAAT (376 bp enhancer with −264/+20 hAAT) was used to treat acute intermittent porphyria (AIP) (rAAV2/5-PBGD) [151,154].

Nevertheless, VTX-801 gene therapy was terminated due to poor efficacy of the doses tested, despite promising results in preclinical studies [151]. Despite some improvement in patients' quality of life, including a significant reduction in depression and anxiety, the rAAV2/5-PBGD therapy also proved to be not enough effective, as ALA (delta-aminolevulinic acid) and PBG (porphobilinogen) levels did not decrease even in the high-dose vector group with 1.8×10^{13} vg/kg [154]. To increase the therapy's efficacy, two copies of the ADRES (ALAS Drug-Responsive Enhancing Sequence) enhancer were added to the EalbAAT promoter, that significantly improved the promoter's activity in the presence of several porphyrinogenic stimuli. This modification of the promoter made it possible to achieve the same level of PBGD activity with 10 times lower viral load [247]. The same group of authors also obtained a hyperfunctional mutant PBGD variant that provided protection in AIP mice against PB-induced attack [248].

An interesting observation was that the addition of several copies of HNF3 binding sites does not enhance the hAAT promoter that is in contrast to mAlb. The previously characterized ApoE/HCR1 (hepatic control region 1) enhancer region was used to enhance the hAAT promoter. hAAT promoter and HCR1 enhancer combination subsequently be-

https://doi.org/10.3390/cells15010014

came the most popular for providing liver-specific expression. The HCR1 region, which together with HCR2 control the expression of all E/C-I/C-IV/C-II cluster genes, was characterized into regulatory regions of varying lengths and enhancer activities [249,250]. The enhancer activity is driven by the presence of multiple binding sites for HNF3, HNF4, and C/EBP [251–253]. Originally, researchers used several copies of the 154 bp HCR1 enhancer region (PvuII-ApaI region) to enhance the −347/+56 hAAT promoter (Figure 2D). Interestingly, the greater promoter activity was attributed to the HCR1 enhancer in reverse orientation as the positioning of the enhancer downstream of the transgene coding sequence [47,254–257]. The use of a greater number of copies of the 154 bp (4 and 8) enhancer in the adenoviral (Ad) expression vector leads to alanine aminotransferase elevations and faster loss APOA1 transgene expression loss [255]. Nevertheless, immune tolerance to transgene was observed in mice using Ad and AAV vectors with the 4xApoE-hAAT promoter [258,259].

A key factor that advances the use of the hAAT promoter was the observation that the complete 711 bp HCR1 enhancer locus combined with 408 bp hAAT fragment resulted in a four-fold higher level of hFIX expression compared with four copies of the 154 bp enhancer [260]. Subsequently, the derived 1.1 kb HCR1-hAAT promoter was used in AAV and LV vectors in preclinical trials for the treatment of hemophilia B. It is important to note that liver-specific expression resulted in the immune tolerance induction in the animals [259,261–263]. However, the HCR1-hAAT promoter became most widespread as a variant in which the HCR1 enhancer region was reduced to ~320 bp containing full functional LCR activities and maintaining a similar promoter activity (Figure 2E) [260,264]. The promoter was at least twice more active than mTTR enhancer/promoter by normalized expression level [265]. Currently, the 730 bp HCR1-hAAT promoter is used for liver-specific expression in a various AAV-mediated gene therapies for the treatment of hemophilia B (Beqvez, idanacogene elaparvovec), Fabry disease (ST-920), Gaucher disease (FLT201, promoter titled as FRE76), Pompe disease (SPK-3006, vanglusagene ensiparvovec), Crigler–Najja syndrome (GNT003) and phenylketonuria (BMN 307; NGGT002) [157,162,165]. SPK-3006 gene therapy was terminated for strategic reasons while BMN 307 was discontinued based on preclinical results, where 6 out of 7 mice receiving the highest dose (2e14 vg/kg) developed tumors in liver necropsy with evidence for integration of portions of the AAV vector into the genome. This observation was most likely due to the extremely high dose of the vector [159].

The LSP1 (liver-specific promoter 1) promoter was obtained by the addition of another copy of the HCR1 enhancer, which enhances the HCR1-hAAT promoter by 2–3 times (Figure 2F) [20,266–270]. LSP1 promoter was used in preclinical studies of ornithine transcarbamylase (OTC) deficiency in mice and for the expression of piggyBac transposase for editing OTC-deficient patient-derived primary human hepatocytes [271–273]. However, LSP1 has not found clinical application despite its identical activity to TBG promoter [227].

Based on the capability of the AMBP (alpha-1-microglobulin/bikunin precursor) to enhance the hAAT promoter to the same degree as HCR1, two copies of the 160 bp AMBP enhancer were added to the 890 bp hAAT promoter to obtain the DC172 promoter, which has five times greater activity compared to previously reported DC190 (Figure 2G) [266,274,275]. An additional modification of DC172 involved inserting copies of the HCR1 enhancer; however, the promoter failed to be enhanced by the 154 bp HCR1 enhancer regardless of the copy number. This confirms the results of previous studies that AMBP and HCR1 enhancers do not have a combinatorial effect and are most effective when acting separately [256,266]. Copies of the 774 bp HCR1 enhancer can enhance the DC172 promoter by 1.5–2 times, but this significantly increases the size of the already long promoter, that limits its application [256]. The derived promoter was used in a more high-capacity Ad

https://doi.org/10.3390/cells15010014

vector for regression and stabilization of advanced murine atherosclerotic lesions [276]. At the same time, the ~1.2 kb DC172 promoter was used for AAV-mediated expression of glucocerebrosidase in a mouse model of Gaucher disease, that results in the immune tolerance induction [275].

However, the great length of the promoters 730 bp HCR1-hAAT, ~1 kb LSP1, ~1.2 kbp DC172 limited their use in a number of gene therapies with larger CDS and for use in scAAV vectors. Based on analysis of TF binding sites, the HCR1-hAAT promoter was truncated to a 448 bp LP1 promoter during the development of the scAAV8 vector for hemophilia B therapy (Figure 2H) [174]. The shortened length of the promoter enabled its use in preclinical studies of AAV-mediated gene therapy for hemophilia A [277–279]. Some attempts were made to modify the LP1 promoter through nucleotide alteration to include HNF1 and HNF4 binding sites, but this provided a minor effect on promoter activity [280]. Due to the lower efficiency of assembling oversized AAV vectors such for hemophilia A gene therapy, further shortening of the promoter was required. Researchers from University College London obtained a 252 bp HLP (hybrid liver promoter) promoter during the development of hemophilia A gene therapy with a new FVIII-V3 variant which has increased expression efficacy (Figure 2I) [183]. HLP promoter design based on the results of the early pilot work mentioned above focusing on the characterization of the promoter functional regions [238]. It was reported that shortening hAAT promoter from −261 to −208 resulted in significant activity reduction, but no change was observed with further shortening to −137. Therefore, it was decided to combine the DE1 with TSE regulatory regions (Figure 2B) and add the first 34 bp of the 192 bp HCR1 enhancer used for LP1. The LP1 promoter was used in gene therapy for the treatment of hemophilia B (AMT-061, Hemgenix; scAAV2/8-LP1-hFIXco) and phenylketonuria (HMI-102 and HMI-103 are currently terminated), while HLP is used for the treatment of hemophilia A (BMN 270, Roctavian; GO-8) [171,176,177]. HMI-102 was discontinued due to the greater promise of HMI-103 therapy, designed to integrate PAH cDNA into the PAH locus via homology-directed repair (HDR). Despite encouraging preliminary results, HMI-103 was also terminated, presumably for economic reasons [281,282].

Despite reports of identical potency of LP1 and HLP promoters in AAV vectors in in vivo studies, in vitro studies on Huh7 cells demonstrated that the activity of various HCR1-hAAT promoter variants decreased with promoter length, and the HLP promoter provided the lowest FVIII expression [172]. The low activity of the HLP promoter is also confirmed in a large-scale study of enhancer and promoter combinations, where it is denoted as E01.A1AT [203]. To enhance HLP promoter the 34 bp HCR1 enhancer was replaced with 117 bp HCR1 [283]. The derived 335 bp in length HLP2 (FRE1) promoter was only 1.5 times less active than HCR1-hAAT (Figure 2J) [187]. HLP2 was used in gene therapies for Fabry disease (FLT190) and hemophilia B (FLT180a), which are currently terminated [186,188]. FLT180a was terminated after 10 patients had been enrolled because of changes to the clinical development plan and recruitment difficulties due to the COVID-19 pandemic (NCT03369444) while development of FLT190 in Fabry disease was paused to focus company resources on advancing FLT201 (NCT04040049).

HLP2 promoter underwent further modifications to provide fully identical to HCR1-hAAT activity. The 119 bp FRE72 promoter was obtained exclusively from the −130/+44 hAAT region by introduction of an internal deletion −50/+4 (Figure 2K). The obtained promoter had 1.5–2 times greater activity compared to its ancestor HLP2 promoter, despite the deletion of the most proximal regions involving the TATA box and TSS, which is quite unexpected in accordance with current concepts about the critical role of the basal promoter region. A similar approach with the proximal elements removal is observed for the new 139 bp Em-hAATsh promoter, which contains multiple deletions

https://doi.org/10.3390/cells15010014

within the −155/+33 hAAT promoter region. In a similar manner the most proximal region was deleted which may indicate the low significance of the −40/+22 region of the basal promoter, additionally considering the extremely low activity of the −85/+9 hAAT promoter [284]. hAATsh consists of the following hAAT regions: −155/−141, −123/−106, −77/−41, +23/+33, and in combination with a synthetic enhancer, it results in a 10-fold activity increase compared to the HLP promoter (Figure 2L; Supplementary Material ZS802 Em-hAATsh) [190]. There were no reports yet on the use of the FRE72 promoter in clinical trials, however Em-hAATsh is presumably already utilized in the ZS802 hemophilia A gene therapy.

Thus, despite the long-standing combined use of the hAAT promoter and HCR1 enhancer, their application is becoming more widespread with an emphasis on reducing the promoter length. The 34 bp HCR1 enhancer is being investigated for the enhancing of other promoters, such as TBG and mTTR, while the AAT region is also used as an enhancer for the mTTR and Alb (*Xenopus laevis*) promoters [203,241]. The CRM8 region was further modified by aligning orthologs (CRMSBS2) or by returning 4 nucleotides to the pan-ortholog consensus HS-CRM8 sequence to the human consensus sequence within or proximal to the predicted transcription factor-binding sites [210,285].

The hAAT promoter offers a compelling example in which a truncated form outperforms the full-length form. The observed effect may be attributed to the presence of an upstream silencer and to differences in transcription factor-binding site density across promoter regions. A low number of such sites may indicate that the DNA remains in a nucleosome-associated, compact configuration, as transcription factors may be insufficient to displace the nucleosome [67,135].

6. Promoters Based on TTR Gene

Transthyretin, formerly known as prealbumin, is a highly conserved plasma transport protein expressed mainly in the liver, as well as in the choroidal plexus and retinal pigment epithelium. Notably, hepatic expression can be suppressed during the APR, since cytokines prevent hepatocyte TFs from binding to the promoter, whereas suppression does not occur in the choroidal plexus, indicating distinct regulatory mechanisms in these tissues [285,286].

An early study focusing on the TTR promoter was conducted in 1986 by Costa et al. The researchers discovered an extremely high homology between the human and mouse *TTR* gene in the region 290n to the cap site after cDNA and the promoter region were sequenced. It was suggested that the promoter region necessary for TTR expression is located within 190 bp upstream of the cap site, since this region possesses the highest homology at 84%. Through deletion analysis, it was found that the NcoI-SstI enhancer element (−2150/−1600) is necessary for high liver-specific expression from the −329 promoter (Figure 3A) [287]. The activity of the −3000/−329 enhancer in antisense orientation was comparable to that in the native orientation when used to enhance both native and heterologous promoters. The same researchers succeeded in more accurately localizing the TTR enhancer region within −1.96/−1.86 kb (−1879/−1780 bp according to PubMed) and the core promoter region within −202 (Figure 3B). The identified enhancer region was able to enhance hepatocyte-specific expression from the heterologous β-globin promoter by 9-fold and contained 4 TF binding sites from hepatocyte nuclear extracts: two C/EBP, AP-1, and HNF4 sites. In addition, it was demonstrated that the mTTR core promoter and the mAAT enhancer region bind to similar TFs, such as HNF1, FOXA, C/EBP, and AP-1 [242,243]. Coordinated interaction between HNF1A, FOXA1/2, HNF4A, and HNF6A factors is necessary to provide core promoter activity, since mutation in a single site has a dramatic negative effect on promoter activity [285].

https://doi.org/10.3390/cells15010014

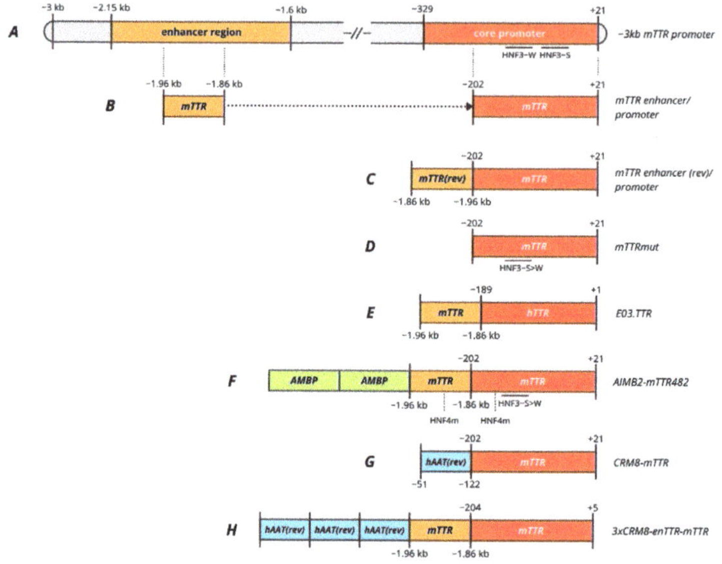

Figure 3. Promoters based on TTR gene. (**A**)—−3 kb murine TTR promoter regions. (**B**)—mTTR enhancer/promoter composed of mTTR promoter with 100 bp mTTR enhancer. (**C**)—mTTR enhancer (rev)/promoter (or enTTR-mTTR) composed of mTTR promoter with 100 bp mTTR enhancer in antisense orientation. (**D**)—mTTRmut, minimal mTTR promoter with HNF3 site mutated. (**E**)—E03.TTR composed of hTTR promoter with 100 bp mTTR. (**F**)—AIMB2-mTTR482 composed of mTTR promoter with HNF3 and HNF4 sites mutated, 100 bp mTTR enhancer with HNF4 site mutated and two copies of modified 162 bp AMBP enhancer. (**G**)—CRM8-mTTR composed of mTTR promoter with hAAT promoter in antisense orientation as enhancer. (**H**)—3xCRM8-enTTR-mTTR composed of mTTR promoter, mTTR enhancer and three copies of hAAT promoter in antisense orientation as enhancer. HNF3-W, HNF3 weak affinity binding site; HNF3-S, HNF3 strong affinity binding site; HNF3-S>W, substitution of HNF3-W site for HNF3-S site; HNF4m, modified HNF4 site for more affinity.

The derived ~320 bp mTTR promoter, consisting of a −202 promoter and a 100 bp enhancer, was compared with a −3 kbp promoter to endogenous expression upon the integration into the mouse genome. The 320 bp promoter provided full expression levels in the liver, but increased general expression in the brain outside the choroid plexus, while the −3 kb promoter provided low expression levels in the brain precisely restricting expression within the choroid plexus [288]. The 320 bp mTTR promoter was further compared with other liver-specific promoters in AAV vectors for factor IX expression including 190 bp mAlb, 730 bp HCR-hAAT, 845 bp LSP and 450 bp −219/+21 hFIX with the site 5 as an enhancer [289]. A scAAV vector with an mTTR promoter provided higher levels of FIX expression in vivo [265]. A similar enTTR-mTTR promoter, but with the mTTR enhancer in the antisense orientation, was used in gene therapies for the treatment of hemophilia A (TAK-754/BAX 888) and hemophilia B (AskBio009/BAX 335) (Figure 3C) [195,197]. BAX-335 was terminated due to rapid loss of transgene expression due to the use of a CpG-enriched sequence, which increases the immunogenicity of the vector and leads to the elimination of transduced cells, as it was also observed for DTX101 gene therapy [290]. However, a rapid decline in transgene expression within the first year was also observed for CpG-depleted TAK-754 gene therapy. The loss of transgene expression was not associated with liver dysfunction, and the cause remains unclear [291].

The 100 bp mTTR (antisense) enhancer was modified by random ligation of binding sites of various hepatocyte-specific transcription factors DPB, C/EBP, HNF1, HNF3, HNF4,

https://doi.org/10.3390/cells15010014

and HNF6 to create the Enh1mTTR (ET) promoter [292]. ET promoter was used in lentiviral-based preclinical studies of hemophilia B therapy in mice and dogs, which resulted in stable but weak expression level of cFIX [293]. A shortened version of the promoter was subsequently used in the development of AAV5-mediated ANB-002 gene therapy for hemophilia B [200]. −202 core mTTR promoter was used in pilot studies of AAV-mediated gene therapy for hemophilia A, where the key requirement for the promoter was its minimal length with sufficient activity [287,294,295]. It was also found that the most important regions of the proximal mTTR promoter seemed to be the HNF4 and HNF3-S (strong affinity) binding sites, since mutations in these regions significantly reduced promoter activity even in the presence of an enhancer. The enhancer was able to compensate for mutations in the HNF1 and HNF3-W sites (weak affinity) sites. Furthermore, the conversion of the HNF3-W site to the HNF3-S site increased promoter activity 1.2-fold in the presence of a 100 bp enhancer and 1.8-fold in its absence (Figure 3D) [296]. The mTTR mut promoter with the mentioned mutation was subsequently used for factor VIII expression in AAV-mediated gene therapies SPK-8011 and NGGT003 for hemophilia A without an enhancer region due to AAV vector capacity limitations. This mutation was reported to increase FVIII expression in vivo in mice by 4-fold compared to the wild-type promoter [193]. Spark suspended phase 3 trials of SPK-8011 (NCT06297486) despite stable but low levels of FVIII activity (mild HA) over 4 years, which amounted to mean 7.4% [297]. The company stated its intention to use the enhanced function FVIII variant to improve the effectiveness of gene therapy. mTTR mut promoter was used in other preclinical studies of gene therapy for hemophilia A [298].

It should be noted that no distal enhancer element in human TTR similar to that in mice has been characterized, despite high homology in the proximal promoter region. The human TTR core promoter has poor activity without the 100 bp mTTR enhancer [299]. Various enhancers assessment was performed to increase activity that resulted in the generation of E03.TTR promoter, which represents a combination of the 100 bp mTTR enhancer with the 190 bp hTTR core promoter (Figure 3E) [203,300]. The E03.TTR promoter was used in gene therapies for hemophilia A DTX201 (BAY2599023) and Wilson's disease UX701 [203,300]. DTX201 gene therapy provided promising results in clinical trials, achieving stable FVIII expression for at least 23 months despite the high content of CpG motifs [301].

Human and mouse TTR promoters were compared with 448 bp LP1 and 152 bp mFibr (modified human fibrinogen beta promoter) promoters in a factor VIII expression cassette. The LP1 and mTTR promoters, but not hTTR, provided the highest expression of factor VIII, that was confirmed by in vivo expression during hydrodynamic injection with plasmids. mTTR and LP1 promoters provided higher and more stable expression of FVIII in oversized vectors [302].

Due to its short length and compatibility with LP1 activity, mTTR was modified for use in gene therapies with large coding sequences. Nambiar et al. created the mTTR202opt promoter by replacing the HNF3 site with a variant with more affinity, as used for SPK-8011 and NGGT003, and replacing the HNF4 binding site in the proximal region of the promoter. Next, a modified mTTR enhancer was added to the resulting promoter, in which the HNF4 binding site was replaced by analogy with the proximal promoter, resulting in the mTTR482 promoter. mTTR202opt and mTTR482 provided more than twice the expression of FVIII in vivo compared to the WT-202 mTTR promoter [207]. Two copies of the 162 bp modified AMBP enhancer (to make it more affine to HNF3 and HNF4 factors) were added to mTTR482 to obtain the mAlMB2-mTTR482 promoter, which was used in SAR444836 gene therapy for phenylketonuria [207] (Figure 3F). Interestingly, the combination of the TTR promoter with the truncated AMBP enhancer led to the formation of FVIII-specific neutralization antibodies, that may raise some concerns in a number of therapies with high immunogenicity of the secreted transgene [203].

Several approaches were developed to use other genes as enhancers to boost the mTTR promoter. It was demonstrated that the proximal region −202/−70 hAAT can act as an enhancer for the heterologous βb-globin promoter, which could enhance transcription more than the 100 bp mTTR enhancer. In the study by Chuah et al., researchers used computational methods to search for highly conserved cis-acting regulatory modules (CRMs) in highly expressed liver genes. The regions they found were used as enhancers for the mouse TTR promoter, including the hTTR promoter region denoted as CRM10 [241]. The 71 bp CRM8, which is a part of the previously characterized tissue-specific element of the hAAT promoter region, was able to enhance the mTTR promoter 7–10 times and provided liver-specific expression in non-human primates delivered by the scAAV9 vector (Figure 3G) [141,241]. As a result of further modification, the hsCRM8-mTTR promoter, also known as HSh-TTR or HHS4-TTR, proved to be 2–3 times more active than the modified LP1 in vivo [280]. The Hsh-TTR promoter was used in preclinical studies of gene therapy for hemophilia B in non-human primates, as well as in the study of various FIX variants obtained through ancestral sequence reconstruction [303–305].

Independently of this modification the CRMSBS2-TTRm promoter was obtained by aligning the *SERPINA1* gene sequence (using ENCODE Alignment) and selecting low-conservative variants. The promoter is used in clinical trials of SB-525 (giroctocogene fitelparvovec) gene therapy for hemophilia A [209,211]. Interestingly, up to half of patients experienced supraphysiological FVIII activity levels, requiring anticoagulant therapy until physiological activity levels were restored. This may be related to both high promoter activity and high vector doses [306].

The use of a larger number of copies of CRM8 can significantly enhance the mTTR promoter, but presence of an mTTR enhancer is still necessary to provide maximum TTR promoter activity, that is an obstacle to its use in gene therapies with a large transgene coding sequence, as in the case of hemophilia A [196,307]. Three copies of an unmodified CRM8 fragment were used to enhance the enTTR-mTTR promoter, as in the case of VGB-R04 gene therapy for hemophilia B [213]. However, TAK-748 therapy utilizing the same promoter was terminated prior to the start of trials due to expectations of low therapeutic efficacy [196,212].

7. Promoters Based on Albumin (Alb) Gene

Albumin is the most abundant plasma protein (40–60% of total protein, 35–52 g/L in human plasma), synthesized exclusively by hepatocytes. Among its principal functions, albumin maintains blood oncotic pressure and transports various ligands, including fatty acids and hormones [308,309]. The exceptionally high level of liver-specific albumin expression—accounting for up to 15% of total hepatic protein synthesis—has prompted extensive investigation into its promoter as a powerful tool for achieving tissue-specific gene delivery in the context of gene therapy [310,311]. Building on the robust endogenous synthesis of albumin, various strategies involving zinc-finger nuclease (ZFN)-mediated integration of therapeutic transgenes directly into the albumin intron were pursued (SB-913, SB-318, SB-FIX). Nevertheless, these strategies demonstrated insufficient clinical efficacy in vivo, leading to a reorientation of research approaches toward the use of transcriptional regulation via the albumin promoter (without genomic integration) in AAV-mediated gene therapy systems [312–314].

Pilot studies suggested that the region of the albumin promoter (PAlb) lies between −31 and −213 and contains six potential cis-acting elements, divided into distal (DE) and proximal (PE) [309,315,316]. The regulatory regions of the promoter include the

https://doi.org/10.3390/cells15010014

TATA box, CCAAT box, PEI, PEII, DEI, DEII, and DEIII (Figure 4A). The TFs responsible for transcriptional activity of the albumin promoter include HNF1, C/EBP, DBP and nuclear factors Y and 1 (NF-Y and NF-1). It has been established that promoter regions are highly conserved in rat, mouse, and human *albumin* genes. The characteristic pattern of DNA–protein interactions observed in these promoters is preserved throughout evolution, reflecting the presence of hepatocyte-specific transcription factors that have been conserved over time [317].

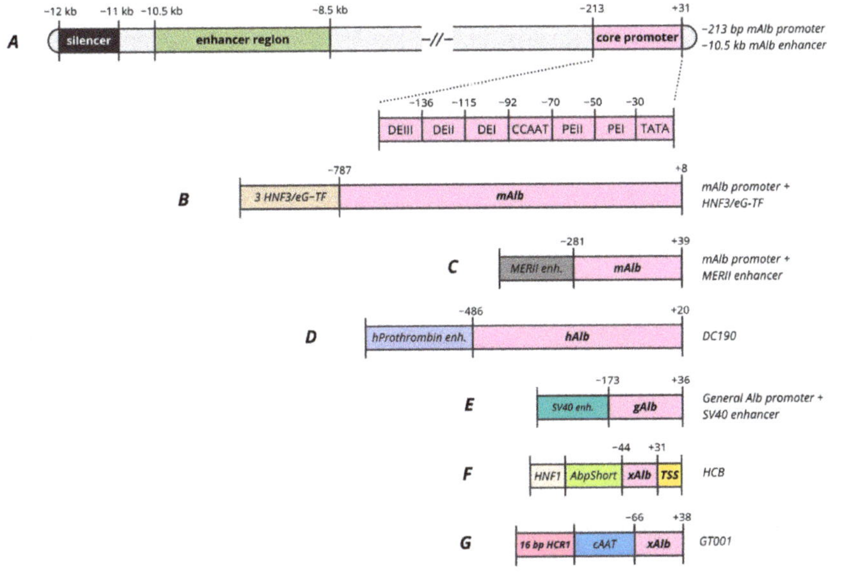

Figure 4. Promoters based on Alb gene. (**A**)—−213 bp albumin promoter and −10.5 kb enhancer regions. (**B**)—mAlb promoter composed of mAlb promoter with 3 copies of HNF3/eG-TF. (**C**)—mAlb promoter composed of mAlb promoter with MERII enhancer. (**D**)—DC190 promoter composed of hAlb promoter with hProthrombin enhancer. (**E**)—gAlb promoter composed of general Alb promoter with SV40 enhancer. (**F**)—HCB promoter composed of xAlb promoter with consensus TSS, AbpShort and HNF1. (**G**)—GT001 vector composed of xAlb promoter, cAAT and 16 bp HCR1.

Considering the functional significance of transcription factor-binding sites, it is noted that deletion of the HNF1 site completely blocks both promoter and enhancer activity [318]. Originally, this proximal region was identified as a negative regulatory element (NRE), owing to the presence of a GATC motif susceptible to methylation. Studies employing methylatable bacterial strains for plasmid preparation revealed a threefold reduction in transcription, underscoring the pivotal roles of HNF1 and epigenetic regulation [315]. The NF-Y site mutation reduces promoter activity by half, but distal enhancers can partially compensate for this effect. The C/EBP(GCAA) site mutation at position −124 leads to a 15-fold reduction in transcription, demonstrating the key function of C/EBP factors.

Analysis of 5′-deletions in the rat albumin promoter has revealed that the size of the promoter region is crucial for its function. A minimal construct extending to −84 bp, with or without a 2 kb albumin enhancer and containing the HNF1 and TATA box, exhibits only low-level activity. For full tissue specificity and proper responsiveness to enhancers, a promoter region extending to −170/−175 bp is required [318,319]. Comparison of variants of the human albumin promoter in reporter adenoviral vectors identified fragment −173/+36 as the most active, demonstrating high promoter activity in Hepa1–6 mouse liver cells (Figure 4E). The −247/+36 variant showed the lowest activity, indicating the

https://doi.org/10.3390/cells15010014

presence of negative regulatory elements in the −247/−173 or unfavorable interaction with the SV40 enhancer used in the study [310].

Several studies reported that the albumin promoter, when paired with its native enhancer, yields only modest levels of expression in vivo [141,320]. However, when regulatory elements from the albumin promoter, including modified variants, are combined with enhancer and promoter regions derived from other highly expressed liver genes, the resulting constructs achieve substantially higher levels of transgene expression.

The albumin enhancer (EAlb), located between −8.5 and −10.5 kb upstream of the transcription start site, harbors functional domains—eH-TF, C/EBP, HNF3, and NF-1—that serve as binding sites for liver-specific transcription factors, including HNF4A, C/EBPα, and FOXA2 [321–323]. This enhancer regulates the promoter by interacting with proximal elements: HNF1 directly activates the promoter, while architectural factors such as HMG-I(Y) and NF-Y fine-tune activation through chromatin looping, thereby ensuring robust and stable transgene expression in the liver [318,324,325].

It was identified that the region located further upstream, −12/−11 kb, is a silencer, as it inhibits transcription in the case of both the homologous albumin promoter and the heterologous herpes simplex virus thymidine kinase promoter (HSV-TK promoter) [326]. In the region −10,284/−9904 bp, point mutations in the C/EBP, C/EBP-RF, HNF3, and eH-TF sites impede enhancer function, and deletion of the eH-TF site completely negates activity, indicating the need for cooperative interaction of all these factors for maximum liver-specific expression [326]. Experiments demonstrated that the albumin enhancer Ealb (−10,500/−8500 bp, 370 bp) has limited effectiveness on its own and only functions with robust support, which was employed to construct the EalbAAT promoter mentioned above (Figure 2C) [141,154]. Understanding the critical role of various regulatory elements has allowed researchers to create optimized promoter constructs for the expression of reporter systems and transgenes.

In several studies the modified mouse albumin promoter (−787/+8 bp) was enhanced by three copies of HNF3/eG-TF binding sites its enhancer region (Figure 4B) [327,328]. This ensures high expression of biologically active human factor VIII/hAAT in HepG2 cells and animals, using retroviral and adenoviral vectors [245,329–331].

The minimal mouse albumin promoter (−281/+39 bp), coupled with the α-fetoprotein enhancer (MERII) (Figure 4C), the 5′ intron of factor IX, and the 3′ intron of albumin (3′iALB), achieves stable secretion of alkaline phosphatase in mouse liver, matching the CMV promoter's efficiency for a full year. The inclusion of 3′iALB after the coding region increases expression fivefold compared to variants without it, whereas the distal albumin enhancer in the construct (−10,284/−9904 bp) does not provide long-term activity [320]. The recombinant human liver-specific DC190 promoter contains the human albumin promoter (−486/+20 bp). To enhance transcriptional activity, two tandem copies of the human prothrombin enhancer (−940/−860 bp), a hybrid intron (three-way adenovirus donor leader/mouse immunoglobulin acceptor) and a BGH polyA signaling sequence for transcript stabilization (Figure 4D) [256,332–334]. Using the DC190 promoter in the AAV2/DC190-αgal construct, the expression of α-galactosidase A in mouse livers increased by 15-fold compared to the CMV promoter [335]. The addition of the hAAT intron further increased transgene expression due to splicing optimization and mRNA stabilization.

The promoter of the frog (*Xenopus laevis*) *albumin* gene, containing a 13-nucleotide HP1 element (CNXNNTTINNNNNC) and a TATA box, was used as a minimal liver-specific promoter capable of sustaining expression in hepatocyte cells in vitro. At the same time, proximal NF-Y/CCAAT, HNF1 and other sites are not essential components of the construct. Similar HP1 motifs have been found in the promoters of mouse and human *albumin*, *α-fetoprotein*, *fibrinogen beta chain*, and *α_1-antitrypsin* genes, demonstrating

https://doi.org/10.3390/cells15010014

significant evolutionary stability of this element [336]. The frog Alb promoter was used in the clinical trial of the gene therapy drug for hemophilia A–ASC618 (NCT04676048) as part of the minimal synthetic promoter HCB (HNF1-AbpShort-SynO-TSS, 146 bp). It consists of SynO (41 bp) from the 5′ UTR of X. laevis albumin as the core promoter, AbpShort (61 bp), a truncated α-microglobulin enhancer, a consensus TSS (Figure 4F) and provides 14 times greater FVIII activity compared to the clinically used HLP promoter (252 bp) in the AAV-FVIII [214]. Similarly, in the hemophilia A therapy (GS1191-0445) using the AAV8 vector, a modified frog Alb promoter (−66/+38, xAlb) was combined with elements from dog AAT (cAAT) and human AAT (hAAT), along with a synthetic enhancer (Figure 4G) [218].

8. Promoters Based on AFP Gene

Alpha-fetoprotein (AFP) is a serum glycoprotein, a member of a multigene family that also includes the albumin gene, which is closely linked to it [337]. During embryonic development, AFP is actively expressed in the visceral endoderm of the yolk sac, the fetal liver and the gastrointestinal tract, but after birth, AFP expression in the liver decreases sharply to practically undetectable levels—the concentration of AFP mRNA decreases by 10^4 times 3–4 weeks after birth [324,338,339]. Paradoxically, this limitation makes AFP a perfect tool for cancer-selective therapy. In hepatocellular carcinoma, AFP is reactivated in about 70–80% of cases. This creates a unique opportunity to target tumor cells specifically, without harming healthy liver cells [340–342]. For liver-specific delivery to healthy hepatocytes, hybrid constructs are developed. These combine AFP enhancers with constitutively active liver-specific promoters, allowing the powerful enhancer properties of AFP elements to be utilized while maintaining the ability to express in healthy hepatocytes.

The AFP promoter region can be divided into two main classes of regulatory elements: the proximal tissue-specific region (−85/−52 bp from the transcription start site) and distal enhancers (−7.6/−1.0 kb), both of which are critical for transcription and tissue-specific expression of AFP. The enhancer domain is represented by three functionally independent elements (I: −1.0 to −3.8 kb; II: −3.8 to −5.3 kb; III: −5.3 to −7.6 kb), which have significant functional redundancy and function independently of their position and orientation relative to the promoter. The most active region is the most proximal element I (−3.8/−1.0 kb), which enhances transcription from the HSV TK promoter by 50-fold [343]. In addition to the activating elements, two repressor sites (−1822/−951 and −402/−169 bp) have been identified, the former being more active and capable of blocking the action of enhancers [344]. Distal enhancers need a PCE (promoter-coupling element) to function properly. This element activates HNF1-binding sites and ensures that distant enhancers interact with proximal promoter elements [318]. AFP regulation is characterized by the involvement of classical liver-specific transcription factors, including HNF1, C/EBP and other hepatocyte nuclear factors [214,243]. The minimal proximal promoter AFP mouse (61 bp) was tested in a study to develop a promoter for hemophilia A gene therapy. In a construct containing AbpShort-HP1-AFP-TSS, the minimal promoter demonstrates a significant increase in expression in vitro [214].

9. Promoters Based on TBG Gene

Thyroxine-binding globulin (TBG) is a liver-derived glycoprotein responsible for the transport of thyroid hormones in blood serum. Since TBG is predominantly synthesized in hepatocytes, its promoter has become a key tool for achieving tissue specificity in gene therapy for inherited liver diseases. Minimal promoter activity is observed in non-target tissues, including the spleen, kidneys, and large intestine [149].

The structural and functional organization of the TBG promoter has been analyzed since 1993 [345]. The key HNF1 site at positions −77/−65 bp is the most critical, as its

https://doi.org/10.3390/cells15010014

mutation completely eliminates promoter activity. Additionally, there are a TATA box (+26 bp), C/EBP, an NF-1 site (−175 bp), and several AP-1 and HNF3 sites associated with tissue specificity. TBG promoter variants with different lengths were characterized in order to discover the most active variant. The study tested three additional TBG promoter fragments TBG1 (−253/−26 bp), TBG2 (−635/−26 bp) and TBG3 (−1190/−26 bp) and compared them with the original −435/−26 bp fragment. The shortest fragment, TBG1, showed a strong drop in activity, while extending the promoter to −635 bp slightly increased expression; however, further extension to −1190 bp reduced activity again. The −435/−26 bp fragment proved to be the most effective and practical, providing high transgene expression without unnecessarily increasing construct size [149,345,346].

Based on TBG, the liver-specific promoter (LSP) has been created, combining the −382/+3 bp fragment of TBG, two copies of the AMBP enhancer (α1-microglobulin/bikunin, −2804/−2704) and the 71 bp leader sequence [347]. LSP is actively used in AAV vectors for the treatment of hemophilia A and B. In animal model experiments, it provides stable expression of coagulation factors and correction of the disease [51,256,277,348–350]. A dual-vector system has been created to treat hemophilia A. It separately expresses the heavy and light chains of FVIII using vectors controlled by the LSP (659 bp). This system demonstrates equivalent efficacy to the single-vector approach using a short synthetic 368 bp IGBP (insulin-like growth factor-binding protein) promoter [351].

In clinical studies, AAV8.TBG.hARSB therapy with two copies of 101 bp of the AMBP enhancer and the −474/+3 region of the TBG promoter maintained 38–67% of normal arylsulfatase B (ARSB) activity in patients with mucopolysaccharidosis type VI (MPS VI) during 45 months of follow-up [219]. In the treatment of Pompe disease, ACTUS-101 gene therapy utilized LSP to express α-glucosidase. Safety and sustained increase in GAA levels in patients in phase 1/2 was demonstrated [222]. Hemophilia B (DTX101) therapy with LSP in AAVrh10 reduces the annual bleeding rate by 71% and provides sustained factor IX activity > 5% in 82% of patients after 24 months [221]. However, the primary endpoint of the clinical trials in terms of IX activity level was not achieved, and the therapy was discontinued.

TBG promoter with AMBP enhancer (680 bp) and Kozak-like sequence in the construct for the treatment of ornithine transcarbamylase deficiency (OTCD) (scAAV8.TBG.hOTCco) provides 60% of normal OTC activity on day 28 [225].

10. Synthetic Promoters

Over recent decades, the rapid development of computational approaches analyzing regulatory element conservation and multi-omics data has enabled rational design of fully synthetic regulatory elements. Omics datasets (particularly ChIP-seq for profiling histone modifications and TFBS, ATAC-seq for assessing chromatin accessibility, as well as spatial and single-cell transcriptomics) provide valuable insights into regulatory grammar and the development of synthetic promoters. These datasets allow the identification of natural regulatory elements, evaluation of their activity, and subsequent application to bioinformatic tasks such as classifying sequences as promoter or non-promoter, predicting promoter activity, and generating de novo promoters.

Promoter design typically involves an iterative process of sequence generation, activity prediction, and subsequent optimization. Various deep learning approaches, such as convolutional and recurrent neural networks, are used to identify and enrich synthetic constructs with motifs specific to the target tissue. The current state of these approaches is reviewed by Wang et al.; in this work we focus on tools and synthetic constructs for liver-directed applications [352].

https://doi.org/10.3390/cells15010014

For enhancer identification, a hybrid neural network, Enhancer-CRNN was developed [353]. Trained on publicly available histone modification data, it accurately localizes natural enhancers, including those in the HepG2 cell line.

DeepLiver, a hierarchical convolutional neural network, was trained to predict enhancer activity and hepatocyte zonation using multi-omics data [354]. The model identified key transcription factor motifs, such as HNF4A, C/EBPa, HNF1A, FOXA1, and AP-1, as drivers of enhancer activity, and its predictions strongly correlated with experimental MPRA data, including for synthetic sequences.

The CODA platform generates 200-bp synthetic enhancers for three cell lines (HepG2, K562, and SK-N-SH) based on activity predictions from the Malinois convolutional neural network, followed by sequence optimization [355]. Analysis of the optimized sequences showed enrichment of binding sites characteristic for the target cell types; for HepG2, these included TFBS for HNF1B and HNF4A. Subsequent testing in transgenic zebrafish demonstrated liver expression in 27 out of 36 cases.

Regarding the practical application of synthetic constructs, the Lxp2.1 promoter was designed to ensure selective transcriptional activity in hepatocytes and comprises a minimal promoter supplemented with two enhancer elements. Its structure contains three binding sites for HNF1, two sites each for HNF3 and HNF4A, and one site for SP1. The Lxp2.1 promoter is part of the vector systems BBM-H901, the first gene therapy for hemophilia B approved in China for clinical use, and BBM-H803, a gene therapy for hemophilia A [229]. Notably, during therapeutic development, the arrangement and relative spacing of binding sites have been shown to play a major role in the efficiency of synthetic constructs.

Another example of a synthetic construct is the 54-bp Em enhancer, incorporated in a construct developed for ZS802 hemophilia A gene therapy [189]. This enhancer was used in combination with the human AAT core promoter and contains HNF4A, C/EBPa/b, FOXA2, HNF1A/B, and D site-binding protein (DBP). The AAT core promoter was also used in G6PC_COMP_v1 and G6PC_COMP_v3. The synthetic sequences contain TFBS, which increase expression in hepatocytes up to 20-fold over the LP-1 promoter [172]. The TFBS are organized as consecutive pairs: HNF1/HNF3, HNF3/HNF3, and C/EBP/HNF4, separated by spacers.

Although we cannot precisely determine how the promoters described above were rationally designed, fully synthetic promoters nonetheless constitute an important tool. Their further development may be more effectively aided by design methods based on neural networks.

11. Conclusions and Future Prospects

The importance of the liver as a central metabolic organ cannot be overstated. The large number of secreted proteins, the maintenance of balance between carbohydrate and lipid metabolism, and its special immune status make the liver an attractive target for gene therapy for various diseases. Current strategies for developing promoters for gene therapy build on fundamental principles of the liver's regulatory network. More specifically, these strategies are based on the interaction of transcription factors with natural promoters of tissue-specific and highly expressed genes. Despite significant progress in identifying many specific regulatory elements, our current knowledge does not allow us to precisely and flexibly regulate expression, which is necessary for a number of metabolic diseases. Furthermore, researchers face several challenges in developing liver-directed gene therapies due to inherent limitations of delivery vectors.

Early attempts at liver-directed gene therapy faced significant hurdles related to promoter choice. The use of adenoviral vectors was complicated by the rapid loss of transgene expression when driven by strong ubiquitous promoters, including viral ones [356,357].

https://doi.org/10.3390/cells15010014

Although the use of liver-specific promoters was a shift to reduce the overall immunogenicity of the vector resulted in a more stable and prolonged therapeutic levels of transgene expression, these promoters failed to eliminate the immunogenicity of the adenoviral vector itself, and loss of transgene expression persisted [258,358–362]. Similarly, initial studies using AAV vectors with viral promoters also failed to achieve stable, high-level transgene expression in target tissues like the liver and muscle [33,143,363,364]. A critical turning point was the discovery that ubiquitous promoters increases the risk of inhibitor formation against the expressed transgene, whereas liver-specific promoters facilitate the immune tolerance [259,365]. In 1999 a seminal research study demonstrated the use of a liver-specific LSP in an AAV vector, achieving stable FIX expression in mice at levels up to 10 μg/mL—twice the normal human physiological concentration [347]. Subsequent studies showed that liver-specific promoter expression could induce transgene-specific immune tolerance, underscoring the safety and efficacy of AAV vectors in gene therapy [259,275,366]. Transgene expression in the liver is actively applied in therapies in which the therapeutic protein needs to perform its physiological effect in muscle cells, such as in Pompe disease. Directing expression to the liver serves to induce immune tolerance, since the use of muscle-specific or ubiquitous promoters leads to the formation of transgene-specific neutralizing antibodies, as observed in clinical trials of AAV1-CMV-GAA [367–369].

To enable dual expression in the liver and muscles, new promoters such as LiMP (liver–muscle promoter) have been developed. LiMP combines the previously characterized strong liver-specific HCR1-hAAT promoter with the spC5.12 promoter, which is active in skeletal and cardiac muscle. The promoter was utilized in preclinical trials for the treatment of Pompe disease and MPS IVA (mucopolysaccharidosis type IVA) [370–372]. To date, all approved liver-directed gene therapies are based exclusively on the delivery of AAV vectors. mRNA delivery via lipid nanoparticles (LNPs) is also being actively developed and is currently in clinical trials However, mRNA delivery cannot provide sustained expression and requires regular drug administration [373].

An additional advantage of using liver-specific promoters in AAV vectors is their potency to induce immune tolerance in subjects with pre-existing antibodies, which may be highly relevant in human therapy [51,366,374]. Currently, the presence of neutralizing antibodies to the transgene in patients is the exclusion criteria for gene therapy. However, there are encouraging results suggesting the possibility of eliminating pre-existing neutralizing antibodies with appearance of the transgene therapeutic activity in the same time in humans, as recently demonstrated in clinical trials of BMN 270 gene therapy (NCT04684940) [53]. All these results highlight the critical role of liver-specific promoters in ensuring the efficacy and safety of gene therapy.

Despite greatest frequency of liver-specific promoters to use in tissue-specific gene therapies and the diversity of their variants, the most of liver-specific promoters are based on just three genes including *AAT*, *TTR*, and *TBG* [140]. Up to half of the gene therapies presented in this review utilize promoters based on *AAT* (48.7%), 20.5% on *TTR*, 7.7% on combinations of *AAT* and *TTR*, and 10.3% on *TBG* (Table 1). It should be noted that all three AAV-mediated liver-directed FDA approved gene therapies use liver-specific promoters based on human AAT promoter with ApoE/HCR1 enhancer, including Hemgenix, Beqvez, and Roctavian. Although these promoters proved their effectiveness, new promoter variants based on other genes and/or computational technologies need to be developed to improve the efficacy of gene therapies.

In summary, our data indicate that a universal strategy for designing liver-specific promoters is to adopt the architecture of highly expressed hepatic genes and to reduce it in a rational way. In most cases, such promoters are constructed by truncating natural sequences to minimal functional regions, by varying their length, and by testing expression

https://doi.org/10.3390/cells15010014

constructs in vitro [375]. TF footprinting is an informative tool for rational truncation. Three widely used promoters (AAT, TTR, TBG) contain TFBS that are strongly associated with high expression in vivo and in vitro: HNF1, FOX, CEBP, MyoD, LEF-1, LEF-1/TCFβ, and Tal1β/E47 for AAT; HNF1, CEBP, FOX, LEF-1, LEF-1/TCF, and MyoD for TTR [241]; and HNF1 and HNF3 for TBG [345]. Preference is also given to promoters of genes that show more predictable and stable expression. For example, the ALB promoter has not been widely used, apparently because of its more complex regulatory organization and its dependence on the physiological state of the liver parenchyma, including changes during inflammation and nutritional stress [376,377].

Unfortunately, despite numerous studies investigating or creating promoters based on the *hAAT*, *TTR*, and *TBG* genes, there is no information on a direct comparison of all the promoters mentioned in this review within a single study. Comparing promoter activity based on multiple studies is complicated by different study design approaches. First, various reference promoters can be used to evaluate relative promoter activity, which prevents general normalization to one specific promoter. Second, promoter activity can be assessed both at the level of transgene mRNA and by the amount of reporter or therapeutic gene protein. Third, various approaches can be used for transgene expression, including both plasmid expression vectors and viral vectors, that also impacts expression efficiency. Fourth, and most importantly, the activity of different promoters can be assessed in vitro in various cell lines and in vivo in animal models. All these variables prevent direct comparison of both the activity and safety of promoters based on the results of different studies. However, with the development of sequencing technology, high-throughput methods for screening promoters have become available, such as based on the creation of barcoded libraries, that allow the activity of multiple promoter and enhancer elements to be evaluated in parallel [150,378–380].

It was recently reported that the 2.2 kb GFAP (glial fibrillary acidic protein) promoter provides the highest level of expression in the liver among a variety of strong ubiquitous and liver-specific promoters, that was rather unexpected, since GFAP promoter is known for astrocyte-specific expression in the central nervous system. An apparent limitation of the GFAP promoter is a large length, and the currently existing truncated variants do not provide the same activity [150,381]. It is also important to mention the previously described promoters that have not yet found application in gene therapies, such as the ADH6 (class V alcohol dehydrogenase) promoter, which has comparable activity to hAAT [266,382]; the APOA1 promoter used in Ad vectors [254,358]; the APOC2 promoter, which is more than 6 times more active than APOA1 [255,383]; the IGBP promoter, which in combination with 2 copies of the AMBP enhancer provide similar to the HCR1-hAAT promoter activity [277]; the AHSG (or pp63 for rat, alpha-2-HS-glycoprotein) promoter, which provide greater activity compared to the β actin promoter [384]; the L-PK (liver-type pyruvate kinase) promoter, which has activity similar to that of the mAlb promoter despite its short length [385]; the ALDOB (aldolase B) promoter, which can be enhanced more than 100-fold by enhancer regions within the intron [386]. In the future these promoters after optimization may find an application in liver-directed gene therapy. However, for some promoters, a serious limitation for application is the inducibility from external signals, such as cytokines IL-1 and IL-6 (acute-phase proteins) or hormones. The promoters of *HP* (haptoglobin) [387], *Hpx* (hemopexin) [388], *SERPINA3c* [389], *FGA* (fibrinogen alpha chain) [390,391], *FCB* (fibrinogen beta chain) [392], *PEPCK* (phosphoenolpyruvate carboxykinase) [244,245], *L-PK* [385], and *CRP* (C-reactive protein) [393] genes have weak basal activity without induction, but at the same time can be extremely active in vivo depending on the body conditions, that makes them difficult to control and unsafe for in vivo application. A promising solution to this problem could be the modification/replacement of inducible regulatory sites and

https://doi.org/10.3390/cells15010014

maintenance of always-on high promoter activity without regulation by external signals. However, in practice such modifications often diminish not only inducible activity but also basal promoter activity [387,391,392].

The use of strong liver-specific promoters can overcome barriers to the effective and safe application of liver-directed gene therapies through increased transgene expression and reduced viral load at the same time. The ideal promoter for gene therapy (not only for liver-directed) should provide stable high tissue-specific expression levels with its minimal size. However, in practical application, the choice of promoter is based on the characteristics of the delivery vector. The use of scAAV or transgenes with a large CDS greatly restricts promoter length, which can negatively impact its activity. Moreover, recent studies suggest that the use of strong promoters may be undesirable, despite their specificity. The utilization of a stronger promoter with a reduced vector dose results in a greater metabolic load within a smaller number of transduced cells, which also negatively affects expression stability. The strongest promoters can lead to unfolded protein response (UPR), endoplasmic reticulum (ER) stress, loss of transduced cells, and the formation of neutralizing antibodies to the transgene [34,203,394].

However, it is important to note that these studies were conducted in the context of preclinical studies of AAV-mediated gene therapies for hemophilia A for endogenous expression of factor VIII, which has low secretion efficiency and ability to induce UPR and ER stress [395]. The results may also be influenced by the use of oversized vectors, the AAV manufacturing platform, and the vector serotype [396]. This suggests that the immunogenicity of promoters should primarily be evaluated in the context of the characteristics of the vector used, the expressed transgene, and the route of vector administration. Thus, tightly regulated promoters of special interest could find wide application in the development of gene therapies for metabolic diseases. However, to date, approaches to promoter selection for the treatment of metabolic disorders or coagulation factor deficiencies do not differ, focusing on stable high transgene expression.

A particularly attractive prospect for highly effective gene therapy is the ability to regulate promoter activity in a controlled manner via internal stimuli—for example, to stimulate insulin expression in response to elevated glucose levels or to reduce transgene expression at supraphysiological levels [397]. Such features could significantly expand the list of diseases suitable for gene therapy and increase safety. The use of synthetic promoters remains limited at the current stage of genetic engineering development, although several clinical applications have been presented in this review. The methods for generation of these promoters are not always clearly established, but it can be concluded that they represent a combination of well-known liver-specific TFBS. The most effective sequences and optimal distances between them were experimentally determined for these sites. Attempts to create synthetic promoters using deep learning technologies experience a number of challenges. The key challenge is developing models that can accurately predict the correlations between sequence features and promoter activity. The primary complexity is accounting for spatial regulation, DNA structure and plasticity, and interactions with regulatory proteins. Large amounts of specialized annotated genomic data are required to identify patterns and model the complex spatial organization of regulatory elements. One challenge is the necessity to develop models that can accurately predict the relationships between sequences and their activity levels. The principal difficulty lies in accounting for spatial regulation, DNA structure and plasticity, and interactions with regulatory proteins. Large amounts of specialized annotated genomic data are required to identify patterns and model the complex spatial organization of regulatory elements. In addition, the accumulation of MPRA data is particularly important for model training, as it provides predictive power in the context of synthetic sequences. Training exclusively on natural sequences faces the problem

https://doi.org/10.3390/cells15010014

of limited data volume, which can be partially addressed through data augmentation methods; however, their use may introduce distortions. Future research should focus on developing new architectural solutions aimed not only at identifying specific sites, but also at accounting for their interactions and generating diverse candidate sequences.

In summary, further development in the tissue-specific promoters area requires a comprehensive multidisciplinary approach, involving fundamental research to a deeper understanding of gene expression regulation, including epigenetic mechanisms. This also suggests the need to study aspects of intrahepatic signaling, which plays a key role in inducing cellular response and adaptation. In addition, a critical objective is to create and improve computational architectures capable of effectively extracting and processing information about complex epigenetic and signaling features. The accumulation of relevant data will serve as the platform for developing more accurate models to predict activity and generate synthetic constructs implemented in practical use.

Supplementary Materials: The following supporting information can be downloaded at: https://www.mdpi.com/article/10.3390/cells15010014/s1, Material S1: List of promoters; Material S2: Supplementary Table S1: Gene therapies with liver-specific promoters [151–232].

Author Contributions: Conceptualization, V.A., A.M. and S.G.F.; investigation, V.A., A.I.P., A.A.D. and D.M.; writing—original draft preparation, V.A., A.I.P., A.A.D., D.M., V.B. and J.K.; writing—review and editing, V.A., A.I.P., A.A.D., D.M. and A.M.; visualization, V.A., A.A.D. and D.M.; supervision, S.G.F. and P.Y.V.; project administration, O.N.M. and P.Y.V.; funding acquisition, V.A., S.G.F., O.N.M. and P.Y.V. All authors have read and agreed to the published version of the manuscript.

Funding: This study was supported by the Russian Science Foundation (Grant No. 23-64-00002) (Mityaeva O.N. and Volchkov P.Y.) and by the Ministry of Science and Higher Education of the Russian Federation (agreement 075-03-2023-106/12) (Bogdanov V. and Krupinova J.).

Institutional Review Board Statement: Not applicable.

Informed Consent Statement: Not applicable.

Data Availability Statement: No new data were created or analyzed in this study.

Conflicts of Interest: The authors declare no conflicts of interest.

References

1. Jacobs, F.; Gordts, S.; Muthuramu, I.; De Geest, B. The Liver as a Target Organ for Gene Therapy: State of the Art, Challenges, and Future Perspectives. *Pharmaceuticals* **2012**, *5*, 1372–1392. [CrossRef]
2. Sanderson, S.; Green, A.; Preece, M.A.; Burton, H. The Incidence of Inherited Metabolic Disorders in the West Midlands, UK. *Arch. Dis. Child.* **2006**, *91*, 896–899. [CrossRef]
3. Chuecos, M.A.; Lagor, W.R. Liver Directed Adeno-associated Viral Vectors to Treat Metabolic Disease. *J. Inherit. Metab. Dis.* **2024**, *47*, 22–40. [CrossRef] [PubMed]
4. Aravalli, R.N.; Belcher, J.D.; Steer, C.J. Liver-targeted Gene Therapy: Approaches and Challenges. *Liver Transplant.* **2015**, *21*, 718–737. [CrossRef]
5. Huebert, R.C.; Rakela, J. Cellular Therapy for Liver Disease. *Mayo Clin. Proc.* **2014**, *89*, 414–424. [CrossRef]
6. Lin, L.; Li, H. Analysis of Clinical Trials of New Drugs for Liver Diseases in China. *Drug Des. Dev. Ther.* **2021**, *15*, 3181–3191. [CrossRef] [PubMed]
7. Ramadori, G.; Saile, B.; Ramadori, G.; Saile, B. Mesenchymal Cells in the Liver—One Cell Type or Two? *Liver* **2002**, *22*, 283–294. [CrossRef]
8. Ishibashi, H.; Nakamura, M.; Komori, A.; Migita, K.; Shimoda, S. Liver Architecture, Cell Function, and Disease. *Semin Immunopathol.* **2009**, *31*, 399–409. [CrossRef] [PubMed]
9. Herzog, R.W.; Kaczmarek, R.; High, K.A. Gene Therapy for Hemophilia—From Basic Science to First Approvals of "One-and-Done" Therapies. *Mol. Ther.* **2025**, *33*, 2015–2034. [CrossRef]
10. Nguyen, T.; Ferry, N. Liver Gene Therapy: Advances and Hurdles. *Gene Ther.* **2004**, *11*, S76–S84. [CrossRef]

https://doi.org/10.3390/cells15010014

11. Herrmann, J.; Abriss, B.; Van De Leur, E.; Weiskirchen, S.; Gressner, A.M.; Weiskirchen, R. Comparative Analysis of Adenoviral Transgene Delivery via Tail or Portal Vein into Rat Liver. *Arch. Virol.* **2004**, *149*, 1611–1617. [CrossRef] [PubMed]
12. Sherman, A.; Schlachterman, A.; Cooper, M.; Merricks, E.P.; Raymer, R.A.; Bellinger, D.A.; Herzog, R.W.; Nichols, T.C. Portal Vein Delivery of Viral Vectors for Gene Therapy for Hemophilia. In *Gene Correction*; Storici, F., Ed.; Methods in Molecular Biology; Humana Press: Totowa, NJ, USA, 2014; Volume 1114, pp. 413–426, ISBN 978-1-62703-760-0.
13. Sands, M.S. AAV-Mediated Liver-Directed Gene Therapy. In *Adeno-Associated Virus*; Snyder, R.O., Moullier, P., Eds.; Methods in Molecular Biology; Humana Press: Totowa, NJ, USA, 2012; Volume 807, pp. 141–157, ISBN 978-1-61779-369-1.
14. Braet, F.; Wisse, E. Structural and Functional Aspects of Liver Sinusoidal Endothelial Cell Fenestrae: A Review. *Comp. Hepatol.* **2002**, *1*, 1. [CrossRef]
15. Warren, A.; Le Couteur, D.G.; Fraser, R.; Bowen, D.G.; McCaughan, G.W.; Bertolino, P. T Lymphocytes Interact with Hepatocytes through Fenestrations in Murine Liver Sinusoidal Endothelial Cells. *Hepatology* **2006**, *44*, 1182–1190. [CrossRef]
16. Snoeys, J.; Lievens, J.; Wisse, E.; Jacobs, F.; Duimel, H.; Collen, D.; Frederik, P.; De Geest, B. Species Differences in Transgene DNA Uptake in Hepatocytes after Adenoviral Transfer Correlate with the Size of Endothelial Fenestrae. *Gene Ther.* **2007**, *14*, 604–612. [CrossRef]
17. Bucher, N.L.; Swaffield, M.N. The Rate of Incorporation of Labeled Thymidine Into the Deoxyribonucleic Acid of Regenerating Rat Liver in Relation to the Amount of Liver Excised. *Cancer Res.* **1964**, *24*, 1611–1625.
18. Rozga, J. Hepatocyte Proliferation in Health and in Liver Failure. *Med. Sci. Monit.* **2002**, *8*, RA32–RA38.
19. Xu, L.; Haskins, M.E.; Melniczek, J.R.; Gao, C.; Weil, M.A.; O'Malley, T.M.; O'Donnell, P.A.; Mazrier, H.; Ellinwood, N.M.; Zweigle, J.; et al. Transduction of Hepatocytes after Neonatal Delivery of a Moloney Murine Leukemia Virus Based Retroviral Vector Results in Long-Term Expression of β-Glucuronidase in Mucopolysaccharidosis VII Dogs. *Mol. Ther.* **2002**, *5*, 141–153. [CrossRef] [PubMed]
20. Cunningham, S.C.; Dane, A.P.; Spinoulas, A.; Alexander, I.E. Gene Delivery to the Juvenile Mouse Liver Using AAV2/8 Vectors. *Mol. Ther.* **2008**, *16*, 1081–1088. [CrossRef] [PubMed]
21. Arii, S.; Imamura, M. Physiological Role of Sinusoidal Endothelial Cells and Kupffer Cells and Their Implication in the Pathogenesis of Liver Injury. *J. Hepato-Biliary-Pancreat. Surg.* **2000**, *7*, 40–48. [CrossRef]
22. Wolff, G.; Worgall, S.; Van Rooijen, N.; Song, W.R.; Harvey, B.G.; Crystal, R.G. Enhancement of in Vivo Adenovirus-Mediated Gene Transfer and Expression by Prior Depletion of Tissue Macrophages in the Target Organ. *J. Virol.* **1997**, *71*, 624–629. [CrossRef]
23. Fitzpatrick, Z.; Leborgne, C.; Barbon, E.; Masat, E.; Ronzitti, G.; Van Wittenberghe, L.; Vignaud, A.; Collaud, F.; Charles, S.; Simon Sola, M.; et al. Influence of Pre-Existing Anti-Capsid Neutralizing and Binding Antibodies on AAV Vector Transduction. *Mol. Ther.-Methods Clin. Dev.* **2018**, *9*, 119–129. [CrossRef] [PubMed]
24. Mingozzi, F.; Maus, M.V.; Hui, D.J.; Sabatino, D.E.; Murphy, S.L.; Rasko, J.E.J.; Ragni, M.V.; Manno, C.S.; Sommer, J.; Jiang, H.; et al. $CD8^+$ T-Cell Responses to Adeno-Associated Virus Capsid in Humans. *Nat. Med.* **2007**, *13*, 419–422. [CrossRef]
25. Huttner, N.A.; Girod, A.; Perabo, L.; Edbauer, D.; Kleinschmidt, J.A.; Büning, H.; Hallek, M. Genetic Modifications of the Adeno-Associated Virus Type 2 Capsid Reduce the Affinity and the Neutralizing Effects of Human Serum Antibodies. *Gene Ther.* **2003**, *10*, 2139–2147. [CrossRef]
26. Mével, M.; Bouzelha, M.; Leray, A.; Pacouret, S.; Guilbaud, M.; Penaud-Budloo, M.; Alvarez-Dorta, D.; Dubreil, L.; Gouin, S.G.; Combal, J.P.; et al. Chemical Modification of the Adeno-Associated Virus Capsid to Improve Gene Delivery. *Chem. Sci.* **2020**, *11*, 1122–1131. [CrossRef]
27. Gan, C.; Leng, M.; Liu, Y.; Zheng, Z.; He, S.; Qiao, W.; Xiao, L.; Xiao, Y.; Ye, J.; Zhou, L.; et al. The Combination of rAAV Pseudo-Lipid Nanoparticle and Triamcinolone Acetonide Enables Multi-Administration to Liver. *Mol. Ther. Methods Clin. Dev.* **2025**, *33*, 101399. [CrossRef]
28. Pan, T.; Yang, C.-H.; Zhao, K.; Sun, Y.-L.; Rao, Z.-Y.; Qu, W.-Q.; Zhang, S.-M.; Jin, X.-K.; Yan, Y.; Zhang, X.-Z. Biomimetic Artificial Enveloped Viral Vectors: Overcoming Immune Barriers for Re-Administration and Long-Term Gene Therapy. *Cell Biomater.* **2025**, *1*, 100143. [CrossRef]
29. Papadakis, E.; Nicklin, S.; Baker, A.; White, S. Promoters and Control Elements: Designing Expression Cassettes for Gene Therapy. *Curr. Gene Ther.* **2004**, *4*, 89–113. [CrossRef] [PubMed]
30. Bell, P.; Wang, L.; Chen, S.-J.; Yu, H.; Zhu, Y.; Nayal, M.; He, Z.; White, J.; Lebel-Hagan, D.; Wilson, J.M. Effects of Self-Complementarity, Codon Optimization, Transgene, and Dose on Liver Transduction with AAV8. *Hum. Gene Ther. Methods* **2016**, *27*, 228–237. [CrossRef]
31. Martinelli, I.; De Stefano, V.; Mannucci, P.M. Inherited Risk Factors for Venous Thromboembolism. *Nat. Rev. Cardiol.* **2014**, *11*, 140–156. [CrossRef]

https://doi.org/10.3390/cells15010014

32. Herzog, R.W.; Mount, J.D.; Arruda, V.R.; High, K.A.; Lothrop, C.D. Muscle-Directed Gene Transfer and Transient Immune Suppression Result in Sustained Partial Correction of Canine Hemophilia B Caused by a Null Mutation. *Mol. Ther.* **2001**, *4*, 192–200. [CrossRef]
33. Herzog, R.W.; Yang, E.Y.; Couto, L.B.; Hagstrom, J.N.; Elwell, D.; Fields, P.A.; Burton, M.; Bellinger, D.A.; Read, M.S.; Brinkhous, K.M.; et al. Long-Term Correction of Canine Hemophilia B by Gene Transfer of Blood Coagulation Factor IX Mediated by Adeno-Associated Viral Vector. *Nat. Med.* **1999**, *5*, 56–63. [CrossRef]
34. Fong, S.; Handyside, B.; Sihn, C.-R.; Liu, S.; Zhang, L.; Xie, L.; Murphy, R.; Galicia, N.; Yates, B.; Minto, W.C.; et al. Induction of ER Stress by an AAV5 BDD FVIII Construct Is Dependent on the Strength of the Hepatic-Specific Promoter. *Mol. Ther.-Methods Clin. Dev.* **2020**, *18*, 620–630. [CrossRef]
35. Kumar, S.R.P.; Hoffman, B.E.; Terhorst, C.; De Jong, Y.P.; Herzog, R.W. The Balance between CD8+ T Cell-Mediated Clearance of AAV-Encoded Antigen in the Liver and Tolerance Is Dependent on the Vector Dose. *Mol. Ther.* **2017**, *25*, 880–891. [CrossRef]
36. Basner-Tschakarjan, E.; Mingozzi, F. Cell-Mediated Immunity to AAV Vectors, Evolving Concepts and Potential Solutions. *Front. Immunol.* **2014**, *5*, 350. [CrossRef]
37. Niederkorn, J.Y. See No Evil, Hear No Evil, Do No Evil: The Lessons of Immune Privilege. *Nat. Immunol.* **2006**, *7*, 354–359. [CrossRef] [PubMed]
38. Forrester, J.V.; Xu, H.; Lambe, T.; Cornall, R. Immune Privilege or Privileged Immunity? *Mucosal Immunol.* **2008**, *1*, 372–381. [CrossRef]
39. Crispe, I.N.; Giannandrea, M.; Klein, I.; John, B.; Sampson, B.; Wuensch, S. Cellular and Molecular Mechanisms of Liver Tolerance. *Immunol. Rev.* **2006**, *213*, 101–118. [CrossRef]
40. Dent, D.M.; Hickman, R.; Uys, C.J.; Saunders, S.; Terblanche, J. The Natural History of Liver Allo- and Autotransplantation in the Pig. *Br. J. Surg.* **1971**, *58*, 407–413. [CrossRef] [PubMed]
41. Tiegs, G.; Lohse, A.W. Immune Tolerance: What Is Unique about the Liver. *J. Autoimmun.* **2010**, *34*, 1–6. [CrossRef] [PubMed]
42. Zhou, J.; Peng, H.; Li, K.; Qu, K.; Wang, B.; Wu, Y.; Ye, L.; Dong, Z.; Wei, H.; Sun, R.; et al. Liver-Resident NK Cells Control Antiviral Activity of Hepatic T Cells via the PD-1-PD-L1 Axis. *Immunity* **2019**, *50*, 403–417.e4. [CrossRef]
43. Sharpe, A.H.; Wherry, E.J.; Ahmed, R.; Freeman, G.J. The Function of Programmed Cell Death 1 and Its Ligands in Regulating Autoimmunity and Infection. *Nat. Immunol.* **2007**, *8*, 239–245. [CrossRef]
44. Hoffman, B.E.; Martino, A.T.; Sack, B.K.; Cao, O.; Liao, G.; Terhorst, C.; Herzog, R.W. Nonredundant Roles of IL-10 and TGF-β in Suppression of Immune Responses to Hepatic AAV-Factor IX Gene Transfer. *Mol. Ther.* **2011**, *19*, 1263–1272. [CrossRef] [PubMed]
45. Cao, O.; Dobrzynski, E.; Wang, L.; Nayak, S.; Mingle, B.; Terhorst, C.; Herzog, R.W. Induction and Role of Regulatory CD4+CD25+ T Cells in Tolerance to the Transgene Product Following Hepatic in Vivo Gene Transfer. *Blood* **2007**, *110*, 1132–1140. [CrossRef] [PubMed]
46. Fields, P.A.; Arruda, V.R.; Armstrong, E.; Chu, K.; Mingozzi, F.; Hagstrom, J.N.; Herzog, R.W.; High, K.A. Risk and Prevention of Anti-Factor IX Formation in AAV-Mediated Gene Transfer in the Context of a Large Deletion of F9. *Mol. Ther.* **2001**, *4*, 201–210. [CrossRef]
47. Mount, J.D.; Herzog, R.W.; Tillson, D.M.; Goodman, S.A.; Robinson, N.; McCleland, M.L.; Bellinger, D.; Nichols, T.C.; Arruda, V.R.; Lothrop, C.D.; et al. Sustained Phenotypic Correction of Hemophilia B Dogs with a Factor IX Null Mutation by Liver-Directed Gene Therapy. *Blood* **2002**, *99*, 2670–2676. [CrossRef] [PubMed]
48. Markusic, D.M.; Hoffman, B.E.; Perrin, G.Q.; Nayak, S.; Wang, X.; LoDuca, P.A.; High, K.A.; Herzog, R.W. Effective Gene Therapy for Haemophilic Mice with Pathogenic Factor IX Antibodies. *EMBO Mol. Med.* **2013**, *5*, 1698–1709. [CrossRef]
49. Zheng, M.; Tian, Z. Liver-Mediated Adaptive Immune Tolerance. *Front. Immunol.* **2019**, *10*, 2525. [CrossRef]
50. LoDuca, P.; Hoffman, B.; Herzog, R. Hepatic Gene Transfer as a Means of Tolerance Induction to Transgene Products. *Curr. Gene Ther.* **2009**, *9*, 104–114. [CrossRef]
51. Finn, J.D.; Ozelo, M.C.; Sabatino, D.E.; Franck, H.W.G.; Merricks, E.P.; Crudele, J.M.; Zhou, S.; Kazazian, H.H.; Lillicrap, D.; Nichols, T.C.; et al. Eradication of Neutralizing Antibodies to Factor VIII in Canine Hemophilia A after Liver Gene Therapy. *Blood* **2010**, *116*, 5842–5848. [CrossRef]
52. Young, G.; Park, Y.-S.; Ozelo, M.C.; Chou, S.-C.; Li, M.; Imtiaz, U.; Chavele, K.-M. European Association for Haemophilia and Allied Disorders (EAHAD) 2024 SLAM Session 6. Available online: https://medical.biomarin.com/en-us/wp-content/uploads/sites/2/2024/02/EAHAD-2024-GENEr8-INH.pdf?v=1.1 (accessed on 14 November 2025).
53. Kaczmarek, R.; Samelson-Jones, B.J.; Herzog, R.W. Immune Tolerance Induction by Hepatic Gene Transfer: First-in-Human Evidence. *Mol. Ther.* **2024**, *32*, 863–864. [CrossRef]
54. Haberle, V.; Stark, A. Eukaryotic Core Promoters and the Functional Basis of Transcription Initiation. *Nat. Rev. Mol. Cell Biol.* **2018**, *19*, 621–637. [CrossRef] [PubMed]

https://doi.org/10.3390/cells15010014

55. Hernandez-Garcia, C.M.; Finer, J.J. Identification and Validation of Promoters and Cis-Acting Regulatory Elements. *Plant Sci.* **2014**, *217–218*, 109–119. [CrossRef]
56. Beagrie, R.A.; Scialdone, A.; Schueler, M.; Kraemer, D.C.A.; Chotalia, M.; Xie, S.Q.; Barbieri, M.; De Santiago, I.; Lavitas, L.-M.; Branco, M.R.; et al. Complex Multi-Enhancer Contacts Captured by Genome Architecture Mapping. *Nature* **2017**, *543*, 519–524. [CrossRef] [PubMed]
57. Pedersen, A.G.; Baldi, P.; Chauvin, Y.; Brunak, S. The Biology of Eukaryotic Promoter Prediction—A Review. *Comput. Chem.* **1999**, *23*, 191–207. [CrossRef]
58. Artemyev, V.; Gubaeva, A.; Paremskaia, A.I.; Dzhioeva, A.A.; Deviatkin, A.; Feoktistova, S.G.; Mityaeva, O.; Volchkov, P.Y. Synthetic Promoters in Gene Therapy: Design Approaches, Features and Applications. *Cells* **2024**, *13*, 1963. [CrossRef]
59. Fassler, J.S.; Gussin, G.N. [1] Promoters and Basal Transcription Machinery in Eubacteria and Eukaryotes: Concepts, Definitions, and Analogies. In *Methods in Enzymology*; Elsevier: Amsterdam, The Netherlands, 1996; Volume 273, pp. 3–29, ISBN 978-0-12-182174-6.
60. Powell, S.K.; Rivera-Soto, R.; Gray, S.J. Viral Expression Cassette Elements to Enhance Transgene Target Specificity and Expression in Gene Therapy. *Discov. Med.* **2015**, *19*, 49–57.
61. Gingrich, J.R.; Roder, J. Inducible Gene Expression in the Nervous System of Transgenic Mice. *Annu. Rev. Neurosci.* **1998**, *21*, 377–405. [CrossRef]
62. Shlyueva, D.; Stampfel, G.; Stark, A. Transcriptional Enhancers: From Properties to Genome-Wide Predictions. *Nat. Rev. Genet.* **2014**, *15*, 272–286. [CrossRef]
63. Spitz, F. Gene Regulation at a Distance: From Remote Enhancers to 3D Regulatory Ensembles. *Semin. Cell Dev. Biol.* **2016**, *57*, 57–67. [CrossRef]
64. Calo, E.; Wysocka, J. Modification of Enhancer Chromatin: What, How, and Why? *Mol. Cell* **2013**, *49*, 825–837. [CrossRef] [PubMed]
65. Huminiecki, Ł.; Horbańczuk, J. Can We Predict Gene Expression by Understanding Proximal Promoter Architecture? *Trends Biotechnol.* **2017**, *35*, 530–546. [CrossRef]
66. Natoli, G.; Andrau, J.-C. Noncoding Transcription at Enhancers: General Principles and Functional Models. *Annu. Rev. Genet.* **2012**, *46*, 1–19. [CrossRef]
67. Long, H.K.; Prescott, S.L.; Wysocka, J. Ever-Changing Landscapes: Transcriptional Enhancers in Development and Evolution. *Cell* **2016**, *167*, 1170–1187. [CrossRef]
68. Lin, T.; Huang, C.; Tong, C.; Xu, W.; Wang, H.; Wu, S.; Zhang, W.; Li, Y.; Liu, H.; Shao, C.; et al. Regulation of Essential Hepatocyte Functions and Identity by Super-Enhancers in Health and Disease. *bioRxiv* **2025**, bioRxiv:2025.06.04.657826. [CrossRef]
69. Chen, Y.; Yao, B.; Zhu, Z.; Yi, Y.; Lin, X.; Zhang, Z.; Shen, G. A Constitutive Super-Enhancer: Homologous Region 3 of Bombyx Mori Nucleopolyhedrovirus. *Biochem. Biophys. Res. Commun.* **2004**, *318*, 1039–1044. [CrossRef]
70. Whyte, W.A.; Orlando, D.A.; Hnisz, D.; Abraham, B.J.; Lin, C.Y.; Kagey, M.H.; Rahl, P.B.; Lee, T.I.; Young, R.A. Master Transcription Factors and Mediator Establish Super-Enhancers at Key Cell Identity Genes. *Cell* **2013**, *153*, 307–319. [CrossRef]
71. Soutourina, J. Transcription Regulation by the Mediator Complex. *Nat. Rev. Mol. Cell Biol.* **2018**, *19*, 262–274. [CrossRef] [PubMed]
72. Allen, B.L.; Taatjes, D.J. The Mediator Complex: A Central Integrator of Transcription. *Nat. Rev. Mol. Cell Biol.* **2015**, *16*, 155–166. [CrossRef]
73. Sabari, B.R.; Dall'Agnese, A.; Boija, A.; Klein, I.A.; Coffey, E.L.; Shrinivas, K.; Abraham, B.J.; Hannett, N.M.; Zamudio, A.V.; Manteiga, J.C.; et al. Coactivator Condensation at Super-Enhancers Links Phase Separation and Gene Control. *Science* **2018**, *361*, eaar3958. [CrossRef] [PubMed]
74. Tang, F.; Yang, Z.; Tan, Y.; Li, Y. Super-Enhancer Function and Its Application in Cancer Targeted Therapy. *npj Precis. Oncol.* **2020**, *4*, 2. [CrossRef] [PubMed]
75. Preissl, S.; Fang, R.; Huang, H.; Zhao, Y.; Raviram, R.; Gorkin, D.U.; Zhang, Y.; Sos, B.C.; Afzal, V.; Dickel, D.E.; et al. Single-Nucleus Analysis of Accessible Chromatin in Developing Mouse Forebrain Reveals Cell-Type-Specific Transcriptional Regulation. *Nat. Neurosci.* **2018**, *21*, 432–439. [CrossRef]
76. Shin, H.Y.; Willi, M.; Yoo, K.H.; Zeng, X.; Wang, C.; Metser, G.; Hennighausen, L. Hierarchy within the Mammary STAT5-Driven Wap Super-Enhancer. *Nat. Genet.* **2016**, *48*, 904–911. [CrossRef]
77. Suzuki, H.I.; Young, R.A.; Sharp, P.A. Super-Enhancer-Mediated RNA Processing Revealed by Integrative MicroRNA Network Analysis. *Cell* **2017**, *168*, 1000–1014.e15. [CrossRef] [PubMed]
78. Boeva, V. Analysis of Genomic Sequence Motifs for Deciphering Transcription Factor Binding and Transcriptional Regulation in Eukaryotic Cells. *Front. Genet.* **2016**, *7*, 24. [CrossRef] [PubMed]

https://doi.org/10.3390/cells15010014

79. Liu, J.; Yu, Z.; Xiao, Y.; Meng, Q.; Wang, Y.; Chang, W. Coordination of FOXA2 and SIRT6 Suppresses the Hepatocellular Carcinoma Progression through ZEB2 Inhibition. *Cancer Manag. Res.* **2018**, *10*, 391–402. [CrossRef]
80. Wolfrum, C.; Stoffel, M. Coactivation of Foxa2 through Pgc-1β Promotes Liver Fatty Acid Oxidation and Triglyceride/VLDL Secretion. *Cell Metab.* **2006**, *3*, 99–110. [CrossRef]
81. Zhou, B.-R.; Feng, H.; Huang, F.; Zhu, I.; Portillo-Ledesma, S.; Shi, D.; Zaret, K.S.; Schlick, T.; Landsman, D.; Wang, Q.; et al. Structural Insights into the Cooperative Nucleosome Recognition and Chromatin Opening by FOXA1 and GATA4. *Mol. Cell* **2024**, *84*, 3061–3079.e10. [CrossRef]
82. Hansen, J.L.; Loell, K.J.; Cohen, B.A. A Test of the Pioneer Factor Hypothesis Using Ectopic Liver Gene Activation. *Elife* **2022**, *11*, e73358. [CrossRef]
83. Odom, D.T.; Zizlsperger, N.; Gordon, D.B.; Bell, G.W.; Rinaldi, N.J.; Murray, H.L.; Volkert, T.L.; Schreiber, J.; Rolfe, P.A.; Gifford, D.K.; et al. Control of Pancreas and Liver Gene Expression by HNF Transcription Factors. *Science* **2004**, *303*, 1378–1381. [CrossRef] [PubMed]
84. Thakur, A.; Park, K.; Cullum, R.; Fuglerud, B.M.; Khoshnoodi, M.; Drissler, S.; Stephan, T.L.; Lotto, J.; Kim, D.; Gonzalez, F.J.; et al. HNF4A Guides the MLL4 Complex to Establish and Maintain H3K4me1 at Gene Regulatory Elements. *Commun. Biol.* **2024**, *7*, 144. [CrossRef]
85. Cai, S.; Lu, S.; Liu, L.; Zhang, C.Z.; Yun, J. Increased Expression of Hepatocyte Nuclear Factor 4 Alpha Transcribed by Promoter 2 Indicates a Poor Prognosis in Hepatocellular Carcinoma. *Ther. Adv. Gastroenterol.* **2017**, *10*, 761–771. [CrossRef]
86. Fang, P.; Wilson, E.R.; Larsen, S.N.; Orellana, W.A.; Hall, M.A.; Stubben, C.; Kabir, A.H.; Affolter, K.; Moffitt, R.A.; Zhang, X.; et al. Differential Control of Growth and Identity by HNF4α Isoforms in Pancreatic Ductal Adenocarcinoma. *Mol. Cancer Res.* **2025**, *23*, 936–952. [CrossRef]
87. Chandra, V.; Huang, P.; Potluri, N.; Wu, D.; Kim, Y.; Rastinejad, F. Multidomain Integration in the Structure of the HNF-4α Nuclear Receptor Complex. *Nature* **2013**, *495*, 394–398. [CrossRef] [PubMed]
88. Lambert, É.; Babeu, J.-P.; Simoneau, J.; Raisch, J.; Lavergne, L.; Lévesque, D.; Jolibois, É.; Avino, M.; Scott, M.S.; Boudreau, F.; et al. Human Hepatocyte Nuclear Factor 4-α Encodes Isoforms with Distinct Transcriptional Functions. *Mol. Cell. Proteom.* **2020**, *19*, 808–827. [CrossRef] [PubMed]
89. Wang, J.-C.; Stafford, J.M.; Granner, D.K. SRC-1 and GRIP1 Coactivate Transcription with Hepatocyte Nuclear Factor 4. *J. Biol. Chem.* **1998**, *273*, 30847–30850. [CrossRef]
90. Kouketsu, T.; Monma, R.; Miyairi, Y.; Sawatsubashi, S.; Shima, H.; Igarashi, K.; Sugawara, A.; Yokoyama, A. IRF2BP2 Is a Novel HNF4α Co-Repressor: Its Role in Gluconeogenic Gene Regulation via Biochemically Labile Interaction. *Biochem. Biophys. Res. Commun.* **2022**, *615*, 81–87. [CrossRef] [PubMed]
91. Nedumaran, B.; Hong, S.; Xie, Y.-B.; Kim, Y.-H.; Seo, W.-Y.; Lee, M.-W.; Lee, C.H.; Koo, S.-H.; Choi, H.-S. DAX-1 Acts as a Novel Corepressor of Orphan Nuclear Receptor HNF4α and Negatively Regulates Gluconeogenic Enzyme Gene Expression. *J. Biol. Chem.* **2009**, *284*, 27511–27523. [CrossRef]
92. Qu, M.; Duffy, T.; Hirota, T.; Kay, S.A. Nuclear Receptor HNF4A Transrepresses CLOCK: BMAL1 and Modulates Tissue-Specific Circadian Networks. *Proc. Natl. Acad. Sci. USA* **2018**, *115*, E12305–E12312. [CrossRef]
93. Qu, M.; Qu, H.; Jia, Z.; Kay, S.A. HNF4A Defines Tissue-Specific Circadian Rhythms by Beaconing BMAL1::CLOCK Chromatin Binding and Shaping the Rhythmic Chromatin Landscape. *Nat. Commun.* **2021**, *12*, 6350. [CrossRef]
94. Mendel, D.B.; Hansen, L.P.; Graves, M.K.; Conley, P.B.; Crabtree, G.R. HNF-1 Alpha and HNF-1 Beta (vHNF-1) Share Dimerization and Homeo Domains, but Not Activation Domains, and Form Heterodimers in Vitro. *Genes Dev.* **1991**, *5*, 1042–1056. [CrossRef]
95. Hansen, S.K.; Párrizas, M.; Jensen, M.L.; Pruhova, S.; Ek, J.; Boj, S.F.; Johansen, A.; Maestro, M.A.; Rivera, F.; Eiberg, H.; et al. Genetic Evidence That HNF-1α–Dependent Transcriptional Control of HNF-4α Is Essential for Human Pancreatic β Cell Function. *J. Clin. Investig.* **2002**, *110*, 827–833. [CrossRef]
96. Haque, E.; Teeli, A.S.; Winiarczyk, D.; Taguchi, M.; Sakuraba, S.; Kono, H.; Leszczyński, P.; Pierzchała, M.; Taniguchi, H. HNF1A POU Domain Mutations Found in Japanese Liver Cancer Patients Cause Downregulation of HNF4A Promoter Activity with Possible Disruption in Transcription Networks. *Genes* **2022**, *13*, 413. [CrossRef]
97. Hu, M.; Huang, X.; Han, X.; Ji, L. Loss of HNF1α Function Contributes to Hepatocyte Proliferation and Abnormal Cholesterol Metabolism via Downregulating miR-122: A Novel Mechanism of MODY3. *Diabetes Metab. Syndr. Obes.* **2020**, *13*, 627–639. [CrossRef] [PubMed]
98. Tan, J.; Xu, J.; Wei, G.; Zhang, L.; Sun, L.; Wang, G.; Li, F.; Jiang, F. HNF1 α Controls Liver Lipid Metabolism and Insulin Resistance via Negatively Regulating the SOCS-3-STAT3 Signaling Pathway. *J. Diabetes Res.* **2019**, *2019*, 5483946. [CrossRef]
99. Liu, M.; Liu, L.; Guo, H.; Fan, X.; Liu, T.; Xu, C.; He, Z.; Song, Y.; Gao, L.; Shao, S.; et al. Dominant-Negative HNF1α Mutant Promotes Liver Steatosis and Inflammation by Regulating Hepatic Complement Factor D. *iScience* **2023**, *26*, 108018. [CrossRef] [PubMed]

https://doi.org/10.3390/cells15010014

100. Peixoto-Barbosa, R.; Reis, A.F.; Giuffrida, F.M.A. Update on Clinical Screening of Maturity-Onset Diabetes of the Young (MODY). *Diabetol. Metab. Syndr.* **2020**, *12*, 50. [CrossRef]
101. Qian, H.; Deng, X.; Huang, Z.-W.; Wei, J.; Ding, C.-H.; Feng, R.-X.; Zeng, X.; Chen, Y.-X.; Ding, J.; Qiu, L.; et al. An HNF1α-Regulated Feedback Circuit Modulates Hepatic Fibrogenesis via the Crosstalk between Hepatocytes and Hepatic Stellate Cells. *Cell Res.* **2015**, *25*, 930–945. [CrossRef] [PubMed]
102. Araújo, T.G.; De Oliveira, A.G.; Vecina, J.F.; Marin, R.M.; Franco, E.S.; Abdalla Saad, M.J.; De Sousa Maia, M.B. *Parkinsonia aculeata (Caesalpineaceae)* Improves High-Fat Diet-Induced Insulin Resistance in Mice through the Enhancement of Insulin Signaling and Mitochondrial Biogenesis. *J. Ethnopharmacol.* **2016**, *183*, 95–102. [CrossRef]
103. Hu, Y.; Luo, Z.; Wang, M.; Wu, Z.; Liu, Y.; Cheng, Z.; Sun, Y.; Xiong, J.-W.; Tong, X.; Zhu, Z.; et al. Prox1a Promotes Liver Growth and Differentiation by Repressing Cdx1b Expression and Intestinal Fate Transition in Zebrafish. *J. Genet. Genom.* **2025**, *52*, 66–77. [CrossRef]
104. Dudas, J.; Elmaouhoub, A.; Mansuroglu, T.; Batusic, D.; Tron, K.; Saile, B.; Papoutsi, M.; Pieler, T.; Wilting, J.; Ramadori, G. Prospero-Related Homeobox 1 (Prox1) Is a Stable Hepatocyte Marker during Liver Development, Injury and Regeneration, and Is Absent from "Oval Cells". *Histochem. Cell Biol.* **2006**, *126*, 549–562. [CrossRef]
105. Meng, F. LYVE1 and PROX1 in the Reconstruction of Hepatic Sinusoids after Partial Hepatectomy in Mice. *Folia Morphol.* **2017**, *76*, 239–245. [CrossRef]
106. Lim, B.; Kamal, A.; Gomez Ramos, B.; Adrian Segarra, J.M.; Ibarra, I.L.; Dignas, L.; Kindinger, T.; Volz, K.; Rahbari, M.; Rahbari, N.; et al. Active Repression of Cell Fate Plasticity by PROX1 Safeguards Hepatocyte Identity and Prevents Liver Tumorigenesis. *Nat. Genet.* **2025**, *57*, 668–679. [CrossRef]
107. Lee, T.K.W.; Ma, S. PROX1: A Key Regulator of Hepatocyte Identity and Tumorigenesis. *Cancer Res.* **2025**, *85*, 1957–1959. [CrossRef] [PubMed]
108. Papoutsi, M.; Dudas, J.; Becker, J.; Tripodi, M.; Opitz, L.; Ramadori, G.; Wilting, J. Gene Regulation by Homeobox Transcription Factor Prox1 in Murine Hepatoblasts. *Cell Tissue Res.* **2007**, *330*, 209–220. [CrossRef] [PubMed]
109. Song, K.-H.; Li, T.; Chiang, J.Y.L. A Prospero-Related Homeodomain Protein Is a Novel Co-Regulator of Hepatocyte Nuclear Factor 4α That Regulates the Cholesterol 7α-Hydroxylase Gene. *J. Biol. Chem.* **2006**, *281*, 10081–10088. [CrossRef]
110. Li, S.; Ou, C.; Zhang, J.; Zeng, M.; Liang, K.; Peng, Q.; Gao, Y. The Effect of FOXA3 Overexpression on Hepatocyte Differentiation and Liver Regeneration in a Fah cKO Mouse Model. *Cell. Mol. Gastroenterol. Hepatol.* **2025**, *19*, 101438. [CrossRef]
111. Li, G.; Zhu, L.; Guo, M.; Wang, D.; Meng, M.; Zhong, Y.; Zhang, Z.; Lin, Y.; Liu, C.; Wang, J.; et al. Characterisation of Forkhead Box Protein A3 as a Key Transcription Factor for Hepatocyte Regeneration. *JHEP Rep.* **2023**, *5*, 100906. [CrossRef]
112. Fan, J.Y.; Dama, G.; Liu, Y.L.; Guo, W.Y.; Lin, J.T. Combinational Overexpression of *Foxa3* and *Hnf4a* Enhance the Proliferation and Prolong the Functional Maintenance of Primary Hepatocytes. *Mol. Biol.* **2023**, *57*, 661–669. [CrossRef]
113. Cereghini, S. Liver-Enriched Transcription Factors and Hepatocyte Differentiation. *FASEB J.* **1996**, *10*, 267–282. [CrossRef]
114. Li, Z.; White, P.; Tuteja, G.; Rubins, N.; Sackett, S.; Kaestner, K.H. Foxa1 and Foxa2 Regulate Bile Duct Development in Mice. *J. Clin. Investig.* **2009**, *119*, 1537–1545. [CrossRef] [PubMed]
115. Tomofuji, K.; Kondo, J.; Onuma, K.; Coppo, R.; Horie, H.; Oyama, K.; Miyoshi, E.; Fukumitsu, K.; Ishii, T.; Hatano, E.; et al. Hepatocyte Differentiation from Mouse Liver Ductal Organoids by Transducing 4 Liver-Specific Transcription Factors. *Hepatol. Commun.* **2023**, *7*, e0134. [CrossRef]
116. Warren, I.; Moeller, M.M.; Guiggey, D.; Chiang, A.; Maloy, M.; Ogoke, O.; Groth, T.; Mon, T.; Meamardoost, S.; Liu, X.; et al. FOXA1/2 Depletion Drives Global Reprogramming of Differentiation State and Metabolism in a Human Liver Cell Line and Inhibits Differentiation of Human Stem Cell-derived Hepatic Progenitor Cells. *FASEB J.* **2023**, *37*, e22652. [CrossRef]
117. Tian, M.; Gao, W.; Ma, S.; Cao, H.; Zhang, Y.; An, F.; Qi, J.; Yang, Z. Role of HNF6 in Liver Homeostasis and Pathophysiology. *Mol. Med.* **2025**, *31*, 48. [CrossRef] [PubMed]
118. Iyaguchi, D.; Yao, M.; Watanabe, N.; Nishihira, J.; Tanaka, I. DNA Recognition Mechanism of the ONECUT Homeodomain of Transcription Factor HNF-6. *Structure* **2007**, *15*, 75–83. [CrossRef] [PubMed]
119. Xue, X.; Li, Z.; Zhao, J.; Zhao, Z.; Li, Z.; Li, Y.; Liu, Y.; He, H. Advances in the Relationship between AP-1 and Tumorigenesis, Development and Therapy Resistance. *Discov. Oncol.* **2025**, *16*, 61. [CrossRef]
120. Leu, J.I.; Crissey, M.A.S.; Leu, J.P.; Ciliberto, G.; Taub, R. Interleukin-6-Induced STAT3 and AP-1 Amplify Hepatocyte Nuclear Factor 1-Mediated Transactivation of Hepatic Genes, an Adaptive Response to Liver Injury. *Mol. Cell. Biol.* **2001**, *21*, 414–424. [CrossRef] [PubMed]
121. Merchant, S.; Korbelik, M. Upregulation of Genes for C-Reactive Protein and Related Pentraxin/Complement Proteins in Photodynamic Therapy-Treated Human Tumor Cells: Enrolment of PI3K/Akt and AP-1. *Immunobiology* **2013**, *218*, 869–874. [CrossRef]

https://doi.org/10.3390/cells15010014

122. Yelins'ka, A.M.; Akimov, O.Y.; Kostenko, V.O. Role of AP-1 Transcriptional Factor in Development of Oxidative and Nitrosative Stress in Periodontal Tissues during Systemic Inflammatory Response. *Ukr. Biochem. J.* **2019**, *91*, 80–85. [CrossRef]
123. Yu, X.; Wang, Y.; Song, Y.; Gao, X.; Deng, H. AP-1 Is a Regulatory Transcription Factor of Inflammaging in the Murine Kidney and Liver. *Aging Cell* **2023**, *22*, e13858. [CrossRef]
124. Bakiri, L.; Hasenfuss, S.C.; Wagner, E.F. A FATal AP-1 Dimer Switch in Hepatosteatosis. *Cell Cycle* **2014**, *13*, 1218–1219. [CrossRef]
125. Bernt, C.; Vennegeerts, T.; Beuers, U.; Rust, C. The Human Transcription Factor AP-1 Is a Mediator of Bile Acid-Induced Liver Cell Apoptosis. *Biochem. Biophys. Res. Commun.* **2006**, *340*, 800–806. [CrossRef] [PubMed]
126. Shaulian, E.; Karin, M. AP-1 in Cell Proliferation and Survival. *Oncogene* **2001**, *20*, 2390–2400. [CrossRef]
127. Müller, C.; Muck, J.S.; Ustyantsev, K.; Kortman, G.; Hartung, J.; Berezikov, E.; Calkhoven, C.F. Enhanced C/EBPα Function Extends Healthspan and Lifespan in the African Turquoise Killifish. *Aging Cell* **2025**, *24*, e70211. [CrossRef] [PubMed]
128. Tolosano, E.; Altruda, F. Hemopexin: Structure, Function, and Regulation. *DNA Cell Biol.* **2002**, *21*, 297–306. [CrossRef]
129. Mischoulon, D.; Rana, B.; Bucher, N.L.R.; Farmer, S.R. Growth-Dependent Inhibition of CCAAT Enhancer-Binding Protein (C/EPBα) Gene Expression during Hepatocyte Proliferation in the Regenerating Liver and in Culture. *Mol. Cell. Biol.* **1992**, *12*, 2553–2560. [CrossRef]
130. Park, J.-C.; Jeong, W.-J.; Seo, S.H.; Choi, K.-Y. WDR76 Mediates Obesity and Hepatic Steatosis via HRas Destabilization. *Sci. Rep.* **2019**, *9*, 19676. [CrossRef]
131. He, L.; Cao, J.; Meng, S.; Ma, A.; Radovick, S.; Wondisford, F.E. Activation of Basal Gluconeogenesis by Coactivator P300 Maintains Hepatic Glycogen Storage. *Mol. Endocrinol.* **2013**, *27*, 1322–1332. [CrossRef] [PubMed]
132. Hou, C.; Lu, S.; Su, Y.; Ding, D.; Tao, L.; Wang, M.; Wang, Y.; Liu, X. C/EBP-α Induces Autophagy by Binding to Beclin1 through Its Own Acetylation Modification in Activated Hepatic Stellate Cells. *Exp. Cell Res.* **2021**, *405*, 112721. [CrossRef]
133. Ramji, D.P.; Vitelli, A.; Tronche, F.; Cortese, R.; Ciliberto, G. The Two C/EBP Isoforms, IL6DBP/NFIL6 and CEBP6δ/NFIL63, Are Induced by IL6β to Promote Acute Phase Gene Transcription via Different Mechanisms. *Nucl. Acids Res.* **1993**, *21*, 289–294. [CrossRef]
134. Li, J.; Ge, Y.; Chai, Y.; Kou, C.; Sun, T.; Liu, J.; Zhang, H. THSR Mediated MiR374b Targeting C/EBP β/FOXO1 to Accelerate Thyroid Stimulating Hormone-Induced Hepatic Steatosis. *Hepatic Med. Évid. Res.* **2024**, *16*, 91–104. [CrossRef]
135. Kim, S.; Wysocka, J. Deciphering the Multi-Scale, Quantitative Cis-Regulatory Code. *Mol. Cell* **2023**, *83*, 373–392. [CrossRef]
136. Borst, E.; Messerle, M. Development of a Cytomegalovirus Vector for Somatic Gene Therapy. *Bone Marrow Transplant.* **2000**, *25*, S80–S82. [CrossRef]
137. Arad, U.; Zeira, E.; El-Latif, M.A.; Mukherjee, S.; Mitchell, L.; Pappo, O.; Galun, E.; Oppenheim, A. Liver-Targeted Gene Therapy by SV40-Based Vectors Using the Hydrodynamic Injection Method. *Hum. Gene Ther.* **2005**, *16*, 361–371. [CrossRef]
138. Wills, K.N.; Maneval, D.C.; Menzel, P.; Harris, M.P.; Sutjipto, S.; Vaillancourt, M.-T.; Huang, W.-M.; Johnson, D.E.; Anderson, S.C.; Wen, S.F.; et al. Development and Characterization of Recombinant Adenoviruses Encoding Human P53 for Gene Therapy of Cancer. *Hum. Gene Ther.* **1994**, *5*, 1079–1088. [CrossRef]
139. Rettinger, S.D.; Kennedy, S.C.; Wu, X.; Saylors, R.L.; Hafenrichter, D.G.; Flye, M.W.; Ponder, K.P. Liver-Directed Gene Therapy: Quantitative Evaluationof Promoter Elements by Using in Vivo Retroviral Transduction. *Proc. Natl. Acad. Sci. USA* **1994**, *91*, 1460–1464. [CrossRef]
140. Kuzmin, D.A.; Shutova, M.V.; Johnston, N.R.; Smith, O.P.; Fedorin, V.V.; Kukushkin, Y.S.; Van Der Loo, J.C.M.; Johnstone, E.C. The Clinical Landscape for AAV Gene Therapies. *Nat. Rev. Drug Discov.* **2021**, *20*, 173–174. [CrossRef] [PubMed]
141. Kramer, M.G.; Barajas, M.; Razquin, N.; Berraondo, P.; Rodrigo, M.; Wu, C.; Qian, C.; Fortes, P.; Prieto, J. In Vitro and in Vivo Comparative Study of Chimeric Liver-Specific Promoters. *Mol. Ther.* **2003**, *7*, 375–385. [CrossRef]
142. Qin, L.; Ding, Y.; Pahud, D.R.; Chang, E.; Imperiale, M.J.; Bromberg, J.S. Promoter Attenuation in Gene Therapy: Interferon-γ and Tumor Necrosis Factor-α Inhibit Transgene Expression. *Hum. Gene Ther.* **1997**, *8*, 2019–2029. [CrossRef] [PubMed]
143. Nakai, H.; Herzog, R.W.; Hagstrom, J.N.; Walter, J.; Kung, S.-H.; Yang, E.Y.; Tai, S.J.; Iwaki, Y.; Kurtzman, G.J.; Fisher, K.J.; et al. Adeno-Associated Viral Vector-Mediated Gene Transfer of Human Blood Coagulation Factor IX Into Mouse Liver. *Blood* **1998**, *91*, 4600–4607. [CrossRef] [PubMed]
144. Pannell, D.; Ellis, J. Silencing of Gene Expression: Implications for Design of Retrovirus Vectors. *Rev. Med. Virol.* **2001**, *11*, 205–217. [CrossRef]
145. Alhaji, S.Y.; Ngai, S.C.; Abdullah, S. Silencing of Transgene Expression in Mammalian Cells by DNA Methylation and Histone Modifications in Gene Therapy Perspective. *Biotechnol. Genet. Eng. Rev.* **2019**, *35*, 1–25. [CrossRef]
146. Lu, D. Epigenetic Modification Enzymes: Catalytic Mechanisms and Inhibitors. *Acta Pharm. Sin. B* **2013**, *3*, 141–149. [CrossRef]
147. Chanda, D.; Hensel, J.; Higgs, J.; Grover, R.; Kaza, N.; Ponnazhagan, S. Effects of Cellular Methylation on Transgene Expression and Site-Specific Integration of Adeno-Associated Virus. *Genes* **2017**, *8*, 232. [CrossRef]
148. Sack, B.K.; Herzog, R.W. Evading the Immune Response upon in Vivo Gene Therapy with Viral Vectors. *Curr. Opin. Mol. Ther.* **2009**, *11*, 493–503.

https://doi.org/10.3390/cells15010014

149. Yan, Z.; Yan, H.; Ou, H. Human Thyroxine Binding Globulin (TBG) Promoter Directs Efficient and Sustaining Transgene Expression in Liver-Specific Pattern. *Gene* **2012**, *506*, 289–294. [CrossRef]
150. Becker, J.; Domenger, C.; Choksi, P.; Krämer, C.; Baumgartl, C.; Maiakovska, O.; Kim, J.-J.; Weinmann, J.; Huber, G.; Schmidt, F.; et al. Identification of a Robust Promoter in Mouse and Human Hepatocytes by in Vivo Biopanning of a Barcoded AAV Library. *Mol. Ther.* **2025**, *33*, 3881–3901. [CrossRef]
151. Murillo, O.; Collantes, M.; Gazquez, C.; Moreno, D.; Hernandez-Alcoceba, R.; Barberia, M.; Ecay, M.; Tamarit, B.; Douar, A.; Ferrer, V.; et al. High Value of 64Cu as a Tool to Evaluate the Restoration of Physiological Copper Excretion after Gene Therapy in Wilson's Disease. *Mol. Ther.-Methods Clin. Dev.* **2022**, *26*, 98–106. [CrossRef]
152. Hodge, T.S.; Torres, E.V. Pharmaceutical Compositions Containing Adeno-Associated Viral Vector. U.S. Patent 18/551,038, 19 September 2022.
153. Aldabe, R.; Lantero, A.; Aseguinolaza, G.G. Gene Therapy Vectors Comprising s/Mar Sequences. 2019. Available online: https://patents.google.com/patent/WO2019219649A1 (accessed on 14 November 2025).
154. D'Avola, D.; López-Franco, E.; Sangro, B.; Pañeda, A.; Grossios, N.; Gil-Farina, I.; Benito, A.; Twisk, J.; Paz, M.; Ruiz, J.; et al. Phase I Open Label Liver-Directed Gene Therapy Clinical Trial for Acute Intermittent Porphyria. *J. Hepatol.* **2016**, *65*, 776–783. [CrossRef] [PubMed]
155. Pena, M.S.R.; Petry, H.; Twisk, J.; Deventer, S.J.H.V.; Ruiz, E.C.S.; Ramirez, A.T. Alanine-Glyoxylate Aminotransferase Therapeutics. 2010. Available online: https://patents.google.com/patent/WO2010087709A1 (accessed on 14 November 2025).
156. Romá, A.F.; Aseguinolaza, G.G.; Pena, M.S.R.; Rodriguez, M.A.P.; Twisk, J.; Valtueña, J.M.P.; Petry, H.; Deventer, S.J.H.V. Porphobilinogen Deaminase Gene Therapy. 2010. Available online: https://patents.google.com/patent/WO2010036118A1 (accessed on 14 November 2025).
157. George, L.A.; Sullivan, S.K.; Giermasz, A.; Rasko, J.E.J.; Samelson-Jones, B.J.; Ducore, J.; Cuker, A.; Sullivan, L.M.; Majumdar, S.; Teitel, J.; et al. Hemophilia B Gene Therapy with a High-Specific-Activity Factor IX Variant. *N. Engl. J. Med.* **2017**, *377*, 2215–2227. [CrossRef] [PubMed]
158. Wang, L.; Wilson, J.M. Gene Therapy for Treating Hemophilia b. 2017. Available online: https://patents.google.com/patent/US11191847B2 (accessed on 14 November 2025).
159. BioMarin Corporate. U.S. FDA Placed a Clinical Hold on BMN 307 Phearless Phase 1/2 Gene Therapy Study in Adults with PKU Based on Interim Pre-Clinical Study Findings—BioMarin Corporate. Available online: https://www.biomarin.com/news/press-releases/u-s-fda-placed-a-clinical-hold-on-bmn-307-phearless-phase-1-2-gene-therapy-study-in-adults-with-pku-based-on-interim-pre-clinical-study-findings/ (accessed on 13 November 2025).
160. Wright, J.B.; Sookiasian, D.L.; Martin, T.B.S.; Francone, O.; Seymour, A.B. Adeno-Associated Virus Compositions for Restoring Pah Gene Function and Methods of Use Thereof. U.S. Patent 12,076,420, 3 September 2024.
161. Wen, S.; Liu, Y.; Guo, Y.; Jiang, L. Optimized PAH gene and Expression Cassette and Use Thereof. 2024. Available online: https://patents.google.com/patent/WO2024094044A1 (accessed on 14 November 2025).
162. Yasuda, M.; Huston, M.W.; Pagant, S.; Gan, L.; Martin, S.S.; Sproul, S.; Richards, D.; Ballaron, S.; Hettini, K.; Ledeboer, A.; et al. AAV2/6 Gene Therapy in a Murine Model of Fabry Disease Results in Supraphysiological Enzyme Activity and Effective Substrate Reduction. *Mol. Ther.-Methods Clin. Dev.* **2020**, *18*, 607–619. [CrossRef] [PubMed]
163. Meyer, K.E.; Lorget, F.; Prawdzik, G.; Falaleeva, M.; Cao, L.; Falese, L.; Huston, M.W.; Hettini, K.; Lu, Y.; Ledeboer, A. A 3-Month Gene Therapy Single-Dose IV Administration Pharmacology and Safety Study with ST-920 (Isaralgagene Civaparvovec) for Fabry Disease in Mice. *Mol. Genet. Metab.* **2024**, *141*, 107963. [CrossRef]
164. Passalacqua, C.; Huston, M.; Souberbielle, B. Methods for Use of Viral Vector Constructs for the Treatment of Fabry Disease. U.S. Patent 18/707,085, 8 May 2025.
165. D'Antiga, L.; Beuers, U.; Ronzitti, G.; Brunetti-Pierri, N.; Baumann, U.; Di Giorgio, A.; Aronson, S.; Hubert, A.; Romano, R.; Junge, N.; et al. Gene Therapy in Patients with the Crigler–Najjar Syndrome. *N. Engl. J. Med.* **2023**, *389*, 620–631. [CrossRef] [PubMed]
166. Mingozzi, F.; Colella, P. Hybrid Regulatory Elements. 2019. Available online: https://patents.google.com/patent/WO2019154939A1 (accessed on 14 November 2025).
167. Bella, T.L.; Ronzitti, G.; Siauve, J. Hybrid Aav Vector Enhancing Transgene Expression in the Liver. 2024. Available online: https://patents.google.com/patent/WO2024079249A1 (accessed on 14 November 2025).
168. Anguela, X.; Armour, S.; Nordin, J. Codon-Optimized Acid Alpha-Glucosidase Expression Cassettes and Methods of Using Same. U.S. Patent 12,084,693, 10 September 2024.
169. Vanglusagene Ensiparvovec. Available online: https://gsrs.ncats.nih.gov/ginas/app/ui/substances/Y47XZ2EW2J (accessed on 13 November 2025).
170. Comper, F.; Miranda, C.J.; Liou, B.; Dodev, T.; Jeyakumar, J.M.; Canavese, M.; Cocita, C.; Khoshrou, K.; Tiscornia, G.; Chisari, E.; et al. FLT201, a Novel Liver-Directed AAV Gene Therapy Candidate for Gaucher Disease Type 1. *Mol. Ther.* **2025**, *33*, 3789–3807. [CrossRef]

https://doi.org/10.3390/cells15010014

171. Pipe, S.W.; Leebeek, F.W.G.; Recht, M.; Key, N.S.; Castaman, G.; Miesbach, W.; Lattimore, S.; Peerlinck, K.; Van Der Valk, P.; Coppens, M.; et al. Gene Therapy with Etranacogene Dezaparvovec for Hemophilia B. *N. Engl. J. Med.* **2023**, *388*, 706–718. [CrossRef]
172. Lubelski, J.; Plessis, D.J.F.D.; Liu, Y.P.; Brake, O.T.; Gonzalez, J.M.I.; Fraser, R.; Roberts, M. Liver-Specific Viral Promoters and Methods of Using the Same. U.S. Patent 12,319,924, 3 June 2025.
173. Reiss, U.M.; Davidoff, A.M.; Tuddenham, E.G.D.; Chowdary, P.; McIntosh, J.H.; Riddell, A.; Pie, A.; Batty, P.; Calvert, J.C.M.; Mangles, S.; et al. Stable Therapeutic Transgenic FIX Levels for More Than 10 Years in Subjects with Severe Hemophilia B Who Received scAAV2/8-LP1-Hfixco Adeno-Associated Virus Gene Therapy. *Blood* **2023**, *142*, 1056. [CrossRef]
174. Nathwani, A.C.; Gray, J.T.; Ng, C.Y.C.; Zhou, J.; Spence, Y.; Waddington, S.N.; Tuddenham, E.G.D.; Kemball-Cook, G.; McIntosh, J.; Boon-Spijker, M.; et al. Self-Complementary Adeno-Associated Virus Vectors Containing a Novel Liver-Specific Human Factor IX Expression Cassette Enable Highly Efficient Transduction of Murine and Nonhuman Primate Liver. *Blood* **2006**, *107*, 2653–2661. [CrossRef] [PubMed]
175. Nathwani, A.C.; Reiss, U.M.; Tuddenham, E.G.D.; Rosales, C.; Chowdary, P.; McIntosh, J.; Della Peruta, M.; Lheriteau, E.; Patel, N.; Raj, D.; et al. Long-Term Safety and Efficacy of Factor IX Gene Therapy in Hemophilia B. *N. Engl. J. Med.* **2014**, *371*, 1994–2004. [CrossRef]
176. Reiss, U.M.; Davidoff, A.M.; Tuddenham, E.G.D.; Chowdary, P.; McIntosh, J.; Muczynski, V.; Journou, M.; Simini, G.; Ireland, L.; Mohamed, S.; et al. Sustained Clinical Benefit of AAV Gene Therapy in Severe Hemophilia B. *N. Engl. J. Med.* **2025**, *392*, 2226–2234. [CrossRef]
177. Ahmed, S.S.; Rubin, H.; Wang, M.; Faulkner, D.; Sengooba, A.; Dollive, S.N.; Avila, N.; Ellsworth, J.L.; Lamppu, D.; Lobikin, M.; et al. Sustained Correction of a Murine Model of Phenylketonuria Following a Single Intravenous Administration of AAVHSC15-PAH. *Mol. Ther.-Methods Clin. Dev.* **2020**, *17*, 568–580. [CrossRef]
178. Iles-Somaratne, J.N.; Cohn, G.M.; Glyman, S.A.; Tzianabos, A.O. Improved Gene Therapy Methods 2022. Available online: https://patents.google.com/patent/WO2022099301A1 (accessed on 14 November 2025).
179. Shioshita, G. P008: pheEDIT: A Phase 1, Open-Label, Dose-Escalation Safety and Efficacy Gene Editing Study Evaluating HMI-103 in Adults with Classical PKU. *Genet. Med. Open* **2023**, *1*, 100018. [CrossRef]
180. Chowdary, P.; Reiss, U.M.; Tuddenham, E.G.D.; Batty, P.; McIntosh, J.H.; Radulescu, V.C.; Chang, E.; Laffan, M.A.; Riddell, A.; Calvert, J.C.M.; et al. GO-8: Stable Expression of Factor VIII over 5 Years Following Adeno-Associated Gene Transfer in Subjects with Hemophilia a Using a Novel Human Factor VIII Variant. *Blood* **2023**, *142*, 3624. [CrossRef]
181. Nathwani, A.; Ward, N.; Thrasher, A.; Tuddenham, E.; Mcvey, J.; Gray, J.; Davidoff, A. Codon-Optimized Factor vi i i Variants and Synthetic Liver-Specific Promoter. 2011. Available online: https://patents.google.com/patent/US9393323B2 (accessed on 14 November 2025).
182. Rangarajan, S.; Walsh, L.; Lester, W.; Perry, D.; Madan, B.; Laffan, M.; Yu, H.; Vettermann, C.; Pierce, G.F.; Wong, W.Y.; et al. AAV5–Factor VIII Gene Transfer in Severe Hemophilia A. *N. Engl. J. Med.* **2017**, *377*, 2519–2530. [CrossRef] [PubMed]
183. McIntosh, J.; Lenting, P.J.; Rosales, C.; Lee, D.; Rabbanian, S.; Raj, D.; Patel, N.; Tuddenham, E.G.D.; Christophe, O.D.; McVey, J.H.; et al. Therapeutic Levels of FVIII Following a Single Peripheral Vein Administration of rAAV Vector Encoding a Novel Human Factor VIII Variant. *Blood* **2013**, *121*, 3335–3344. [CrossRef]
184. Vettermann, C. Methods of Treating Anti-Aav Seropositive Hemophilia Patients. 2024. Available online: https://patents.google.com/patent/WO2024238591A2 (accessed on 14 November 2025).
185. Chowdary, P.; Shapiro, S.; Makris, M.; Evans, G.; Boyce, S.; Talks, K.; Dolan, G.; Reiss, U.; Phillips, M.; Riddell, A.; et al. Phase 1–2 Trial of AAVS3 Gene Therapy in Patients with Hemophilia B. *N. Engl. J. Med.* **2022**, *387*, 237–247. [CrossRef] [PubMed]
186. Peyvandi, F.; Garagiola, I. Clinical Advances in Gene Therapy Updates on Clinical Trials of Gene Therapy in Haemophilia. *Haemophilia* **2019**, *25*, 738–746. [CrossRef]
187. Kia, A.; Corbau, R. Transcription Regulatory Elements. 2021. Available online: https://patents.google.com/patent/WO2021084277A2 (accessed on 14 November 2025).
188. Jeyakumar, J.M.; Kia, A.; Tam, L.C.S.; McIntosh, J.; Spiewak, J.; Mills, K.; Heywood, W.; Chisari, E.; Castaldo, N.; Verhoef, D.; et al. Preclinical Evaluation of FLT190, a Liver-Directed AAV Gene Therapy for Fabry Disease. *Gene Ther.* **2023**, *30*, 487–502. [CrossRef]
189. Dong, B.; Zhang, B.; Ye, J.; Xiao, L.; Zheng, Z.; Yang, L.; Liu, Y. Engineered Human Fviii with Enhanced Secretion Ability and Clotting Activity. U.S. Patent 17/774,450, 1 August 2024.
190. Dong, B.; Yang, L. Engineered Liver-Specific Core Promoters and Their Applications. 2024. Available online: https://patents.google.com/patent/WO2024230802A1 (accessed on 14 November 2025).
191. Cao, Q.; Wen, S.; Wang, Q.; Liu, Y.; Jiang, L. Coagulation factor fviii protein variant, expression vector and use 2024. Available online: https://patents.google.com/patent/WO2024255448A1 (accessed on 14 November 2025).
192. Anguela, X. Factor Viii (Fviii) Gene Therapy Methods. U.S. Patent 18/803,700, 6 March 2025.

193. Elkouby, L.; Armour, S.M.; Toso, R.; DiPietro, M.; Davidson, R.J.; Nguyen, G.N.; Willet, M.; Kutza, S.; Silverberg, J.; Frick, J.; et al. Preclinical Assessment of an Optimized AAV-FVIII Vector in Mice and Non-Human Primates for the Treatment of Hemophilia A. *Mol. Ther.-Methods Clin. Dev.* **2022**, *24*, 20–29. [CrossRef]
194. Anguela, X.; Shen, S.H. Cpg Reduced Factor Viii Variants, Compositions and Methods and Uses for Treatment of Hemostasis Disorders. U.S. Patent 11,168,124, 9 November 2021.
195. Konkle, B.A.; Walsh, C.E.; Escobar, M.A.; Josephson, N.C.; Young, G.; Von Drygalski, A.; McPhee, S.W.J.; Samulski, R.J.; Bilic, I.; De La Rosa, M.; et al. BAX 335 Hemophilia B Gene Therapy Clinical Trial Results: Potential Impact of CpG Sequences on Gene Expression. *Blood* **2021**, *137*, 763–774. [CrossRef]
196. Horling, F.; Lengler, J.; Falkner, F.; Rottensteiner, H.; Scheiflinger, F. Viral Vectors Encoding Recombinant Fix with Increased Expression for Gene Therapy of Hemophilia b. U.S. Patent 10,842,853, 24 November 2020.
197. Lengler, J.; Weiller, M.; Horling, F.; Mayrhofer, J.; Schuster, M.; Falkner, F.G.; Gil-Farina, I.; Klugmann, M.; Scheiflinger, F.; Hoellriegl, W.; et al. Preclinical Development of TAK-754, a High-Performance AAV8-Based Vector Expressing Coagulation Factor VIII. *Mol. Ther. Methods Clin. Dev.* **2025**, *33*, 101424. [CrossRef]
198. Falkner, F.-G.; Horling, F.; Lengler, J.; Rottensteiner, H.; Scheiflinger, F. Viral Vectors Encoding Recombinant Fviii Variants with Increased Expression for Gene Therapy of Hemophilia a. U.S. Patent 10,189,888, 29 January 2019.
199. Rottensteiner, H.; Hoellriegl, W. Gene Therapy of Hemophilia a Using Viral Vectors Encoding Recombinant Fviii Variants with Increased Expression. 2021. Available online: https://patents.google.com/patent/WO2021119357A2/en (accessed on 14 November 2025).
200. Zozulya, N.; Ptushkin, V.; Davydkin, I.; Ivanov, V.; Korobkin, A.; Zorenko, V.; Uss, A.; Fatenkova, E.; Yudina, N.; Linkova, Y.; et al. Arvenacogene Sanparvovec, A Novel Option for Hemophilia B Gene Therapy: First-in-Human Findings. Available online: https://library.ehaweb.org/eha/2025/eha2025-congress/4161329/nadezhda.zozulya.arvenacogene.sanparvovec.a.novel.option.for.hemophilia.b.gene.html?f=menu=6*browseby=8*sortby=2*media=3*ce_id=2882*ot_id=31584*marker=5844*featured=19595 (accessed on 13 November 2025).
201. Prokofyev, A.V.; Gershovich, P.M.; Strelkova, A.N.; Spirina, N.A.; Shugaeva, T.E.; Morozov, D.V. Codon-Optimized Nucleic Acid Encoding the Fix Protein. U.S. Patent 18/280,338, 7 March 2024.
202. Pipe, S.W.; Arruda, V.R.; Lange, C.; Kitchen, S.; Eichler, H.; Wadsworth, S. Characteristics of BAY 2599023 in the Current Treatment Landscape ofHemophilia A Gene Therapy. *Curr. Gene Ther.* **2023**, *23*, 81–95. [CrossRef] [PubMed]
203. Greig, J.A.; Wang, Q.; Reicherter, A.L.; Chen, S.-J.; Hanlon, A.L.; Tipper, C.H.; Clark, K.R.; Wadsworth, S.; Wang, L.; Wilson, J.M. Characterization of Adeno-Associated Viral Vector-Mediated Human Factor VIII Gene Therapy in Hemophilia A Mice. *Hum. Gene Ther.* **2017**, *28*, 392–402. [CrossRef]
204. Wang, L.; Wilson, J.M.; Sidrane, J.A. Gene Therapy for Treating Hemophilia a. 2017. Available online: https://patents.google.com/patent/WO2017180857A1 (accessed on 14 November 2025).
205. Cataldo, J.; Allen, J.; Sankoh, S.; Weiss, K.; Askari, F. eP140: A Novel, Double-Blind Placebo-Controlled Seamless Phase 1/2/3 AAV9 Gene Therapy Study for Wilson Disease. *Genet. Med.* **2022**, *24*, S86. [CrossRef]
206. Livingston, C.; Wadsworth, S. Gene Therapy Constructs for Treating Wilson Disease. U.S. Patent 12,338,450, 24 June 2025.
207. Nambiar, B.; Cornell Sookdeo, C.; Berthelette, P.; Jackson, R.; Piraino, S.; Burnham, B.; Nass, S.; Souza, D.; O'Riordan, C.R.; Vincent, K.A.; et al. Characteristics of Minimally Oversized Adeno-Associated Virus Vectors Encoding Human Factor VIII Generated Using Producer Cell Lines and Triple Transfection. *Hum. Gene Ther. Methods* **2017**, *28*, 23–38. [CrossRef]
208. Kyostio-Moore, S.; Manavalan, P. Generation of Improved Human Pah for Treatment of Severe Pku by Liver-Directed Gene Replacement Therapy. 2020. Available online: https://patents.google.com/patent/WO2020077250A1 (accessed on 14 November 2025).
209. Leavitt, A.D.; Konkle, B.A.; Stine, K.C.; Visweshwar, N.; Harrington, T.J.; Giermasz, A.; Arkin, S.; Fang, A.; Plonski, F.; Yver, A.; et al. Giroctocogene Fitelparvovec Gene Therapy for Severe Hemophilia A: 104-Week Analysis of the Phase 1/2 Alta Study. *Blood* **2024**, *143*, 796–806. [CrossRef]
210. Riley, B.E. Liver-Specific Constructs, Factor Viii Expression Cassettes and Methods of Use Thereof. 2017. Available online: https://patents.google.com/patent/WO2017074526A1 (accessed on 14 November 2025).
211. Conner, J.E.; Crawford, L.A.; Damitz, R.; Davis, B.M.; Hodge, C.M.; Kimmel, M.L., II; Willard, T.Q.; Ramsey, P.; Thorne, D.J.; Young, A.L. Improved Pharmaceutical Compositions Containing Adeno-Associated Viral Vector. U.S. Patent 18/258,027, 29 February 2024.
212. Weiller, M.; Wang, H.; Coulibaly, S.; Schuster, M.; Rottensteiner, H.; Sun, K.; Chuah, M.K.; Vandendriessche, T.; Scheiflinger, F.; Höllriegl, W. Evaluation of the Human Factor IX Gene Therapy Vector TAK-748 in Hemophilia: Results from Non-Clinical Studies in Factor IX Knockout Mice and Rhesus Monkeys. *Blood* **2019**, *134*, 4633. [CrossRef]
213. Lu, G.; Jiabao, H.; Wei, L.; Shin-Shay, T.; Bin, L.; Xi, Z.; Zhao, X.P. Recombinant Adeno-Associated Virus Vector and Method for Treating or Preventing Hemophilia b. 2023. Available online: https://patents.google.com/patent/WO2023280323A1 (accessed on 14 November 2025).

https://doi.org/10.3390/cells15010014

214. Brown, H.C.; Zakas, P.M.; George, S.N.; Parker, E.T.; Spencer, H.T.; Doering, C.B. Target-Cell-Directed Bioengineering Approaches for Gene Therapy of Hemophilia A. *Mol. Ther.-Methods Clin. Dev.* **2018**, *9*, 57–69. [CrossRef]
215. Doering, C.B.; Spencer, T.H.; Brown, H.C. Recombinant Promoters and Vectors for Protein Expression in Liver and Use Thereof. U.S. Patent 10,058,624, 28 August 2018.
216. Brown, H. Nucleic Acid and Amino Acid Sequences Encoding High-Level Expressor Factor Viii Polypeptides and Methods of Use. 2022. Available online: https://patents.google.com/patent/WO2022165390A1 (accessed on 14 November 2025).
217. Yu, T.; Yang, Z.; Zhang, J.; Shen, L.; Rao, Y. Isolated Nucleic Acid Molecule and Use Thereof. 2023. Available online: https://patents.google.com/patent/WO2023078220A1 (accessed on 14 November 2025).
218. Yu, T.; Li, Y.; Yang, Z.; Zhang, J.; Liu, Y.; Zhou, E.; Dai, X.; Rao, Y. Isolated Nucleic Acid Molecule and Application Thereof. U.S. Patent 17/788,713, 5 October 2023.
219. Brunetti-Pierri, N.; Ferla, R.; Ginocchio, V.M.; Rossi, A.; Fecarotta, S.; Romano, R.; Parenti, G.; Yildiz, Y.; Zancan, S.; Pecorella, V.; et al. Liver-Directed Adeno-Associated Virus–Mediated Gene Therapy for Mucopolysaccharidosis Type VI. *NEJM Evid.* **2022**, *1*, EVIDoa2200052. [CrossRef]
220. Auricchio, A.; Alliegro, M.; Ferla, R. Combined Therapy for Mucopolysaccharidosis Type vi (Maroteaux-Lamy-Syndrome). 2018. Available online: https://patents.google.com/patent/WO2018046737A1 (accessed on 14 November 2025).
221. Rosenberg, M. First Patient Dosed with Gene Therapy in Phase 1/2 Study of ACTUS-101 in Patients with Pompe Disease. AskBio 2019. Available online: https://www.askbio.com/first-patient-dosed-with-gene-therapy-in-phase-1-2-study-of-actus-101-in-patients-with-pompe-disease/ (accessed on 14 November 2025).
222. Hopkins, S.; Smith, E.C. Therapeutic Adeno-Associated Virus for Treating Pompe Disease with Long Term Cessation of Gaa Enzyme Replacement Therapy. U.S. Patent 18/840,822, 22 May 2025.
223. Pipe, S.; Poma, A.; Rajasekhar, A.; Everington, T.; Sankoh, S.; Allen, J.; Cataldo, J.; Crombez, E. Gene Therapy for Hemophilia B: Results from the Phase 1/2 101HEMB01/02 Studies. *Blood Adv.* **2025**, *9*, 2980–2987. [CrossRef] [PubMed]
224. Wang, L.; Wilson, J.M. Compositions Useful in Treatment of Otc Deficency. U.S. Patent 9,890,365, 13 February 2018.
225. Wang, L.; Warzecha, C.C.; Kistner, A.; Chichester, J.A.; Bell, P.; Buza, E.L.; He, Z.; Pampena, M.B.; Couthouis, J.; Sethi, S.; et al. Prednisolone Reduces the Interferon Response to AAV in Cynomolgus Macaques and May Increase Liver Gene Expression. *Mol. Ther.-Methods Clin. Dev.* **2022**, *24*, 292–305. [CrossRef]
226. Wang, L.; Bell, P.; Morizono, H.; He, Z.; Pumbo, E.; Yu, H.; White, J.; Batshaw, M.L.; Wilson, J.M. AAV Gene Therapy Corrects OTC Deficiency and Prevents Liver Fibrosis in Aged OTC-Knock out Heterozygous Mice. *Mol. Genet. Metab.* **2017**, *120*, 299–305. [CrossRef] [PubMed]
227. Wang, L.; Wang, H.; Morizono, H.; Bell, P.; Jones, D.; Lin, J.; McMenamin, D.; Yu, H.; Batshaw, M.L.; Wilson, J.M. Sustained Correction of OTC Deficiency in Spfash Mice Using Optimized Self-Complementary AAV2/8 Vectors. *Gene Ther.* **2012**, *19*, 404–410. [CrossRef]
228. Hu, H.; Li, D.; Xu, Y.; Chen, C.; Wang, T.; Yuan, L.; Wang, X.; Du, Z.; Jiang, W.; Wu, X.; et al. Improved Human Coagulation Factor Viii Gene Expression Cassette and Use Thereof. 2024. Available online: https://patents.google.com/patent/WO2024060463A1 (accessed on 14 November 2025).
229. Xue, F.; Li, H.; Wu, X.; Liu, W.; Zhang, F.; Tang, D.; Chen, Y.; Wang, W.; Chi, Y.; Zheng, J.; et al. Safety and Activity of an Engineered, Liver-Tropic Adeno-Associated Virus Vector Expressing a Hyperactive Padua Factor IX Administered with Prophylactic Glucocorticoids in Patients with Haemophilia B: A Single-Centre, Single-Arm, Phase 1, Pilot Trial. *Lancet Haematol.* **2022**, *9*, e504–e513. [CrossRef]
230. Weinstein, D.A.; Derks, T.G.; Rodriguez-Buritica, D.F.; Ahmad, A.; Couce, M.; Mitchell, J.J.; Riba-Wolman, R.; Mount, M.; Sallago, J.B.; Ross, K.M.; et al. Safety and Efficacy of DTX401, an AAV8-Mediated Liver-Directed Gene Therapy, in Adults with Glycogen Storage Disease Type I a (GSDIa). *J. Inher Metab. Disea* **2025**, *48*, e70014. [CrossRef]
231. Tipper, C.; Clark, K.R.; Wadsworth, S. Methods and Compositions for Treating Glycogen Storage Diseases. 2020. Available online: https://patents.google.com/patent/WO2020132115A1 (accessed on 14 November 2025).
232. Liefhebber, J.M.P.; Brasser, G.; Spronck, E.A.; Ottenhoff, R.; Paerels, L.; Ferraz, M.J.; Schwarz, L.K.; Efthymiopoulou, N.; Kuo, C.-L.; Montenegro-Miranda, P.S.; et al. Preclinical Efficacy and Safety of Adeno-Associated Virus 5 Alpha-Galactosidase: A Gene Therapy for Fabry Disease. *Mol. Ther.-Methods Clin. Dev.* **2024**, *32*, 101375. [CrossRef]
233. Janciauskiene, S.; Welte, T. Well-Known and Less Well-Known Functions of Alpha-1 Antitrypsin. Its Role in Chronic Obstructive Pulmonary Disease and Other Disease Developments. *Ann. Am. Thorac. Soc.* **2016**, *13*, S280–S288. [CrossRef]
234. Matamala, N.; Martínez, M.T.; Lara, B.; Pérez, L.; Vázquez, I.; Jimenez, A.; Barquín, M.; Ferrarotti, I.; Blanco, I.; Janciauskiene, S.; et al. Alternative Transcripts of the *SERPINA1* Gene in Alpha-1 Antitrypsin Deficiency. *J. Transl. Med.* **2015**, *13*, 211. [CrossRef]
235. Morgan, K.; Chappell, S.; Guetta-Baranés, T.; Morley, S.; Kalsheker, N. The *Alpha-1-Antitrypsin* Gene Promoter in Human A549 Lung Derived Cells, and a Novel Transcription Initiation Site. *Int. J. Biochem. Cell Biol.* **2009**, *41*, 1157–1164. [CrossRef] [PubMed]
236. Long, G.L.; Chandra, T.; Woo, S.L.C.; Davie, E.W.; Kurachi, K. Complete Sequence of the cDNA for Human. *Alpha. 1-Antitrypsin* and the Gene for the S Variant. *Biochemistry* **1984**, *23*, 4828–4837. [CrossRef] [PubMed]

https://doi.org/10.3390/cells15010014

237. Ciliberto, G.; Dente, L.; Cortese, R. Cell-Specific Expression of a Transfected Human *A1-Antitrypsin* Gene. *Cell* **1985**, *41*, 531–540. [CrossRef]
238. De Simone, V.; Ciliberto, G.; Hardon, E.; Paonessa, G.; Palla, F.; Lundberg, L.; Cortese, R. Cis- and Trans-Acting Elements Responsible for the Cell-Specific Expression of the Human *Alpha 1-Antitrypsin* Gene. *EMBO J.* **1987**, *6*, 2759–2766. [CrossRef] [PubMed]
239. Morgan, K.; Scobie, G.; Marsters, P.; Kalsheker, N.A. Mutation in an *A1-Antitrypsin* Enhancer Results in an Interleukin-6 Deficient Acute-Phase Response Due to Loss of Cooperativity between Transcription Factors. *Biochim. Biophys. Acta (BBA)-Mol. Basis Dis.* **1997**, *1362*, 67–76. [CrossRef]
240. Morgan, K.; Kalsheker, N.A. Regulation of the Serine Proteinase Inhibitor (*SERPIN*) Gene A1-Antitrypsin: A Paradigm for Other SERPINs. *Int. J. Biochem. Cell Biol.* **1997**, *29*, 1501–1511. [CrossRef]
241. Chuah, M.K.; Petrus, I.; De Bleser, P.; Le Guiner, C.; Gernoux, G.; Adjali, O.; Nair, N.; Willems, J.; Evens, H.; Rincon, M.Y.; et al. Liver-Specific Transcriptional Modules Identified by Genome-Wide In Silico Analysis Enable Efficient Gene Therapy in Mice and Non-Human Primates. *Mol. Ther.* **2014**, *22*, 1605–1613. [CrossRef] [PubMed]
242. Samadani, U.; Costa, R.H. The Transcriptional Activator Hepatocyte Nuclear Factor 6 Regulates Liver Gene Expression. *Mol. Cell. Biol.* **1996**, *16*, 6273–6284. [CrossRef]
243. Costa, R.H.; Grayson, D.R.; Darnell, A.E. Multiple Hepatocyte-Enriched Nuclear Factors Function in the Regulation of *Transthyretin* and *A1-Antitrypsin* Genes. *Mol. Cell. Biol.* **1989**, *9*, 1415–1425. [CrossRef]
244. Ponder, K.P.; Dunbar, R.P.; Wilson, D.R.; Darlington, G.J.; Woo, S.L.C. Evaluation of Relative Promoter Strength in Primary Hepatocytes Using Optimized Lipofection. *Hum. Gene Ther.* **1991**, *2*, 41–52. [CrossRef]
245. Hafenrichter, D.; Wu, X.; Rettinger, S.; Kennedy, S.; Flye, M.; Ponder, K. Quantitative Evaluation of Liver-Specific Promoters from Retroviral Vectors after in Vivo Transduction of Hepatocytes. *Blood* **1994**, *84*, 3394–3404. [CrossRef] [PubMed]
246. Zhao, F.; Liang, S.; Zhou, Y.; Wang, Y.; Yan, H.; Wang, X.; Wang, H.; Du, J.; Zhan, L. Evaluation of Hepatitis B Virus Promoters for Sustained Transgene Expression in Mice by Bioluminescence Imaging. *Virus Res.* **2010**, *149*, 162–166. [CrossRef] [PubMed]
247. Serrano-Mendioroz, I.; Sampedro, A.; Alegre, M.; Enríquez De Salamanca, R.; Berraondo, P.; Fontanellas, A. An Inducible Promoter Responsive to Different Porphyrinogenic Stimuli Improves Gene Therapy Vectors for Acute Intermittent Porphyria. *Hum. Gene Ther.* **2018**, *29*, 480–491. [CrossRef]
248. Serrano-Mendioroz, I.; Sampedro, A.; Serna, N.; De Salamanca, R.E.; Sanz-Parra, A.; Corrales, F.; Berraondo, P.; Millet, O.; Fontanellas, A. Bioengineered PBGD Variant Improves the Therapeutic Index of Gene Therapy Vectors for Acute Intermittent Porphyria. *Hum. Mol. Genet.* **2018**, *27*, 3688–3696. [CrossRef]
249. Shih, S.-J.; Allan, C.; Grehan, S.; Tse, E.; Moran, C.; Taylor, J.M. Duplicated Downstream Enhancers Control Expression of the Human Apolipoprotein E Gene in Macrophages and Adipose Tissue. *J. Biol. Chem.* **2000**, *275*, 31567–31572. [CrossRef] [PubMed]
250. Chen, H.P.; Lin, A.; Bloom, J.S.; Khan, A.H.; Park, C.C.; Smith, D.J. Screening Reveals Conserved and Nonconserved Transcriptional Regulatory Elements Including an E3/E4 Allele-Dependent APOE Coding Region Enhancer. *Genomics* **2008**, *92*, 292–300. [CrossRef]
251. Dang, Q.; Walker, D.; Taylor, S.; Allan, C.; Chin, P.; Fan, J.; Taylor, J. Structure of the Hepatic Control Region of the Human Apolipoprotein E/C-I Gene Locus. *J. Biol. Chem.* **1995**, *270*, 22577–22585. [CrossRef]
252. Shachter, N.; Zhu, Y.; Walsh, A.; Breslow, J.; Smith, J. Localization of a Liver-Specific Enhancer in the Apolipoprotein E/C-I/C-II Gene Locus. *J. Lipid Res.* **1993**, *34*, 1699–1707. [CrossRef] [PubMed]
253. Allan, C.M.; Taylor, S.; Taylor, J.M. Two Hepatic Enhancers, HCR.1 and HCR.2, Coordinate the Liver Expression of the Entire Human Apolipoprotein E/C-I/C-IV/C-II Gene Cluster. *J. Biol. Chem.* **1997**, *272*, 29113–29119. [CrossRef] [PubMed]
254. Okuyama, T.; Huber, R.M.; Bowling, W.; Pearline, R.; Kennedy, S.C.; Flye, M.W.; Ponder, K.P. Liver-Directed Gene Therapy: A Retroviral Vector with a Complete LTR and the ApoE Enhancer-α_1-Antitrypsin Promoter Dramatically Increases Expression of Human α_1-Antitrypsin In Vivo. *Hum. Gene Ther.* **1996**, *7*, 637–645. [CrossRef]
255. Van Linthout, S.; Collen, D.; De Geest, B. Effect of Promoters and Enhancers on Expression, Transgene DNA Persistence, and Hepatotoxicity After Adenoviral Gene Transfer of Human Apolipoprotein A-I. *Hum. Gene Ther.* **2002**, *13*, 829–840. [CrossRef]
256. Jacobs, F.; Snoeys, J.; Feng, Y.; Van Craeyveld, E.; Lievens, J.; Armentano, D.; Cheng, S.H.; De Geest, B. Direct Comparison of Hepatocyte-Specific Expression Cassettes Following Adenoviral and Nonviral Hydrodynamic Gene Transfer. *Gene Ther.* **2008**, *15*, 594–603. [CrossRef]
257. Le, M.; Okuyama, T.; Cai, S.-R.; Kennedy, S.C.; Bowling, W.M.; Flye, M.W.; Ponder, K.P. Therapeutic Levels of Functional Human Factor X in Rats After Retroviral-Mediated Hepatic Gene Therapy. *Blood* **1997**, *89*, 1254–1259. [CrossRef]
258. De Geest, B.R.; Van Linthout, S.A.; Collen, D. Humoral Immune Response in Mice against a Circulating Antigen Induced by Adenoviral Transfer Is Strictly Dependent on Expression in Antigen-Presenting Cells. *Blood* **2003**, *101*, 2551–2556. [CrossRef]
259. Mingozzi, F.; Liu, Y.-L.; Dobrzynski, E.; Kaufhold, A.; Liu, J.H.; Wang, Y.; Arruda, V.R.; High, K.A.; Herzog, R.W. Induction of Immune Tolerance to Coagulation Factor IX Antigen by in Vivo Hepatic Gene Transfer. *J. Clin. Investig.* **2003**, *111*, 1347–1356. [CrossRef]

https://doi.org/10.3390/cells15010014

260. Miao, C.H.; Ohashi, K.; Patijn, G.A.; Meuse, L.; Ye, X.; Thompson, A.R.; Kay, M.A. Inclusion of the Hepatic Locus Control Region, an Intron, and Untranslated Region Increases and Stabilizes Hepatic Factor IX Gene Expression in Vivo but Not in Vitro. *Mol. Ther.* **2000**, *1*, 522–532. [CrossRef]
261. Manno, C.S.; Pierce, G.F.; Arruda, V.R.; Glader, B.; Ragni, M.; Rasko, J.J.E.; Ozelo, M.C.; Hoots, K.; Blatt, P.; Konkle, B.; et al. Successful Transduction of Liver in Hemophilia by AAV-Factor IX and Limitations Imposed by the Host Immune Response. *Nat. Med.* **2006**, *12*, 342–347. [CrossRef]
262. Cooper, M.; Nayak, S.; Hoffman, B.E.; Terhorst, C.; Cao, O.; Herzog, R.W. Improved Induction of Immune Tolerance to Factor IX by Hepatic AAV-8 Gene Transfer. *Hum. Gene Ther.* **2009**, *20*, 767–776. [CrossRef]
263. Vandendriessche, T.; Thorrez, L.; Acosta-Sanchez, A.; Petrus, I.; Wang, L.; Ma, L.; De Waele, L.; Iwasaki, Y.; Gillijns, V.; Wilson, J.M.; et al. Efficacy and Safety of Adeno-associated Viral Vectors Based on Serotype 8 and 9 vs. Lentiviral Vectors for Hemophilia B Gene Therapy. *J. Thromb. Haemost.* **2007**, *5*, 16–24. [CrossRef] [PubMed]
264. Davidoff, A.M.; Gray, J.T.; Ng, C.Y.C.; Zhang, Y.; Zhou, J.; Spence, Y.; Bakar, Y.; Nathwani, A.C. Comparison of the Ability of Adeno-Associated Viral Vectors Pseudotyped with Serotype 2, 5, and 8 Capsid Proteins to Mediate Efficient Transduction of the Liver in Murine and Nonhuman Primate Models. *Mol. Ther.* **2005**, *11*, 875–888. [CrossRef] [PubMed]
265. Wu, Z.; Sun, J.; Zhang, T.; Yin, C.; Yin, F.; Van Dyke, T.; Samulski, R.J.; Monahan, P.E. Optimization of Self-Complementary AAV Vectors for Liver-Directed Expression Results in Sustained Correction of Hemophilia B at Low Vector Dose. *Mol. Ther.* **2008**, *16*, 280–289. [CrossRef]
266. Gehrke, S.; Jérôme, V.; Müller, R. Chimeric Transcriptional Control Units for Improved Liver-Specific Transgene Expression. *Gene* **2003**, *322*, 137–143. [CrossRef]
267. Paulk, N.K.; Pekrun, K.; Zhu, E.; Nygaard, S.; Li, B.; Xu, J.; Chu, K.; Leborgne, C.; Dane, A.P.; Haft, A.; et al. Bioengineered AAV Capsids with Combined High Human Liver Transduction In Vivo and Unique Humoral Seroreactivity. *Mol. Ther.* **2018**, *26*, 289–303. [CrossRef] [PubMed]
268. Dane, A.P.; Cunningham, S.C.; Graf, N.S.; Alexander, I.E. Sexually Dimorphic Patterns of Episomal rAAV Genome Persistence in the Adult Mouse Liver and Correlation With Hepatocellular Proliferation. *Mol. Ther.* **2009**, *17*, 1548–1554. [CrossRef]
269. Cabanes-Creus, M.; Westhaus, A.; Navarro, R.G.; Baltazar, G.; Zhu, E.; Amaya, A.K.; Liao, S.H.Y.; Scott, S.; Sallard, E.; Dilworth, K.L.; et al. Attenuation of Heparan Sulfate Proteoglycan Binding Enhances In Vivo Transduction of Human Primary Hepatocytes with AAV2. *Mol. Ther.-Methods Clin. Dev.* **2020**, *17*, 1139–1154. [CrossRef]
270. Cabanes-Creus, M.; Navarro, R.G.; Zhu, E.; Baltazar, G.; Liao, S.H.Y.; Drouyer, M.; Amaya, A.K.; Scott, S.; Nguyen, L.H.; Westhaus, A.; et al. Novel Human Liver-Tropic AAV Variants Define Transferable Domains That Markedly Enhance the Human Tropism of AAV7 and AAV8. *Mol. Ther.-Methods Clin. Dev.* **2022**, *24*, 88–101. [CrossRef]
271. Cunningham, S.C.; Spinoulas, A.; Carpenter, K.H.; Wilcken, B.; Kuchel, P.W.; Alexander, I.E. AAV2/8-Mediated Correction of OTC Deficiency Is Robust in Adult but Not Neonatal Spfash Mice. *Mol. Ther.* **2009**, *17*, 1340–1346. [CrossRef]
272. Cunningham, S.C.; Siew, S.M.; Hallwirth, C.V.; Bolitho, C.; Sasaki, N.; Garg, G.; Michael, I.P.; Hetherington, N.A.; Carpenter, K.; De Alencastro, G.; et al. Modeling Correction of Severe Urea Cycle Defects in the Growing Murine Liver Using a Hybrid Recombinant Adeno-associated Virus/piggyBac Transposase Gene Delivery System. *Hepatology* **2015**, *62*, 417–428. [CrossRef]
273. Ginn, S.L.; Amaya, A.K.; Liao, S.H.Y.; Zhu, E.; Cunningham, S.C.; Lee, M.; Hallwirth, C.V.; Logan, G.J.; Tay, S.S.; Cesare, A.J.; et al. Efficient in Vivo Editing of OTC-Deficient Patient-Derived Primary Human Hepatocytes. *JHEP Rep.* **2020**, *2*, 100065. [CrossRef] [PubMed]
274. Rouet, P.; Raguenez, G.; Tronche, F.; Yaniv, M.; N'Guyen, C.; Salier, J.P. A Potent Enhancer Made of Clustered Liver-Specific Elements in the Transcription Control Sequences of Human Alpha 1-Microglobulin/Bikunin Gene. *J. Biol. Chem.* **1992**, *267*, 20765–20773. [CrossRef]
275. McEachern, K.A.; Nietupski, J.B.; Chuang, W.; Armentano, D.; Johnson, J.; Hutto, E.; Grabowski, G.A.; Cheng, S.H.; Marshall, J. AAV8-mediated Expression of Glucocerebrosidase Ameliorates the Storage Pathology in the Visceral Organs of a Mouse Model of Gaucher Disease. *J. Gene Med.* **2006**, *8*, 719–729. [CrossRef] [PubMed]
276. Van Craeyveld, E.; Gordts, S.C.; Nefyodova, E.; Jacobs, F.; De Geest, B. Regression and Stabilization of Advanced Murine Atherosclerotic Lesions: A Comparison of LDL Lowering and HDL Raising Gene Transfer Strategies. *J. Mol. Med.* **2011**, *89*, 555–567. [CrossRef] [PubMed]
277. Sabatino, D.E.; Lange, A.M.; Altynova, E.S.; Sarkar, R.; Zhou, S.; Merricks, E.P.; Franck, H.G.; Nichols, T.C.; Arruda, V.R.; Kazazian, H.H., Jr. Efficacy and Safety of Long-Term Prophylaxis in Severe Hemophilia A Dogs Following Liver Gene Therapy Using AAV Vectors. *Mol. Ther.* **2011**, *19*, 442–449. [CrossRef]
278. Nguyen, G.N.; George, L.A.; Siner, J.I.; Davidson, R.J.; Zander, C.B.; Zheng, X.L.; Arruda, V.R.; Camire, R.M.; Sabatino, D.E. Novel Factor VIII Variants with a Modified Furin Cleavage Site Improve the Efficacy of Gene Therapy for Hemophilia A. *J. Thromb. Haemost.* **2017**, *15*, 110–121. [CrossRef]

https://doi.org/10.3390/cells15010014

279. Siner, J.I.; Iacobelli, N.P.; Sabatino, D.E.; Ivanciu, L.; Zhou, S.; Poncz, M.; Camire, R.M.; Arruda, V.R. Minimal Modification in the Factor VIII B-Domain Sequence Ameliorates the Murine Hemophilia A Phenotype. *Blood* **2013**, *121*, 4396–4403. [CrossRef] [PubMed]
280. Brown, H.C.; Doering, C.B.; Herzog, R.W.; Ling, C.; Markusic, D.M.; Spencer, H.T.; Srivastava, A.; Srivastava, A. Development of a Clinical Candidate AAV3 Vector for Gene Therapy of Hemophilia B. *Hum. Gene Ther.* **2020**, *31*, 1114–1123. [CrossRef] [PubMed]
281. Thomas, H.; Carlisle, R.C. Progress in Gene Therapy for Hereditary Tyrosinemia Type 1. *Pharmaceutics* **2025**, *17*, 387. [CrossRef]
282. Homology Medicines, Inc. Homology Medicines Announces Plan to Evaluate Strategic Options for the Company and Its Genetic Medicines Programs, Including HMI-103 Gene Editing Candidate for PKU. Available online: https://www.globenewswire.com/news-release/2023/07/27/2712687/0/en/Homology-Medicines-Announces-Plan-to-Evaluate-Strategic-Options-for-the-Company-and-its-Genetic-Medicines-Programs-including-HMI-103-Gene-Editing-Candidate-for-PKU.html (accessed on 10 December 2025).
283. Nathwani, A.; Raj, D. Fabry Disease Gene Therapy. U.S. Patent 11,103,596, 31 August 2021.
284. Dong, B.; Yang, L. Engineered Liver-Specific Enhancers and Their Applications. U.S. Patent 18/919,331, 17 April 2025.
285. Wang, Z.; Burke, P.A. Hepatocyte Nuclear Factor-4α Interacts with Other Hepatocyte Nuclear Factors in Regulating *Transthyretin* Gene Expression. *FEBS J.* **2010**, *277*, 4066–4075. [CrossRef]
286. Sanguinetti, C.; Minniti, M.; Susini, V.; Caponi, L.; Panichella, G.; Castiglione, V.; Aimo, A.; Emdin, M.; Vergaro, G.; Franzini, M. The Journey of Human Transthyretin: Synthesis, Structure Stability, and Catabolism. *Biomedicines* **2022**, *10*, 1906. [CrossRef] [PubMed]
287. Costa, R.H.; Lai, E.; Darnell, J.E. Transcriptional Control of the Mouse *Prealbumin* (*Transthyretin*) Gene: Both Promoter Sequences and a Distinct Enhancer Are Cell Specific. *Mol. Cell. Biol.* **1986**, *6*, 4697–4708. [CrossRef]
288. Yan, C.; Costa, R.H.; Darnell, J.E.; Chen, J.D.; Van Dyke, T.A. Distinct Positive and Negative Elements Control the Limited Hepatocyte and Choroid Plexus Expression of Transthyretin in Transgenic Mice. *EMBO J.* **1990**, *9*, 869–878. [CrossRef]
289. Hoag, H.; Gore, J.; Barry, D.; Mueller, C.R. Gene Therapy Expression Vectors Based on the Clotting Factor IX Promoter. *Gene Ther.* **1999**, *6*, 1584–1589. [CrossRef]
290. Muhuri, M.; Levy, D.I.; Schulz, M.; McCarty, D.; Gao, G. Durability of Transgene Expression after rAAV Gene Therapy. *Mol. Ther.* **2022**, *30*, 1364–1380. [CrossRef]
291. Chapin, J.; Álvarez Román, M.T.; Ayash-Rashkovsky, M.; Diogo, D.; Kenniston, J.; Lopez-Jaime, F.; Maggiore, C.; Mingot-Castellano, M.; Rajavel, K.; Rauch, A.; et al. A Phase 1/2 Safety and Efficacy Study of TAK-754 Gene Therapy: The Challenge of Achieving Durable Factor VIII Expression in Haemophilia A Clinical Trials. *Haemophilia* **2025**, *31*, 108–117. [CrossRef]
292. Vigna, E.; Amendola, M.; Benedicenti, F.; Simmons, A.D.; Follenzi, A.; Naldini, L. Efficient Tet-Dependent Expression of Human Factor IX in Vivo by a New Self-Regulating Lentiviral Vector. *Mol. Ther.* **2005**, *11*, 763–775. [CrossRef]
293. Cantore, A.; Ranzani, M.; Bartholomae, C.C.; Volpin, M.; Valle, P.D.; Sanvito, F.; Sergi, L.S.; Gallina, P.; Benedicenti, F.; Bellinger, D.; et al. Liver-Directed Lentiviral Gene Therapy in a Dog Model of Hemophilia B. *Sci. Transl. Med.* **2015**, *7*, 277ra28. [CrossRef]
294. Jiang, H.; Lillicrap, D.; Patarroyo-White, S.; Liu, T.; Qian, X.; Scallan, C.D.; Powell, S.; Keller, T.; McMurray, M.; Labelle, A.; et al. Multiyear Therapeutic Benefit of AAV Serotypes 2, 6, and 8 Delivering Factor VIII to Hemophilia A Mice and Dogs. *Blood* **2006**, *108*, 107–115. [CrossRef] [PubMed]
295. Scallan, C.D.; Lillicrap, D.; Jiang, H.; Qian, X.; Patarroyo-White, S.L.; Parker, A.E.; Liu, T.; Vargas, J.; Nagy, D.; Powell, S.K.; et al. Sustained Phenotypic Correction of Canine Hemophilia A Using an Adeno-Associated Viral Vector. *Blood* **2003**, *102*, 2031–2037. [CrossRef]
296. Costa, R.H.; Grayson, D.R. Site-Directed Mutagenesis of Hepatocyte Nuclear Factor (HNF) Binding Sites in the Mouse Transthyretin (TTR) Promoter Reveal Synergistic Interactions with Its Enhancer Region. *Nucleic Acids Res.* **1991**, *19*, 4139–4145. [CrossRef]
297. Croteau, S.E.; Eyster, M.E.; Tran, H.; Ragni, M.V.; Samelson-Jones, B.J.; George, L.; Sullivan, S.; Rasko, J.E.J.; Moormeier, J.; Angchaisuksiri, P.; et al. Long-Term Durable FVIII Expression with Improvements in Bleeding Rates Following AAV-Mediated FVIII Gene Transfer for Hemophilia A: Multiyear Follow-up on the Phase I/II Trial of SPK-8011. *Blood* **2022**, *140*, 1899–1901. [CrossRef]
298. Nguyen, G.N.; Lindgren, J.R.; Seleme, M.C.; Kafle, S.; Zander, C.B.; Zheng, X.L.; Sabatino, D.E. Altered Cleavage of Human Factor VIII at the B-Domain and Acidic Region 3 Interface Enhances Expression after Gene Therapy in Hemophilia A Mice. *J. Thromb. Haemost.* **2023**, *21*, 2101–2113. [CrossRef]
299. Yamamura, K.; Wakasugi, S.; Maeda, S.; Inomoto, T.; Iwanaga, T.; Uehira, M.; Araki, K.; Miyazaki, J.; Shimada, K. Tissue-specific and Developmental Expression of Human *Transthyretin* Gene in Transgenic Mice. *Dev. Genet.* **1987**, *8*, 195–205. [CrossRef] [PubMed]
300. Greig, J.A.; Nordin, J.M.L.; White, J.W.; Wang, Q.; Bote, E.; Goode, T.; Calcedo, R.; Wadsworth, S.; Wang, L.; Wilson, J.M. Optimized Adeno-Associated Viral-Mediated Human Factor VIII Gene Therapy in Cynomolgus Macaques. *Hum. Gene Ther.* **2018**, *29*, 1364–1375. [CrossRef] [PubMed]

https://doi.org/10.3390/cells15010014

301. Pipe, S.W.; Sheehan, J.P.; Coppens, M.; Eichler, H.; Linardi, C.; Wiegmann, S.; Hay, C.R.; Lissitchkov, T. First-in-Human Dose-Finding Study of AAVhu37 Vector-Based Gene Therapy: BAY 2599023 Has Stable and Sustained Expression of FVIII over 2 Years. *Blood* **2021**, *138*, 3971. [CrossRef]
302. Kyostio-Moore, S.; Berthelette, P.; Piraino, S.; Sookdeo, C.; Nambiar, B.; Jackson, R.; Burnham, B.; O'Riordan, C.R.; Cheng, S.H.; Armentano, D. The Impact of Minimally Oversized Adeno-Associated Viral Vectors Encoding Human Factor VIII on Vector Potency in Vivo. *Mol. Ther.-Methods Clin. Dev.* **2016**, *3*, 16006. [CrossRef]
303. Kumar, S.R.P.; Xie, J.; Hu, S.; Ko, J.; Huang, Q.; Brown, H.C.; Srivastava, A.; Markusic, D.M.; Doering, C.B.; Spencer, H.T.; et al. Coagulation Factor IX Gene Transfer to Non-Human Primates Using Engineered AAV3 Capsid and Hepatic Optimized Expression Cassette. *Mol. Ther.-Methods Clin. Dev.* **2021**, *23*, 98–107. [CrossRef]
304. Coyle, C.W.; Knight, K.A.; Brown, H.C.; George, S.N.; Denning, G.; Branella, G.M.; Childers, K.C.; Spiegel, P.C.; Spencer, H.T.; Doering, C.B. Humanization and Functional Characterization of Enhanced Coagulation Factor IX Variants Identified through Ancestral Sequence Reconstruction. *J. Thromb. Haemost.* **2024**, *22*, 633–644. [CrossRef] [PubMed]
305. Knight, K.A.; Coyle, C.W.; Radford, C.E.; Parker, E.T.; Fedanov, A.; Shields, J.M.; Szlam, F.; Purchel, A.; Chen, M.; Denning, G.; et al. Identification of Coagulation Factor IX Variants with Enhanced Activity through Ancestral Sequence Reconstruction. *Blood Adv.* **2021**, *5*, 3333–3343. [CrossRef]
306. Jiang, D.; Wang, M.; Wheeler, A.P.; Croteau, S.E. 2025 Clinical Trials Update on Hemophilia VWD, and Rare Inherited Bleeding Disorders. *Am. J. Hematol.* **2025**, *100*, 666–684. [CrossRef] [PubMed]
307. Chuah, M.; Vandendriessche, T. Optimized Liver-Specific Expression Systems for Fviii and Fix 2016. Available online: https://patents.google.com/patent/WO2016146757A1 (accessed on 14 November 2025).
308. Mazzaferro, E.M.; Edwards, T. Update on Albumin Therapy in Critical Illness. *Vet. Clin. N. Am. Small Anim. Pract.* **2020**, *50*, 1289–1305. [CrossRef]
309. Cereghini, S.; Raymondjean, M.; Carranca, A.G.; Herbomel, P.; Yaniv, M. Factors Involved in Control of Tissue-Specific Expression of *Albumin* Gene. *Cell* **1987**, *50*, 627–638. [CrossRef]
310. Tang, J.; Wu, Q.; Li, Y.; Wu, X.; Wang, Y.; Zhu, L.; Shi, Y.; Bu, H.; Bao, J.; Xie, M. Construction of a General Albumin Promoter Reporter System for Real-Time Monitoring of the Differentiation Status of Functional Hepatocytes from Stem Cells in Mouse, Rat and Human. *Biomed. Rep.* **2017**, *6*, 627–632. [CrossRef] [PubMed]
311. Vanrell, L.; Di Scala, M.; Blanco, L.; Otano, I.; Gil-Farina, I.; Baldim, V.; Paneda, A.; Berraondo, P.; Beattie, S.G.; Chtarto, A.; et al. Development of a Liver-Specific Tet-On Inducible System for AAV Vectors and Its Application in the Treatment of Liver Cancer. *Mol. Ther.* **2011**, *19*, 1245–1253. [CrossRef]
312. Harmatz, P.; Prada, C.E.; Burton, B.K.; Lau, H.; Kessler, C.M.; Cao, L.; Falaleeva, M.; Villegas, A.G.; Zeitler, J.; Meyer, K.; et al. First-in-Human in Vivo Genome Editing via AAV-Zinc-Finger Nucleases for Mucopolysaccharidosis I/II and Hemophilia B. *Mol. Ther.* **2022**, *30*, 3587–3600. [CrossRef]
313. Muenzer, J.; Prada, C.E.; Burton, B.; Lau, H.A.; Ficicioglu, C.; Foo, C.W.P.; Vaidya, S.A.; Whitley, C.B.; Harmatz, P. CHAMPIONS: A Phase 1/2 Clinical Trial with Dose Escalation of SB-913 ZFN-Mediated in Vivo Human Genome Editing for Treatment of MPS II (Hunter Syndrome). *Mol. Genet. Metab.* **2019**, *126*, S104. [CrossRef]
314. Harmatz, P.; Lau, H.A.; Heldermon, C.; Leslie, N.; Foo, C.W.P.; Vaidya, S.A.; Whitley, C.B. EMPOWERS: A Phase 1/2 Clinical Trial of SB-318 ZFN-Mediated in Vivo Human Genome Editing for Treatment of MPS I (Hurler Syndrome). *Mol. Genet. Metab.* **2019**, *126*, S68. [CrossRef]
315. Herbomel, P.; Rollier, A.; Tronche, F.; Ott, M.-O.; Yaniv, M.; Weiss, M.C. The Rat Albumin Promoter Is Composed of Six Distinct Positive Elements within 130 Nucleotides. *Mol. Cell. Biol.* **1989**, *9*, 4750–4758. [CrossRef]
316. Power, S.C.; Cereghini, S.; Rollier, A.; Gannon, F. Isolation and Functional Analysis of the Promoter of the Bovine Serum *Albumin* Gene. *Biochem. Biophys. Res. Commun.* **1994**, *203*, 1447–1456. [CrossRef]
317. Schorpp, M.; Kugler, W.; Wagner, U.; Ryffel, G.U. Hepatocyte-Specific Promoter Element HP1 of the Xenopus Albumin Gene Interacts with Transcriptional Factors of Mammalian Hepatocytes. *J. Mol. Biol.* **1988**, *202*, 307–320. [CrossRef]
318. Vorachek, W.R.; Steppan, C.M.; Lima, M.; Black, H.; Bhattacharya, R.; Wen, P.; Kajiyama, Y.; Locker, J. Distant Enhancers Stimulate the Albumin Promoter through Complex Proximal Binding Sites. *J. Biol. Chem.* **2000**, *275*, 29031–29041. [CrossRef]
319. Gorski, K.; Carneiro, M.; Schibler, U. Tissue-Specific in Vitro Transcription from the Mouse Albumin Promoter. *Cell* **1986**, *47*, 767–776. [CrossRef]
320. Wooddell, C.I.; Reppen, T.; Wolff, J.A.; Herweijer, H. Sustained Liver-specific Transgene Expression from the Albumin Promoter in Mice Following Hydrodynamic Plasmid DNA Delivery. *J. Gene Med.* **2008**, *10*, 551–563. [CrossRef] [PubMed]
321. Herbst, R.S.; Friedman, N.; Darnell, J.E.; Babiss, L.E. Positive and Negative Regulatory Elements in the Mouse Albumin Enhancer. *Proc. Natl. Acad. Sci. USA* **1989**, *86*, 1553–1557. [CrossRef]
322. Feng, R.; Kan, K.; Sticht, C.; Li, Y.; Wang, S.; Liu, H.; Shao, C.; Munker, S.; Niess, H.; Wang, S.; et al. A Hierarchical Regulatory Network Ensures Stable Albumin Transcription under Various Pathophysiological Conditions. *Hepatology* **2022**, *76*, 1673–1689. [CrossRef]

https://doi.org/10.3390/cells15010014

323. Zhang, C.; Liu, D. Transcription Factor Binding Site in Promoter Determines the Pattern of Plasmid-Based Transgene Expression In Vivo. *Pharmaceutics* **2024**, *16*, 544. [CrossRef]
324. Pinkert, C.A.; Ornitz, D.M.; Brinster, R.L.; Palmiter, R.D. An *Albumin* Enhancer Located 10 Kb Upstream Functions along with Its Promoter to Direct Efficient, Liver-Specific Expression in Transgenic Mice. *Genes Dev.* **1987**, *1*, 268–276. 1798. [CrossRef]
325. Xiao, W.; Berta, S.C.; Lu, M.M.; Moscioni, A.D.; Tazelaar, J.; Wilson, J.M. Adeno-Associated Virus as a Vector for Liver-Directed Gene Therapy. *J. Virol.* **1998**, *72*, 10222–10226. [CrossRef] [PubMed]
326. Hu, J.M.; Camper, S.A.; Tilghman, S.M.; Miller, T.; Georgoff, I.; Serra, R.; Isom, H.C. Functional Analyses of Albumin Expression in a Series of Hepatocyte Cell Lines and in Primary Hepatocytes. *Cell Growth Differ.* **1992**, *3*, 577–588. [PubMed]
327. Connelly, S.; Smith, T.A.G.; Dhir, G.; Gardner, J.M.; Mehaffey, M.G.; Zaret, K.S.; McClelland, A.; Kaleko, M. In Vivo Gene Delivery and Expression of Physiological Levels of Functional Human Factor VIII in Mice. *Hum. Gene Ther.* **1995**, *6*, 185–193. [CrossRef] [PubMed]
328. Zaret, K.S.; DiPersio, C.M.; Jackson, D.A.; Montigny, W.J.; Weinstat, D.L. Conditional Enhancement of Liver-Specific Gene Transcription. *Proc. Natl. Acad. Sci. USA* **1988**, *85*, 9076–9080. [CrossRef] [PubMed]
329. Connelly, S.; Andrews, J.L.; Gallo, A.M.; Kayda, D.B.; Qian, J.; Hoyer, L.; Kadan, M.J.; Gorziglia, M.I.; Trapnell, B.C.; McClelland, A.; et al. Sustained Phenotypic Correction of Murine Hemophilia A by In Vivo Gene Therapy. *Blood* **1998**, *91*, 3273–3281. [CrossRef] [PubMed]
330. Hafenrichter, D.G.; Ponder, K.P.; Rettinger, S.D.; Kennedy, S.C.; Wu, X.; Saylors, R.S.; Flye, M.W. Liver-Directed Gene Therapy: Evaluation of Liver Specific Promoter Elements. *J. Surg. Res.* **1994**, *56*, 510–517. [CrossRef] [PubMed]
331. Connelly, S.; Gardner, J.; Lyons, R.; McClelland, A.; Kaleko, M. Sustained Expression of Therapeutic Levels of Human Factor VIII in Mice. *Blood* **1996**, *87*, 4671–4677. [CrossRef]
332. Nietupski, J.B.; Hurlbut, G.D.; Ziegler, R.J.; Chu, Q.; Hodges, B.L.; Ashe, K.M.; Bree, M.; Cheng, S.H.; Gregory, R.J.; Marshall, J.; et al. Systemic Administration of AAV8-α-Galactosidase A Induces Humoral Tolerance in Nonhuman Primates Despite Low Hepatic Expression. *Mol. Ther.* **2011**, *19*, 1999–2011. [CrossRef]
333. Ziegler, R.J.; Cherry, M.; Barbon, C.M.; Li, C.; Bercury, S.D.; Armentano, D.; Desnick, R.J.; Cheng, S.H. Correction of the Biochemical and Functional Deficits in Fabry Mice Following AAV8–Mediated Hepatic Expression of α-Galactosidase A. *Mol. Ther.* **2007**, *15*, 492–500. [CrossRef]
334. Hurlbut, G.D.; Ziegler, R.J.; Nietupski, J.B.; Foley, J.W.; Woodworth, L.A.; Meyers, E.; Bercury, S.D.; Pande, N.N.; Souza, D.W.; Bree, M.P.; et al. Preexisting Immunity and Low Expression in Primates Highlight Translational Challenges for Liver-Directed AAV8-Mediated Gene Therapy. *Mol. Ther.* **2010**, *18*, 1983–1994. [CrossRef]
335. Ziegler, R.J.; Lonning, S.M.; Armentano, D.; Li, C.; Souza, D.W.; Cherry, M.; Ford, C.; Barbon, C.M.; Desnick, R.J.; Gao, G.; et al. AAV2 Vector Harboring a Liver-Restricted Promoter Facilitates Sustained Expression of Therapeutic Levels of α-Galactosidase A and the Induction of Immune Tolerance in Fabry Mice. *Mol. Ther.* **2004**, *9*, 231–240. [CrossRef]
336. Ryffel, G.U.; Kugler, W.; Wagner, U.; Kaling, M. Liver Cell Specific Gene Transcription in Vitro: The Promoter Elements HP1 and TATA Box Are Necessary and Sufficient to Generate a Liver-Specific Promoter. *Nucleic Acids Res.* **1989**, *17*, 939–953. [CrossRef]
337. Mozzi, A.; Forni, D.; Cagliani, R.; Pozzoli, U.; Vertemara, J.; Bresolin, N.; Sironi, M. Albuminoid Genes: Evolving at the Interface of Dispensability and Selection. *Genome Biol. Evol.* **2014**, *6*, 2983–2997. [CrossRef] [PubMed]
338. Hammer, R.E.; Krumlauf, R.; Camper, S.A.; Brinster, R.L.; Tilghman, S.M. Diversity of Alpha-Fetoprotein Gene Expression in Mice Is Generated by a Combination of Separate Enhancer Elements. *Science* **1987**, *235*, 53–58. [CrossRef] [PubMed]
339. Poli, V.; Silengo, L.; Altruda, F.; Cortese, R. The Analysis of the Human Hemopexin Promoter Defines a New Class of Liver-Specific Genes. *Nucleic Acids Res.* **1989**, *17*, 9351–9365.
340. Kanai, F.; Lan, K.H.; Shiratori, Y.; Tanaka, T.; Ohashi, M.; Okudaira, T.; Yoshida, Y.; Wakimoto, H.; Hamada, H.; Nakabayashi, H.; et al. In Vivo Gene Therapy for Alpha-Fetoprotein-Producing Hepatocellular Carcinoma by Adenovirus-Mediated Transfer of Cytosine Deaminase Gene. *Cancer Res.* **1997**, *57*, 461–465.
341. Pin, R.H.; Reinblatt, M.; Fong, Y. Utilizing α-Fetoprotein Expression to Enhance Oncolytic Viral Therapy in Hepatocellular Carcinoma. *Ann. Surg.* **2004**, *240*, 659–666. [CrossRef] [PubMed]
342. Ucdal, M.; Yazarkan, Y.; Sonmez, G.; Celtikci, B.; Balaban, Y. Alpha-Fetoprotein: A Multifaceted Player in Cancer Biology. *Euroasian J. Hepato-Gastroenterol.* **2025**, *15*, 72–82. [CrossRef]
343. Godbout, R.; Ingram, R.; Tilghman, S.M. Multiple Regulatory Elements in the Intergenic Region Between the α-Fetoprotein and Albumin Genes. *Mol. Cell. Biol.* **1986**, *6*, 477–487. [CrossRef]
344. Lazarevich, N.L. Molecular Mechanisms of Alpha-Fetoprotein Gene Expression. *Biochemistry* **2000**, *65*, 117–133.
345. Hayashi, Y.; Mori, Y.; Janssen, O.E.; Sunthornthepvarakul, T.; Weiss, R.E.; Takeda, K.; Weinberg, M.; Seo, H.; Bell, G.I.; Refetoff, S. Human *Thyroxine-Binding Globulin* Gene: Complete Sequence and Transcriptional Regulation. *Mol. Endocrinol.* **1993**, *7*, 1049–1060. [CrossRef] [PubMed]
346. Gao, G.-P.; Alvira, M.R.; Wang, L.; Calcedo, R.; Johnston, J.; Wilson, J.M. Novel Adeno-Associated Viruses from Rhesus Monkeys as Vectors for Human Gene Therapy. *Proc. Natl. Acad. Sci. USA* **2002**, *99*, 11854–11859. [CrossRef] [PubMed]

https://doi.org/10.3390/cells15010014

347. Wang, L.; Takabe, K.; Bidlingmaier, S.M.; Ill, C.R.; Verma, I.M. Sustained Correction of Bleeding Disorder in Hemophilia B Mice by Gene Therapy. *Proc. Natl. Acad. Sci. USA* **1999**, *96*, 3906–3910. [CrossRef]
348. Ill, C.R.; Yang, C.Q.; Bidlingmaier, S.M.; Gonzales, J.N.; Burns, D.S.; Bartholomew, R.M.; Scuderi, P. Optimization of the Human Factor VIII Complementary DNA Expression Plasmid for Gene Therapy of Hemophilia A. *Blood Coagul. Fibrinolysis* **1997**, *8* (Suppl. 2), S23–S30.
349. Wang, L.; Calcedo, R.; Nichols, T.C.; Bellinger, D.A.; Dillow, A.; Verma, I.M.; Wilson, J.M. Sustained Correction of Disease in Naive and AAV2-Pretreated Hemophilia B Dogs: AAV2/8-Mediated, Liver-Directed Gene Therapy. *Blood* **2005**, *105*, 3079–3086. [CrossRef] [PubMed]
350. Wang, L.; Nichols, T.C.; Read, M.S.; Bellinger, D.A.; Verma, I.M. Sustained Expression of Therapeutic Level of Factor IX in Hemophilia B Dogs by AAV-Mediated Gene Therapy in Liver. *Mol. Ther.* **2000**, *1*, 154–158. [CrossRef]
351. Sarkar, R.; Tetreault, R.; Gao, G.; Wang, L.; Bell, P.; Chandler, R.; Wilson, J.M.; Kazazian, H.H. Total Correction of Hemophilia A Mice with Canine FVIII Using an AAV 8 Serotype. *Blood* **2004**, *103*, 1253–1260. [CrossRef]
352. Wang, X.; Xu, K.; Huang, Z.; Lin, Y.; Zhou, J.; Zhou, L.; Ma, F. Accelerating Promoter Identification and Design by Deep Learning. *Trends Biotechnol.* **2025**, *43*, 3071–3087. [CrossRef]
353. Lim, A.; Lim, S.; Kim, S. Enhancer Prediction with Histone Modification Marks Using a Hybrid Neural Network Model. *Methods* **2019**, *166*, 48–56. [CrossRef] [PubMed]
354. Bravo González-Blas, C.; Matetovici, I.; Hillen, H.; Taskiran, I.I.; Vandepoel, R.; Christiaens, V.; Sansores-García, L.; Verboven, E.; Hulselmans, G.; Poovathingal, S.; et al. Single-Cell Spatial Multi-Omics and Deep Learning Dissect Enhancer-Driven Gene Regulatory Networks in Liver Zonation. *Nat. Cell Biol.* **2024**, *26*, 153–167. [CrossRef]
355. Gosai, S.J.; Castro, R.I.; Fuentes, N.; Butts, J.C.; Mouri, K.; Alasoadura, M.; Kales, S.; Nguyen, T.T.L.; Noche, R.R.; Rao, A.S.; et al. Machine-Guided Design of Cell-Type-Targeting Cis-Regulatory Elements. *Nature* **2024**, *634*, 1211–1220. [CrossRef] [PubMed]
356. Smith, T.A.G.; Mehaffey, M.G.; Kayda, D.B.; Saunders, J.M.; Yei, S.; Trapnell, B.C.; McClelland, A.; Kaleko, M. Adenovirus Mediated Expression of Therapeutic Plasma Levels of Human Factor IX in Mice. *Nat. Genet.* **1993**, *5*, 397–402. [CrossRef]
357. Kung, S.-H.; Nathan Hagstrom, J.; Cass, D.; Jen Tai, S.; Lin, H.-F.; Stafford, D.W.; High, K.A. Human Factor IX Corrects the Bleeding Diathesis of Mice With Hemophilia B. *Blood* **1998**, *91*, 784–790. [CrossRef]
358. Geest, B.D.; Linthout, S.V.; Lox, M.; Collen, D.; Holvoet, P. Sustained Expression of Human Apolipoprotein A-I after Adenoviral Gene Transfer in C57BL/6 Mice: Role of Apolipoprotein A-I Promoter, Apolipoprotein A-I Introns, and Human Apolipoprotein E Enhancer. *Hum. Gene Ther.* **2000**, *11*, 101–112. [CrossRef] [PubMed]
359. Ehrhardt, A.; Kay, M.A. A New Adenoviral Helper–Dependent Vector Results in Long-Term Therapeutic Levels of Human Coagulation Factor IX at Low Doses in Vivo. *Blood* **2002**, *99*, 3923–3930. [CrossRef]
360. De Geest, B.; Van Linthout, S.; Collen, D. Sustained Expression of Human Apo A-I Following Adenoviral Gene Transfer in Mice. *Gene Ther.* **2001**, *8*, 121–127. [CrossRef]
361. Connelly, S.; Gardner, J.M.; McClelland, A.; Kaleko, M. High-Level Tissue-Specific Expression of Functional Human Factor VIII in Mice. *Hum. Gene Ther.* **1996**, *7*, 183–195. [CrossRef]
362. Pastore, L.; Morral, N.; Zhou, H.; Garcia, R.; Parks, R.J.; Kochanek, S.; Graham, F.L.; Lee, B.; Beaudet, A.L. Use of a Liver-Specific Promoter Reduces Immune Response to the Transgene in Adenoviral Vectors. *Hum. Gene Ther.* **1999**, *10*, 1773–1781. [CrossRef]
363. Snyder, R.O.; Miao, C.H.; Patijn, G.A.; Spratt, S.K.; Danos, O.; Nagy, D.; Gown, A.M.; Winther, B.; Meuse, L.; Cohen, L.K.; et al. Persistent and Therapeutic Concentrations of Human Factor IX in Mice after Hepatic Gene Transfer of Recombinant AAV Vectors. *Nat. Genet.* **1997**, *16*, 270–276. [CrossRef]
364. Snyder, R.O.; Miao, C.; Meuse, L.; Tubb, J.; Donahue, B.A.; Lin, H.-F.; Stafford, D.W.; Patel, S.; Thompson, A.R.; Nichols, T.; et al. Correction of Hemophilia B in Canine and Murine Models Using Recombinant Adeno-Associated Viral Vectors. *Nat. Med.* **1999**, *5*, 64–70. [CrossRef]
365. Chao, H.; Mao, L.; Bruce, A.T.; Walsh, C.E. Sustained Expression of Human Factor VIII in Mice Using a Parvovirus-Based Vector. *Blood* **2000**, *95*, 1594–1599. [CrossRef]
366. Arruda, V.R.; Samelson-Jones, B.J. Gene Therapy for Immune Tolerance Induction in Hemophilia with Inhibitors. *J. Thromb. Haemost.* **2016**, *14*, 1121–1134. [CrossRef]
367. Sun, B.; Zhang, H.; Franco, L.M.; Brown, T.; Bird, A.; Schneider, A.; Koeberl, D.D. Correction of Glycogen Storage Disease Type II by an Adeno-Associated Virus Vector Containing a Muscle-Specific Promoter. *Mol. Ther.* **2005**, *11*, 889–898. [CrossRef] [PubMed]
368. Eggers, M.; Vannoy, C.H.; Huang, J.; Purushothaman, P.; Brassard, J.; Fonck, C.; Meng, H.; Prom, M.J.; Lawlor, M.W.; Cunningham, J.; et al. Muscle-directed Gene Therapy Corrects Pompe Disease and Uncovers Species-specific GAA Immunogenicity. *EMBO Mol. Med.* **2022**, *14*, e13968. [CrossRef]
369. Corti, M.; Liberati, C.; Smith, B.K.; Lawson, L.A.; Tuna, I.S.; Conlon, T.J.; Coleman, K.E.; Islam, S.; Herzog, R.W.; Fuller, D.D.; et al. Safety of Intradiaphragmatic Delivery of Adeno-Associated Virus-Mediated Alpha-Glucosidase (rAAV1-CMV-hGAA) Gene Therapy in Children Affected by Pompe Disease. *Hum. Gene Ther. Clin. Dev.* **2017**, *28*, 208–218. [CrossRef] [PubMed]

https://doi.org/10.3390/cells15010014

370. Colella, P.; Sellier, P.; Costa Verdera, H.; Puzzo, F.; Van Wittenberghe, L.; Guerchet, N.; Daniele, N.; Gjata, B.; Marmier, S.; Charles, S.; et al. AAV Gene Transfer with Tandem Promoter Design Prevents Anti-Transgene Immunity and Provides Persistent Efficacy in Neonate Pompe Mice. *Mol. Ther.-Methods Clin. Dev.* **2019**, *12*, 85–101. [CrossRef]
371. Sellier, P.; Vidal, P.; Bertin, B.; Gicquel, E.; Bertil-Froidevaux, E.; Georger, C.; Van Wittenberghe, L.; Miranda, A.; Daniele, N.; Richard, I.; et al. Muscle-specific, Liver-detargeted Adeno-Associated Virus Gene Therapy Rescues Pompe Phenotype in Adult and Neonate *Gaa*$^{-/-}$ Mice. *J. Inher Metab. Disea* **2024**, *47*, 119–134. [CrossRef]
372. Khan, S.A.; Álvarez, J.V.; Nidhi, F.N.U.; Benincore-Florez, E.; Tomatsu, S. Evaluation of AAV Vectors with Tissue-Specific or Ubiquitous Promoters in a Mouse Model of Mucopolysaccharidosis Type IVA. *Mol. Ther. Methods Clin. Dev.* **2025**, *33*, 101447. [CrossRef] [PubMed]
373. Baruteau, J.; Brunetti-Pierri, N.; Gissen, P. Liver-directed Gene Therapy for Inherited Metabolic Diseases. *J. Inher Metab. Disea* **2024**, *47*, 9–21. [CrossRef]
374. Crudele, J.M.; Finn, J.D.; Siner, J.I.; Martin, N.B.; Niemeyer, G.P.; Zhou, S.; Mingozzi, F.; Lothrop, C.D.; Arruda, V.R. AAV Liver Expression of FIX-Padua Prevents and Eradicates FIX Inhibitor without Increasing Thrombogenicity in Hemophilia B Dogs and Mice. *Blood* **2015**, *125*, 1553–1561. [CrossRef] [PubMed]
375. Jin, L.; Nawab, S.; Xia, M.; Ma, X.; Huo, Y. Context-dependency of Synthetic Minimal Promoters in Driving Gene Expression: A Case Study. *Microb. Biotechnol.* **2019**, *12*, 1476–1486. [CrossRef] [PubMed]
376. Kaysen, G.A.; Dubin, J.A.; Müller, H.-G.; Mitch, W.E.; Rosales, L.M.; Levin, N.W. The Hemo Study Group Relationships among Inflammation Nutrition and Physiologic Mechanisms Establishing Albumin Levels in Hemodialysis Patients. *Kidney Int.* **2002**, *61*, 2240–2249. [CrossRef]
377. Levitt, D.; Levitt, M. Human Serum Albumin Homeostasis: A New Look at the Roles of Synthesis, Catabolism, Renal and Gastrointestinal Excretion, and the Clinical Value of Serum Albumin Measurements. *IJGM* **2016**, *9*, 229–255. [CrossRef]
378. Zahm, A.M.; Owens, W.S.; Himes, S.R.; Fallon, B.S.; Rondem, K.E.; Gormick, A.N.; Bloom, J.S.; Kosuri, S.; Chan, H.; English, J.G. A Massively Parallel Reporter Assay Library to Screen Short Synthetic Promoters in Mammalian Cells. *Nat. Commun.* **2024**, *15*, 10353. [CrossRef] [PubMed]
379. Degner, K.N.; Bell, J.L.; Jones, S.D.; Won, H. Just a SNP Away: The Future of in Vivo Massively Parallel Reporter Assay. *Cell Insight* **2025**, *4*, 100214. [CrossRef]
380. Inoue, F.; Ahituv, N. Decoding Enhancers Using Massively Parallel Reporter Assays. *Genomics* **2015**, *106*, 159–164. [CrossRef]
381. Brenner, M.; Messing, A. Regulation of GFAP Expression. *ASN Neuro* **2021**, *13*, 1759091420981206. [CrossRef]
382. Zhi, X.; Chan, E.M.; Edenberg, H.J. Tissue-Specific Regulatory Elements in the Human Alcohol Dehydrogenase 6 Gene. *DNA Cell Biol.* **2000**, *19*, 487–497. [CrossRef]
383. Vorgia, P.; Zannis, V.I.; Kardassis, D. A Short Proximal Promoter and the Distal Hepatic Control Region-1 (HCR-1) Contribute to the Liver Specificity of the Human Apolipoprotein C-II Gene. *J. Biol. Chem.* **1998**, *273*, 4188–4196. [CrossRef] [PubMed]
384. Falquerho, L.; Paquereau, L.; Vilarem, M.J.; Galas, S.; Patey, G.; Le Cam, A. Functional Characterization of the Promoter of Pp63, a Gene Encoding a Natural Inhibitor of the Insulin Receptor Tyrosine Kinase. *Nucleic Acids Res.* **1992**, *20*, 1983–1990. [CrossRef]
385. Vaulont, S.; Puzenat, N.; Kahn, A.; Raymondjean, M. Analysis by Cell-Free Transcription of the Liver-Specific Pyruvate Kinase Gene Promoter. *Mol. Cell. Biol.* **1989**, *9*, 4409–4415. [CrossRef]
386. Gregori, C.; Porteu, A.; Lopez, S.; Kahn, A.; Pichard, A.-L. Characterization of the Aldolase B Intronic Enhancer. *J. Biol. Chem.* **1998**, *273*, 25237–25243. [CrossRef]
387. Oliviero, S.; Cortese, R. The Human Haptoglobin Gene Promoter: Interleukin-6-Responsive Elements Interact with a DNA-Binding Protein Induced by Interleukin-6. *EMBO J.* **1989**, *8*, 1145–1151. [CrossRef]
388. Nagae, Y.; Muller-Eberhard, U. Identification of an Interleukin-6 Responsive Element and Characterization of the Proximal Promoter Region of the Rat Hemopexin Gene. *Biochem. Biophys. Res. Commun.* **1992**, *185*, 420–429. [CrossRef] [PubMed]
389. Paquereau, L.; Vilarem, M.J.; Rossi, V.; Rouayrenc, J.F.; Le Cam, A. Regulation of Two Rat Serine-protease Inhibitor Gene Promoters by Somatotropin and Glucocorticoids: Study with Intact Hepatocytes and Cell-free Systems. *Eur. J. Biochem.* **1992**, *209*, 1053–1061. [CrossRef]
390. Fish, R.J.; Neerman-Arbez, M. Fibrinogen Gene Regulation. *Thromb. Haemost.* **2012**, *108*, 419–426. [CrossRef]
391. Hu, C.-H.; Harris, J.E.; Davie, E.W.; Chung, D.W. Characterization of the 5′-Flanking Region of the Gene for the α Chain of Human Fibrinogen. *J. Biol. Chem.* **1995**, *270*, 28342–28349. [CrossRef]
392. Dalmon, J.; Laurent, M.; Courtois, G. The Human β Fibrinogen Promoter Contains a Hepatocyte Nuclear Factor 1-Dependent Interleukin-6-Responsive Element. *Mol. Cell. Biol.* **1993**, *13*, 1183–1193. [CrossRef] [PubMed]
393. Toniatti, C.; Demartis, A.; Monaci, P.; Nicosia, A.; Ciliberto, G. Synergistic Trans-Activation of the Human C-Reactive Protein Promoter by Transcription Factor HNF-1 Binding at Two Distinct Sites. *EMBO J.* **1990**, *9*, 4467–4475. [CrossRef] [PubMed]
394. Lundgren, T.S.; Denning, G.; Stowell, S.R.; Spencer, H.T.; Doering, C.B. Pharmacokinetic Analysis Identifies a Factor VIII Immunogenicity Threshold after AAV Gene Therapy in Hemophilia A Mice. *Blood Adv.* **2022**, *6*, 2628–2645. [CrossRef]

https://doi.org/10.3390/cells15010014

395. Butterfield, J.S.S.; Yamada, K.; Bertolini, T.B.; Syed, F.; Kumar, S.R.P.; Li, X.; Arisa, S.; Piñeros, A.R.; Tapia, A.; Rogers, C.A.; et al. IL-15 Blockade and Rapamycin Rescue Multifactorial Loss of Factor VIII from AAV-Transduced Hepatocytes in Hemophilia A Mice. *Mol. Ther.* **2022**, *30*, 3552–3569. [CrossRef]
396. Handyside, B.; Ismail, A.M.; Zhang, L.; Yates, B.; Xie, L.; Sihn, C.-R.; Murphy, R.; Bouwman, T.; Kim, C.K.; De Angelis, R.; et al. Vector Genome Loss and Epigenetic Modifications Mediate Decline in Transgene Expression of AAV5 Vectors Produced in Mammalian and Insect Cells. *Mol. Ther.* **2022**, *30*, 3570–3586. [CrossRef] [PubMed]
397. Mazloum, A.; Feoktistova, S.G.; Gubaeva, A.; Alsalloum, A.; Mityaeva, O.N.; Kim, A.; Bodunova, N.A.; Woroncow, M.V.; Volchkov, P.Y. Maturity-Onset Diabetes of the Young 10 (MODY10): A Comprehensive Review of Genetics, Clinical Features, and Therapeutic Advances. *IJMS* **2025**, *26*, 8110. [CrossRef] [PubMed]

https://doi.org/10.3390/cells15010014

cells

MDPI

Review

Synthetic Promoters in Gene Therapy: Design Approaches, Features and Applications

Valentin Artemyev [1,2,*], Anna Gubaeva [1], Anastasiia Iu. Paremskaia [1], Amina A. Dzhioeva [2], Andrei Deviatkin [1], Sofya G. Feoktistova [1,*], Olga Mityaeva [1,2,3] and Pavel Yu. Volchkov [1,3,4]

1 Federal Research Center for Innovator and Emerging Biomedical and Pharmaceutical Technologies, 125315 Moscow, Russia; as.gubaeva@gmail.com (A.G.); devyatkin_aa@academpharm.ru (A.D.); mityaeva_on@academpharm.ru (O.M.); vpwwww@gmail.com (P.Y.V.)
2 Moscow Center for Advanced Studies, Kulakova Str. 20, 123592 Moscow, Russia; dzhioeva05@gmail.com
3 Faculty of Fundamental Medicine, Moscow State University, Lomonosovsky Pr., 27, 119991 Moscow, Russia
4 Moscow Clinical Scientific Center N.A. A.S. Loginov, 111123 Moscow, Russia
* Correspondence: artemev.vv@genlab.llc (V.A.); feoktistova_sg@academpharm.ru (S.G.F.)

Abstract: Gene therapy is a promising approach to the treatment of various inherited diseases, but its development is complicated by a number of limitations of the natural promoters used. The currently used strong ubiquitous natural promoters do not allow for the specificity of expression, while natural tissue-specific promoters have lowactivity. These limitations of natural promoters can be addressed by creating new synthetic promoters that achieve high levels of tissue-specific target gene expression. This review discusses recent advances in the development of synthetic promoters that provide a more precise regulation of gene expression. Approaches to the design of synthetic promoters are reviewed, including manual design and bioinformatic methods using machine learning. Examples of successful applications of synthetic promoters in the therapy of hereditary diseases and cancer are presented, as well as prospects for their clinical use.

Keywords: synthetic promoters; gene therapy; promoter design; eukaryotic promoters; gene expression

1. Introduction

Monogenic diseases, although individually rare, collectively affect a significant number of people, creating a substantial global health burden. These disorders, caused by mutations in a single gene, are particularly suited for gene therapy, a promising strategy based on delivering a functional copy of the faulty gene. The estimated prevalence of monogenic diseases is approximately 1.7–5 per 1000 neonates [1]. The cumulative economic impact of these diseases is immense, with lifelong medical care and loss of productivity contributing to significant costs. Gene therapy has already demonstrated its efficacy in treating various rare genetic disorders, with notable examples including Leber's congenital amaurosis [2], spinal muscular atrophy [3], Duchenne muscular dystrophy [4], hemophilia A [5], and hemophilia B [6]. It is important to note that despite the diversity of these diseases, all of these therapies are based on the same underlying platform—adeno-associated viral (AAV) vectors. This is largely attributed to the AAV vector's broad tissue tropism and its favorable safety profile. An AAV vector offers several advantages, including being non-pathogenic, exhibiting a low propensity for genomic integration, and providing sustained transgene expression over extended periods, making it versatile and effective for treating a wide range of genetic disorders [7,8].

AAV vectors are also being explored for cancer treatment, with promising results in preclinical studies [9–11]. Unlike conventional therapies, AAV vector-based strategies can modulate or knock down specific gene expressions, offering targeted anti-tumor approaches. AAV vectors can transduce both dividing and non-dividing cells, making them effective for targeting the heterogeneous cell populations within tumors. For example, in a xenograft mouse model of colorectal cancer, AAV vector-mediated delivery of interferon β (IFNβ) successfully suppressed tumor growth. In glioblastoma, AAV vector-driven IFNβ expression significantly extended survival in an orthotopic mouse model [12]. Additionally, AAV capsids have been engineered to specifically target tumors, such as in the case of Her2-targeting AAV vectors that enhanced specificity and reduced tumor burden more effectively than conventional therapies like Herceptin [13].

Most existing gene therapies are delivered systemically, and high titers are used to achieve effective expression, e.g., for Duchenne myodystrophy therapy, the therapeutic dose is 1.33×10^{14} genome copies/kg, and for hemophilia A, the therapeutic dose is 6×10^{13} genome copies/kg. Potentially high titers can promote off-target transduction to various organs such as the liver, which can lead to hepatotoxicity, immunotoxicity, and, in extreme cases, death [14].

Reducing the viral load in AAV vector-mediated gene therapy is essential for minimizing immune responses, off-target effects, and toxicity while preserving therapeutic efficacy. One approach is engineering AAV capsids with enhanced tissue tropism [15,16] that allows for a more efficient targeting of specific tissues, such as the central nervous system [17] or liver [18,19], thereby reducing the required dose. Additionally, localized delivery methods, such as intravitreal [20,21] or intracerebral injections [22–24], can further reduce the overall viral load by concentrating the therapeutic vector in the affected area, ensuring efficacy with lower systemic exposure. However, local delivery of gene therapy is not always feasible, especially when the disease affects multiple tissues or organs throughout the body. While local delivery can be effective for disorders confined to a specific area, such as the eye or a localized region of the brain, many genetic conditions require a more widespread therapeutic effect. This is particularly true for metabolic diseases [25–29] where enzyme deficiencies or other genetic defects impact various tissues, including the liver, muscles, heart, and nervous system. Additionally, AAV vectors can be designed for systemic delivery but with transgene expression restricted to specific tissues, using tissue-specific or synthetic promoters.

Currently, most gene therapy approaches using adeno-associated viruses (AAVs) employ natural ubiquitous constitutive promoters such as CBA (chicken beta-actin), CMV (cytomegalovirus), or CAG (a synthetic promoter consisting of the CMV enhancer, CBA promoter, and rabbit beta-globin splice acceptor) [30]. While these promoters provide high levels of transgene expression, their use comes with several drawbacks. Unlike tissue-specific promoters, strong constitutive promoters are susceptible to extensive methylation, leading to their inactivation and subsequent suppression of transcription [31]. Additionally, the use of a strong promoter increases the risk of cytotoxicity due to overexpression and/or off-target expression of the transgene [32]. There is also evidence linking the presence of additional cis-regulatory sequences upstream of the transgene in the vector to toxicity in immune-privileged areas like the eyes [33]. Moreover, transgene overexpression may compete with the normal expression of other genes, disrupting cell metabolism and function [30].

Retroviral vectors (including lentiviral vectors) were the first FDA-approved cell-based gene therapy instrument (Kymriah, Novartis Pharmaceuticals Corporation, Philadelphia, PA, USA) [34]. These vectors find their greatest application in ex vivo therapies: stem cell gene therapy for the treatment of primary immunodeficiencies, as well as T-cell editing to create immunotherapies [35,36]. The majority of these ex vivo therapies require high expression efficiency of the transgene integrated into the genome and do not depend on the specificity of expression because the desired cells are isolated from the patient's body and then manipulated in vivo, mediating the re-administration of the edited cells back into the patient's body [35,37]. For this reason, ex vivo gene therapies mediated by lentivirus transduction are currently significantly prevalent using strong constitutive promoters (EF1α, CMV, CAG, PGK, SFFV), and new stronger synthetic promoters are being developed (MND, MCU3) [38,39]. Based on the mentioned general concept of ex vivo therapies, which differ significantly from in vivo therapies with direct administration of the transient vector, examples of synthetic promoters for lentiviral vectors were omitted in this work. However, it is important to note that synthetic promoters for lentiviral therapies are also beginning to gain popularity. The need to use tissue-specific promoters for the physiological regulation of transgene expression to avoid toxicity due to transgene overexpression has been demonstrated in a number of primary immunodeficiency disease models [36]. The creation of synthetic promoters is necessary in ex vivo therapies where expression only in specific cell types is required, such as in the case of chronic granulomatous disease with a number of synthetic promoters [36,40,41]. The use of tissue-specific promoters for lentiviral vectors allows us to restrict the expression in certain cell types after the differentiation from stem cells derived from a patient and transduced with vectors used for myeloid cells (CD11B), macrophages (CD68), megakaryocytes (PF4, hGP6, hGP1BA), B cells (EμB29) or erythrocytes (β-globin promoter) [37,40,42–47]. In vivo therapies using lentiviral vectors with the required specificity of expression and reduced immunogenicity are also under active development [37,38,48–50]. More information on lentiviral promoters, including synthetic promoters, can be found in the review articles by Claire Booth et al. [36], Estera Rintz et al. [37], Benjamin Houghton and Claire Booth [51].

Thus, although ubiquitous promoters are often used for their simplicity and efficiency, they come with disadvantages such as uncontrolled expression and potential toxicity. A more effective solution is to use tissue-specific natural promoters, which allow for more targeted and regulated expression and minimizing off-target effects. Alternatively, the development of new synthetic promoters designed for the precise control of transgene expression could further enhance the safety and efficacy of gene therapies.

All the above limitations necessitate research to design and explore new promoters to realize effective and safe gene therapies for diseases. Therefore, the aim of this review is to highlight the latest scientific advances in the field of synthetic promoter design and application as one of the directions in improving the effectiveness of gene therapies.

2. Natural Promoters

2.1. Structure of the Eukaryotic Natural Promoter

A promoter is a DNA sequence that is recognized by RNA polymerase and serves as a transcription start site. Promoter activity is traditionally understood as a quantification of gene expression in the form of mRNA and protein molecules. Higher levels of mRNA can be produced from a vector with higher promoter activity. In other words, the same quantity of therapeutic protein can be generated using a lower quantity of the vector in the case of higher promoter activity.

The eukaryotic natural promoter consists of a core promoter region, also called the minimal promoter, and proximal and distal promoter regions (Figure 1). The core promoter for RNA polymerase II is generally defined as the set of sequences that is sufficient for the assembly of a pre-initiation complex, and for exactly specifying the point of transcriptional initiation in vitro [52]. The study of core promoters has led to the discovery of the major constituent motifs such as the TATA-box (although the TATA-box is the best known core promoter motif, it is present in only 10–15% of mammalian core promoters [53,54], BRE (TFIIB recognition element) and Inr (initiator)) [55]. In addition, enhancers and silencers are DNA regions that can attract factors that interact with other elements involved in transcription.

Emami et al. [56] showed the existence of different classes of core promoters depending on the presence of certain regulatory elements in them. The type of promoter is defined by the presence of a TATA-box or other motifs (e.g., BRE) that regulates enhancer/promoter interactions. Some enhancers have been shown to mainly affect TATA-dependent promoters, while others prefer promoters with other motifs [57,58]. This specificity of enhancers is related to the selectivity of associated transcription factors. The most common core promoter motif is the initiator Inr, which includes the TSS (transcription start site) and has a consensus sequence [59].

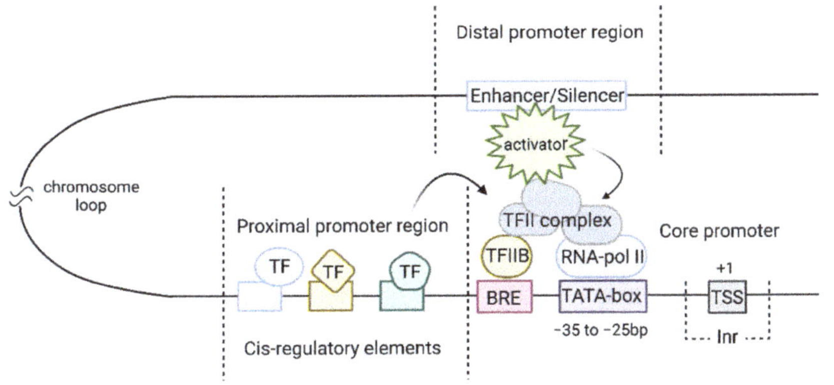

Figure 1. Functional model of transcription initiation at genomic promoters. The eukaryotic natural promoter consists of a core promoter, and proximal and distal promoter regions. The core promoter attracts a set of basic transcription factors (TFs), TFIIA, TFIIB, TFIID, TFIIE, TFIIF and TFIIH, and recruits RNA polymerase II. The proximal promoter region is formed by a set of cis-regulatory elements that bind to additional TFs, enhancing or weakening transcription. The distal region, represented by the enhancer or silencer, is located several kilobases away from the core promoter and mediates regulation of transcription levels by recruiting transcription factors (TFs) and cofactors (COFs). Vertical dashed lines indicate the boundaries between different promoter regions. Arrows show the influence of proximal and distal regions on the core promoter associated with basal TFs.

Transcription initiation of protein-coding genes occurs when RNA polymerase II binds to DNA with the participation of "basal" transcription factors: TFIIA, TFIIB, TFIID, TFIIE, TFIIF and TFIIH. The main basal transcription factor TFIID consists of TBP and TAF (TBP-associated factors recognizing Inr) and is involved in the recognition of core promoters [60]. The basal transcription factor TFIIB binds to the sequences of the BRE motif of the core promoter. The other basal factors bind to the core promoter after TFIIB and TFIID and participate in the initial steps of transcription. The required combination of transcription factors is different for promoters with different regulations. Thus, the set of basal TFs required for the transcription of one promoter cannot ensure transcription from another promoter.

The presence of the core promoter is a sufficient condition for the initiation of transcription [61], but ensures its low basal level [52]. The level of transcription depends on many factors, including chromatin packing density and the location of additional regulatory elements [62]. Some of these elements are close to the core promoter and form the proximal promoter region, while others are located a few kb away and form the distal promoter region.

The proximal promoter region is a several-hundred-nucleotide long sequence formed by cis-regulatory elements. Cis-regulatory elements (CREs) are sequences that bind various transcription factors, which can repress or enhance transcription by binding to DNA. Types, number, and location of cis-regulatory elements in the proximal and distal regions of the promoter affects its activity [63]. The set of CREs associated with a gene, as well as the specificity and availability of TFs that bind to them, determine the promoter as tissue-specific or ubiquitous. Tissue-specific promoters provide targeted expression in specific tissues or cells and are often associated with low levels of activity, whereas ubiquitous promoters provide widespread expression and are associated with higher activity level [64].

Distal promoter elements include enhancer and silencer sequences, which can be identified by chromatin modifications, including histones [65], and by the bidirectional transcription of enhancer RNA (eRNA) [66,67]. Enhancers affect the level of transcription from the core promoter regardless of their location and orientation [68]. Contact between the core promoter and distal enhancer is ensured by the three-dimensional structure of chromatin and determines the promoter activity [69]. Enhancers bind to TFs, which interact directly with DNA, as well as to transcriptional cofactors that affect transcription levels indirectly through TFs [67]. An increase in the level of transcription can be achieved by accelerating initiation by stabilizing the pre-initiator complex through the action of TFs or cofactors attracted by enhancers [70].

Thus, the location of CREs and their availability for binding by TFs in proximal and distal promoter regions are key factors for the regulation of gene expression levels. The identification of natural CREs and the creation of synthetic sequences are urgent tasks of molecular and synthetic biology.

2.2. Modern Approaches for Recognizing CREs

Modern approaches to the creation of synthetic tissue- and cell-specific promoters are based on the identification of naturally occurring unique CREs, which are active in specific cell types. However, the precise boundaries of these promoters are often not well defined, presenting a significant challenge in the fields of synthetic biology and bioinformatics. The identification of promoters can be achieved through both laboratory methods and computational approaches based on genomic data.

Promoters may be identified using the "trapping" technique [71]. The method is based on the random integration of a reporter gene into the genome. Since the reporter gene lacks its own promoter, it will only be expressed if it integrates near an active endogenous promoter. Measuring the expression level of the reporter gene provides insights into the activity of the promoter. A limitation of the classical approach is the frequent integration of the vector outside the promoter, making the process time-consuming. Nevertheless, the approach continues to improve: a recently developed vector with a bicistronic system allows for increased efficiency in promoter detection [72].

Other techniques such as chromatin immunoprecipitation followed by sequencing (ChIP-seq), formaldehyde-assisted isolation of regulatory elements (FAIRE-seq), DNase I hypersensitive site sequencing (DNase-seq), and assay for transposase-accessible chromatin with high-throughput sequencing (ATAC-seq) allow for the detection of open chromatin regions characterized by high transcriptional activity [71]. These methods are based on the ability of active promoters to be less tightly packaged within nucleosomes, making them accessible to specific enzymes and proteins. ChIP-seq helps to identify DNA regions bound by specific transcription factors, such as p300, indicating the presence of active promoter elements [73,74]. FAIRE-seq and DNase-seq isolate DNA regions in an open configuration, which is a hallmark of active promoters [75,76]. ATAC-seq data, when combined with RNA-seq, can provide insights into promoter accessibility and tissue specificity [77]. Moreover, studying open chromatin regions can reveal new enhancers that drive expression in specific cell subtypes [78–80]. For example, chromatin accessibility data have been compiled for different cell types, allowing for the identification of enhancers specific to various cell types [79].

Machine learning (ML) and deep learning (DL) approaches enable the identification of not only elements with known motifs, but also hidden patterns within genomic sequences. Tools developed before 2021 for promoter prediction in prokaryotic and eukaryotic genomes are described by M. Zhang et al. [81]. Often, such models function as classifiers trained to distinguish promoter sequences from non-promoter sequences. For *E. coli*, two tools, one described by Le et al. [82] and the other being iPSW(2L)-PseKNC [83], not only recognize promoters but also predict their strength by classifying them as weak or strong. Among the tools with the highest efficiency and accuracy for predicting promoters of *Homo sapiens* are Depicter and iProEP. Depicter utilizes convolutional neural network (CNN) architecture comprising two one-dimensional convolutional layers, a one-dimensional capsule layer, and a fully connected layer. This network was trained on sequences encoded via one-hot encoding. iProEP is based on the classic supervised ML algorithm, the support vector machine (SVM). Notably, the authors combined structural information with physicochemical properties using the positional correlation scoring function and pseudo-k-tuple nucleotide composition (PseKNC). PseKNC expands the k-mer nucleotide composition concept by incorporating physicochemical properties of nucleotides.

A recent study implemented a hybrid approach in the PromGER tool [84], where the authors utilized various nucleotide characteristics, such as chemical properties, nucleotide density, pseudopotential, and probabilistic sequence distribution. Promoters were modeled as graphs, where nodes represent individual promoter sequences, and edges indicate potential interactions between them. By leveraging graph-embedding methods, the tool extracts both local and global graph features, integrating them into a classification model based on the CatBoost algorithm [85]. The hybrid model DeePromClass [86], which combines layers of convolutional neural networks with long short-term memory (LSTM) layers and regular expression search, demonstrated improved efficiency for identifying human TATA promoters.

The identification of promoters across various genomes and their validation using experimental methods has led to the development of databases. The EPD (The Eukaryotic Promoter Database) and EPDnew [87] are the largest sources of experimentally validated promoter sequences, including 29,598 Homo sapiens promoters. EPDnew also incorporates expanded data obtained through high-throughput technologies such as RNA-seq and ChIP-seq. The TRRD contains information on transcriptional regulatory regions of eukaryotic genes [88], while DBTSS provides transcription start site (TSS) positions in human embryonic and adult tissues as well as in the mouse genome [89]. The FANTOM consortium made a significant contribution by developing The FANTOM5 promoter atlas [90], which includes data from over 1000 human and mouse samples, covering a wide range of primary cells, tissues, and cancer cell lines. In the context of gene therapy, the results on tissue-specific promoters, obtained using the CAGE technology, are particularly useful as they allow for the precise mapping of TSSs by capturing the 5′ ends of mRNA.

Additionally, there are open-access databases dedicated to specific model organisms: CEPDB for *C. elegans*; RegulonDB contains information on 4050 promoter sequences of *E. coli* and other transcriptional regulators [91]; and DBTBS focuses on regulatory elements of *B. subtilis* [92].

Accumulated data on promoter sequences with established effects on gene expression now enable the application of computational approaches. These data can be utilized to predict the functionality of synthetic constructs and to model the interactions between regulatory elements.

3. Synthetic Promoters

Since tissue-specific natural promoters have limitations due to their large size and low activity, and strong ubiquitous promoters lead to off-target expression, new synthetic promoters are being actively developed to achieve high levels of transgene expression in the target tissue in the absence of expression in other tissues. There are different approaches to create synthetic promoters, which can be categorized into two main strategies: (1) the creation of synthetic promoters based on already available sequences (shuffling/substitution/truncation of natural promoter sequences, addition of cis-elements from other promoters, combination of regions of several promoters [93,94]), and (2) the prediction of promoters using computational technologies [95]. CREs of natural promoters are used as building blocks to create synthetic promoters by random ligation or rational design [96]. Thus, the term synthetic promoter refers to a DNA sequence that does not exist in nature, but which allows for the activation of controlled gene expression [94].

3.1. Synthetic Promoters' Manual Design

The main challenge in the creation of synthetic promoters is to find out which elements of the distal/proximal promoter and other UTRs (untranslated regions) determine its activity [97]. The level of enhancer and promoter activity has been shown to be driven by certain typical motifs, with increased promoter activity associated with transcription factor binding, demonstrating a modular system of promoters [98]. Due to the modular promoter system, which allows it to be divided into separate functional elements, it is possible to use the CREs of natural promoters to create new synthetic promoters, while preserving their activity. Thus, a library of various cis-acting elements and core promoters can be created [99].

In its simplest form, a synthetic promoter is a variant of a defined core promoter for the binding of RNA polymerase II and major TFs and a set of upstream CREs specifically conjugated to it. The main advantage of synthetic promoters is their tunability, i.e., the ability to adjust the desired level of expression and specificity to different tissues by selecting CREs [100,101].

One of the main limitations of natural promoters in their great length is solved by removing the space both between different promoter regions and between different CREs (Figure 2A). Often, low-conserved sequences are removed, and the significance of the remaining fragments is confirmed by mutational analysis [102]. For example, the 2.2 kb long human glial fibrillary acidic protein (GFAP) promoter was shortened in several steps to a 681 bp long variant gfaABC1D, which has a 2-fold higher activity compared to the original promoter [103]. However, if the CREs are too close, transcription factors may compete for the binding site, resulting in a decrease in promoter activity and a drop in expression level. A similar problem can occur if there is insufficient distance between the core promoter and the CRE in the proximal region, as TF IIB may compete with other transcription factors for binding to DNA [104]. Therefore, the required number of CREs and the optimal distance between them should be determined experimentally.

Since it is possible to use the core promoters of various genes to create synthetic promoters, it is necessary to characterize the core promoter, i.e., to define its region accurately. The choice of a core promoter for the creation of a synthetic promoter is based on the properties of the original natural promoter, including its activity. Thus, most of the previously reported synthetic mammalian promoters are constructed from a small set of core promoters, the most frequent of which is the minimal cytomegalovirus (minCMV) promoter, and the main disadvantage of which is its high basal expression level. In order to create highly specific promoters, core promoters of tissue-specific genes are increasingly being investigated [109]. In a study to create a liver-specific synthetic promoter, the core promoters SynO (a 41 bp fragment from the *Xenopus laevis* albumin 5′ UTR), which contains an HP1 transcription factor binding site and TATA-box, and a 61 bp fragment of the 5′ UTR of *mouse α-fetoprotein* (AFP) gene were used, to which various binding sites for liver-specific TFs were additionally added. The authors found that with its small size (146 n), the HNF1-AbpShort-SynO-TSS promoter (named HCB) had greater activity compared to the HLP (hybrid liver promoter) promoter (Figure 2B) [110].

The addition of specific CREs to already characterized promoters, including core promoters, makes it possible to increase their activity and tissue specificity of regulated transgene expression. Promoters from different genes obtained by this approach are called chimeric promoters. For example, the chimeric CMVenh/MLC1.5 promoter, consisting of the enhancer region of CMV and part of the MLC1.5 promoter, increases luciferase expression in the myocardium of the mouse's left ventricle 50-fold compared to the CMV promoter, while also changing the expression profile in other tissues [111]. Other authors also found that adding the enhancer region of CMV to gene promoters resulted in greater expression of the target protein [112,113]. Similarly, a highly active constitutive synthetic CAG promoter consisting of the CMV enhancer region, a portion of the CBA promoter, and a rabbit β-globin splicing acceptor was created [114]. The addition of randomly ligated CRE regions of various myogenic TFs to the minimal skeletal muscle α-actin promoter resulted in an SPc5-12 promoter that had up to a 10-fold higher activity than the original α-actin promoter, while maintaining specificity to muscle cells [115].

An alternative way to increase promoter activity is to add CRE repeats in varying amounts upstream of the core promoter, since increasing the number of bound transcription factors increases transcription activity from a promoter obtained using this approach [99–101]. Typically, such promoters consist of tandem repeats of cis-acting elements that bind one or more types of TFs active only in the cell type of interest that are encoded upstream of the core promoter [116]. An example is the anti-cancer 5HRE/hCMVmp promoter, consisting of a minCMV promoter to which five repetitive HRE (hypoxia-responsive element) sequences have been added upstream. The obtained promoter allows inducing gene expression under conditions of low oxygen availability in cells, such as in solid tumors [117]. Later, other researchers used this promoter to express VEGF (vascular endothelial growth factor) and CBD (collagen-binding domain) in heart tissues to improve cardiac function after myocardial infarction [118]. However, at a certain stage, it becomes impossible to increase the copy number of a single CRE because after a certain number of copies, further addition of copies leads to a decrease in transgene expression [119]. Another problem of this approach is the long length of the synthetic promoter obtained in this way, which complicates their application in the development of gene therapies.

Promoters containing combinations of different CREs, compared to promoters containing copies of a single element, tend to show better expression and specificity. This can be explained by two reasons: (1) the presence of several different CREs allows for the use of different signaling and promoter activation pathways; and (2) some CREs are not active by themselves but can form functional complexes with other regulating elements [99].

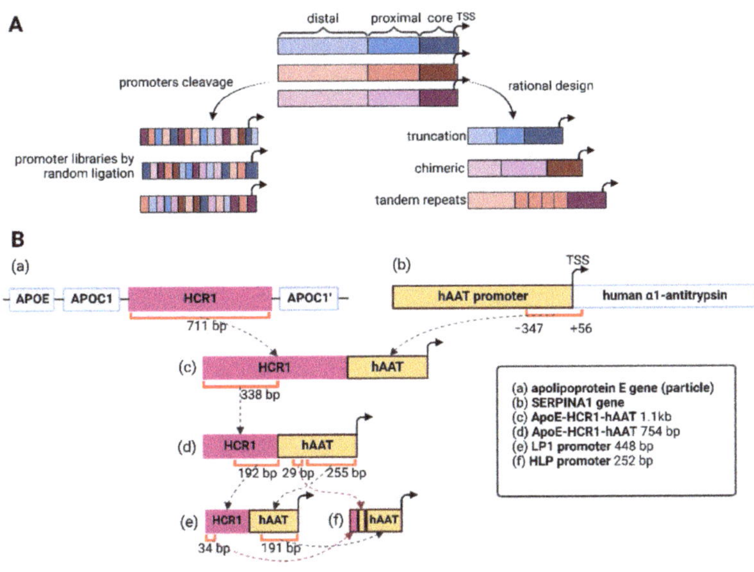

Figure 2. Synthetic promoter design. (**A**) Synthetic promoters' manual design approaches. Design approaches can be divided into rational design methods (truncation, chimeric/hybrid promoters, tandem repeats of cis-regulating elements) and creation of promoter libraries by cleavage and random ligation. (**B**) Creation of a synthetic promoter on the example of HLP. A 711 bp sequence was taken from the HCR1 enhancer region of the apolipoprotein E gene (**a**) and fused to the hAAT promoter region (−347 to +56) of the *SERPINA1* gene (**b**), which resulted in the first ApoE-HCR1-hAAT 1.1 kb promoter (**c**) [105]. The ApoE-HCR1-hAAT promoter was further shortened to 754 bp (**d**) by deleting the last 373 nucleotides from the HCR1 enhancer (338 bp of HCR1 remaining) [106]. The 448 bp LP1 (**e**) promoter was derived from the 192 bp region of the HCR1 enhancer and the 255 bp region of the hAAT promoter [107]. The final HLP (**f**) promoter in these series was obtained by adding a 29 bp region from the ApoE-HCR1-hAAT 754 bp promoter to the 191 bp region of the hAAT and 34 bp region of the HCR1 enhancer of the LP1 promoter [108]. The shaded arrows represent the strategies division for creating promoters while the dashed lines indicate the succession of promoter regions.

Manual design methods have become widespread, and, to date,, all synthetic promoters used in gene therapies have been created using this approach. However, this approach to create synthetic promoters implies a clear understanding of the functioning of individual CREs and is labor-intensive. In recent decades, due to the development of computational technologies, it has become possible both to define functional units of promoters and to create new synthetic promoters based on generative models, which can significantly simplify the design of synthetic promoters and ensure their greater use in the development of new gene therapies.

3.2. Computational Design of Synthetic Promoters

There are two primary approaches to developing CREs using bioinformatics tools. The first approach involves annotating natural non-coding regions of highly expressed genes. Using available information about regulatory motifs, researchers either identify their optimal combinations or enhance the sequence by enriching it with significant motifs while removing potentially insignificant regions. This strategy not only preserves but, in some cases, enhances CRE activity while reducing sequence length, thereby improving its suitability for use in delivery systems This approach has been applied to various cell types, such as pan-neuronal cells [120], as well as different eye tissues, including retinal ganglion cells, bipolar cells, cones, and corneal cells [121,122]. The second approach assumes that naturally occurring CREs may not be optimal for efficient regulation. An innovative approach in the creation of synthetic promoters involves the use of generative algorithms, enabling the design of novel promoters based on patterns learned from large annotated datasets. These models include generative adversarial networks (GANs), variational autoencoders (VAEs), and diffusion models.

The architecture of a generative adversarial network consists of two neural networks, a generator and a discriminator, trained simultaneously in opposition to each other. The discriminator learns to recognize natural promoters by identifying their attributes, while the generator, starting from random or noisy sequences, aims to create new promoters. The discriminator computes the probability that the generator's output resembles the original promoter dataset and sends feedback for noise adjustment. Through this process, the generator learns to produce sequences classified by the discriminator as real promoter sequences. This architecture allows for the creation of unique promoter sequences that differ in nucleotide composition from natural ones, while retaining their functional characteristics.

The first application of GANs in promoter design was for *E. coli* [123], with experimental validation via a reporter system demonstrating that over 70% of the generated promoters were functionally active. Another example is the iPro-GAN tool [124], which can not only generate new sequences but also classify promoters by strength, achieving up to 92% accuracy on independent datasets. However, GAN models face several challenges. First, there can be an imbalance between the generator and discriminator, leading to either overfitting or underfitting of the generator. Second, a critical issue with GANs is mode collapse, where the generator, instead of producing a diverse range of sequences, repeatedly generates similar ones due to a limited sample space successfully selected by the discriminator.

To some extent, the diversity issue is addressed by the DeepSEED framework [125], which allows for the generation of promoter flanking regions based on a seed sequence chosen by the user. DeepSEED was trained on the HACER dataset, which includes enhancers specific to various cell types. During training, HEK293 cell lines were used. Although this approach does not yield more effective TF binding motifs (seed), optimizing the DNA shape (minor groove width, roll, propeller twist, and helix twist) of the flanking regions positively impacts promoter efficiency.

VAEs also consist of two components. The first, the encoder, transforms the input sequence into a more compact representation, while the decoder reconstructs the sequence with similar properties from this compact representation. This approach has been used to design synthetic promoters in cyanobacteria [126]. The encoder can generalize features and detect functionally significant promoter patterns, although excessive generalization may lead to the loss of unique, essential regions. In a recent study, VAEs were used in conjunction with the gradient-based biological sequence optimization method Fast SeqProp [127,128]. Since nucleotide selection is a discrete process, direct gradient computation is not feasible. To address this issue, Fast SeqProp employs the softmax straight-through estimator, which allows for an approximation of the gradients. Additionally, distribution normalization is applied to ensure stability during the training process. In vivo testing on mice and zebrafish demonstrated that the resulting synthetic sequences exhibited greater efficiency and tissue specificity compared to natural variants. To the best of our knowledge, this study represents the first instance where synthetic promoters obtained using generative models have been experimentally tested in mammals.

Promoter generation and strength prediction have also been implemented using transformer architectures in tools like BERT-Promoter and msBERT-Promoter. In BERT-Promoter, Shapley Additive Explanations (SHAP) analysis and a random forest classifier were used for feature selection, while msBERT-Promoter utilized tokenization for feature extraction and soft voting for classification. The training of the msBERT-Promoter model was carried out in multiple stages. In the initial stage, bacterial sequences from RegulonDB were employed, followed by fine-tuning using human sequences.

The diffusion probabilistic model, incorporating Markov chains, has shown the ability to generate a wide variety of promoters while maintaining properties such as GC content, k-mer frequency, and motif preservation. However, no experimental validation has yet confirmed the functional efficacy of these promoters.

To facilitate the use of generative algorithms, a pipeline called GPro has been developed [129], which allows for the integration of various architectures and the use of custom or pre-prepared datasets. The pipeline aims to simplify the integration of advanced architecture into new studies.

Thus, generative models trained on large-scale genomic datasets represent an effort to create novel promoter constructs by learning patterns from existing data. However, these models often assume that the learned patterns will be universally applicable across different tissue types and conditions. This assumption can lead to unpredictable results, as promoter activity in one tissue or condition may significantly differ from its function in another context due to factors such as chromatin accessibility and nucleosome positioning. Moreover, promoters are integral parts of regulatory networks, interacting with transcription factors, enhancers, and other regulatory elements. For example, recent studies by Belokopytova et al. [130] and Zheng et al. [131] proposed models for predicting promoter/enhancer interactions based on epigenetic data.

In conclusion, while generative models offer powerful tools for designing new promoters, the careful consideration of the biological context and validation in appropriate systems is essential to ensure that the promoters function as intended without causing undesirable side effects.

3.3. *Synthetic Promoter Evaluation Methods*

The use of new synthetic promoters obtained by both manual design methods and computational approaches is limited by the need to evaluate their activity in live models and compare them with available natural variants. There are many different experimental methods for promoter characterization, the applicability of which usually depends on the number of constructs under study. The methods also differ depending on the characterization being investigated, e.g., methods for investigating promoter tissue specificity may differ from methods for investigating promoter activity.

When studying the characteristics of a small number of promoters, usually obtained by manual design, it can be used to assess the expression level of a controlled transgene, such as a reporter protein, in different cell cultures and tissues of the body. A classic method for assessing promoter specificity and strength in vivo is the determination of luciferase activity [115]. This method of promoter analysis can be performed in vitro using cell cultures instead of animals [132], as this technique was originally developed [133]. Additionally, RNA can be isolated from organs and the reporter gene expression levels can be assessed by RT-qPCR with normalization to housekeeping gene expression, which allows for the estimation of promoter activity at the level of transgene mRNA synthesis [95,134]. In this way, it becomes possible to assess the activity of the investigated promoter in different tissues, i.e., its specificity. Other reporter systems can be used instead of luciferase [103, 110,135]. The advantage of methods based on the enzymatic activity of expressed reporter genes is that these methods allow for the study of promoter characteristics quantitatively and qualitatively directly at the level of protein expression rather than mRNA synthesis. Despite their simplicity and efficiency, these methods are time-consuming and labor-intensive, which severely limits their applicability to the simultaneous study of multiple promoters. In addition to enzymatic activity, the expression of reporter genes in various tissues can be assessed using antibodies in Western blotting and immunofluorescence [136].

To improve performance, methods for assessing promoter activity based on reporter genes have evolved to utilize promoter libraries. A reporter gene as part of a library with many different promoters is delivered to cells, and the resulting reporter protein signal is detected using flow cytometry [137]. Cells can possibly be sorted by signal level (FACS) and sequenced to identify more active promoters within libraries [116]. A limitation of FACS is the inability to assess the tissue specificity of the promoters under study.

To study the characteristics of multiple promoters, high-throughput screening methods are used, such as the use of vector libraries with unique barcodes after the stop codon or before the start codon of the reporter protein used, which is under the control of the promoter corresponding to the barcode. After sequencing, it becomes possible to estimate the frequency of the occurrence of certain barcodes and, based on this fact, make conclusions about the level of expression of the reporter protein under the corresponding promoter and about the activity of the promoter itself. The described method (also called massively parallel reporter assays) makes it possible to evaluate not only many individual promoters, but also individual short regions of various promoters both in vitro and in vivo [98,137,138].

4. Synthetic Promoter Applications

Despite the widespread use of natural promoters in the production of vectors for gene therapies [30,139], their use is complicated by the disadvantages discussed above. The development of synthetic promoters remains relevant, as evidenced by their growing presence in gene therapy clinical trials in recent years [139].

As previously mentioned, tissue-specific promoters have several advantages over ubiquitous ones, primarily the ability to deliver vectors in lower and safer doses. Most tissue- and cell-specific promoters currently employed in clinical trials are designed to target the CNS, liver, eye, and muscle (Table 1) [30,140]. Currently, optimized versions of natural promoters are predominantly used. They are mainly achieved using the following strategies: (1) shortening of the original promoter by removing insignificant parts of sequence, (2) adding one or several copies of enhancers, and (3) introducing modifications (deletions, substitutions) aimed at increasing promoter activity (Table 1).

Table 1. Tissue- and cell-specific promoters currently used in clinical trials.

Gene of Origin	Promoter	Disease	Specificity	Design	Therapy	Trial	Status
Liver-specific							
TTR	TTR	Hemophilia B	hepatocytes	murine TTR promoter/enhancer	SHP648	NCT04394286, phase 1/2	terminated
		Hemophilia B		TTR promoter/enhancer	AskBio009, Bax335	NCT01687608, phase 1/2	active, not recruiting
		Wilson's Disease		murine TTR enhancer/human TTR promoter + SV40 intron	UX701	NCT04884815, phase 1/2	active, not recruiting
		Hemophilia A		truncated TTR promoter/enhancer	SPK-8011	NCT06297486, phase 3	recruiting
		Hemophilia A		TTR promoter/enhancer	Bax 888, AAV8-BDD-FVIIIopt	NCT03370172, phase 1/2	active, not recruiting

Table 1. *Cont.*

Gene of Origin	Promoter	Disease	Specificity	Design	Therapy	Trial	Status
ApoE, hAAT	LP1	Hemophilia B	hepatocytes	truncated HCR/hAAT hybrid promoter (448 bp)	AMT-061	NCT03569891, phase 3	active, not recruiting
	LP1	Hemophilia B		truncated HCR/hAAT hybrid promoter (448 bp)	scAAV 2/8-LP1-hFIXco	NCT00979238, phase 1	active, not recruiting
	HLP	Hemophilia A		truncated HCR/hAAT hybrid promoter (252 bp)	BMN 270-301	NCT03370913, phase 3	active, not recruiting
	HLP	Severe Crigler Najjar Syndrome		truncated HCR/hAAT hybrid promoter (252 bp)	GNT0003	NCT03466463, not applicable	recruiting
	ApoE/hAAT	Fabry disease		enhancer and HCR from human ApoE gene + hAAT promoter + modified chimeric intron (HBB-IgG)	ST-920	NCT05039866, phase 1/2 follow-up	enrolling by invitation
hAAT	hAAT	Wilson's Disease	hepatocytes	hAAT promoter	VTX-801	NCT04537377, phase 1/2	recruiting
AAT, albumin	EalbAAT	Acute Intermittent Porphyria	hepatocytes	albumin enhancer + AAT promoter	rAAV2/5-PBGD	NCT02082860, phase 1	completed
Albumin	Albumin	MPS I	hepatocytes	albumin promoter	SB-318	NCT02702115, phase 1/2	terminated
		MPS II			SB-913	NCT03041324, phase 1/2	terminated
		Hemophilia B			SB-FIX	NCT02695160, phase 1	terminated
TBG	TBG	MPS VI	hepatocytes	TBG promoter	AAV2/8.TBG.hARSB	NCT03173521, phase 1/2	not recruiting
	TBG	Late-onset OTC Deficiency		TBG promoter	DTX301, scAAV8OTC	NCT02991144, phase 1/2	completed
	LSP	Pompe Disease		α1-microglobulin/bikunin enhancer x2 + TBG promoter	ACTUS-101, AAV2/8LSPhGAA	NCT03533673, phase 1/2	not recruiting
Combination	HCB	Hemophilia A	hepatocytes	combination of TFBS + minimal promoter SynO	ASC618	NCT04676048, phase 1/2	recruiting
Muscle-specific							
MCK	CK8	DMD	skeletal muscles	optimized MCK-enhancer + optimized CK6 (enhancer 2RS5 + proximal MCK promoter)	SGT-001	NCT03368742, phase 1/2	active, not recruiting
					SGT-003	NCT06138639, phase 1/2	recruiting
	MHCK7	Dysferlinopathy		α-MHC-enhancer + optimized CK cassette (enhancer 2RS5 + proximal MCK promoter)	rAAVrh.74.MHCK7.DYSF.DV	NCT02710500, phase 1	completed
		LGMD2E			SRP-9003	NCT06246513, phase 3	recruiting
		DMD			SRP-9001	NCT05881408, phase 3	recruiting
		LGMD2B/R2			SRP-6004	NCT05906251, phase 1	active, not recruiting
	tMCK	DMD		enhancer 2RS5 x3 + proximal MCK promoter	PF-06939926	NCT04281485, phase 3	active, not recruiting
		LGMD2D			SRP-9004	NCT01976091, phase 1/2	completed
						NCT00494195, phase 1	completed
		CMT			scAAV1.tMCK.NTF3	NCT03520751, phase 1/2	suspended (vector has not been produced)
	eMCK	Pompe Disease		MCK promoter/enhancer combination	AT845	NCT04174105, phase 1/2	completed
Des	Des	LGMD2C	skeletal muscles	desmin promoter	AAV1-gamma-sarcoglycan vector injection	NCT01344798, phase 1	completed
		X-Linked Myotubular Myopathy		truncated desmin enhancer/promoter (1.05 kb)	Resamirigene bilparvovec, AT132	NCT03199469, phase 2/3	not recruiting

Table 1. *Cont.*

Gene of Origin	Promoter	Disease	Specificity	Design	Therapy	Trial	Status
		Pompe Disease		desmin promoter	rAAV9-DES-hGAA	NCT02240407, phase 1	completed
hCK	hCK	DMD	skeletal and cardiac muscle	hCK promoter	fordadistrogene movaparvovec	NCT05429372, phase 2	active, not recruiting
Combination	Spc5-12	DMD	skeletal and cardiac muscle	combination of TFBS (SRE, MEF-1, MEF-2, TEF-1) + chicken skeletal a-actin promoter	RGX-202	NCT05693142, phase 1/2	recruiting
		OPMD			BB-301	NCT06185673, phase 1/2	recruiting
Eye-specific							
hRS1	hRS1	XLRS	PRs	native human RS1 promoter	AAV8-scRS/IRBPhRS, RS1 AAV Vector	NCT02317887, phase 1/2	active, not recruiting
hRPE65	hRPE65	LCA	RPE	truncated hRPE65 (1.4 kb)	tgAAG76, rAAV 2/2.hRPE65p.hRPE65	NCT00643747, phase 1/2	completed
	NA65p	LCA10		optimized hRPE65 promoter (757 bp)	AAV2/5-OPTIRPE65	NCT02946879, phase 1/2	completed
hGRK1/GRK1	hGRK1	LCA10	PRs(rods and cones)	hGRK1 promoter	EDIT-101, AAV5.SaCas9 AGN-151587	NCT03872479, phase 1/2	active, not recruiting
	GRK1	X-linked RP		GRK1 promoter	AGTC-501, rAAV2tYF-GRK1-RPGR	NCT06275620, phase 2	enrolling by invitation
hRK	hRK	X-linked RP	PRs (rods and cones)	hPK promoter	Cotoretigene toliparvovec, BIIB112, AAV8-RPGR	NCT03116113, phase 2/3	completed
		X-linked RP		truncated hPK promoter	Botaretigene Sparoparvovec, AAV5-hRKp.RPGR	NCT03252847, phase 1/2	completed
		Autosomal Recessive RP		truncated hPK promoter	AAV2/5-hPDE6B	NCT03252847, phase 1/2	recruiting
		XLRS		hPK promoter	ATSN-201, rAAV.SPR-hGRK1-hRS1syn	NCT05878860, phase 1/2	recruiting
human red opsin	PR1.7	Achromatopsia	PRs (cones)	LCR enhancer fragment + red opsin promoter	(rAAV2tYF-PR1.7-hCNGB3), AGTC-401	NCT02599922, phase 1/2	active, not recruiting
					(rAAV2tYF-PR1.7-hCNGA3), AGTC-402	NCT02935517, phase 1/2	active, not recruiting
Neuron-specific							
hSyn1	hSyn1	MTLE	neurons	hSyn1 promoter	AMT-260, AAV9-hSyn1-miGRIK	NCT06063850, phase 1/2	recruiting
		FTD			AVB-101	NCT06064890, phase 1/2	recruiting
mPGK	mPGK	Sanfilippo Syndrome B	cortical neurons and oligodendrocytes	mPGK promoter	rAAV2/5-hNAGLU	NCT03300453, phase 1/2	completed
MeCP	P546	Batten Disease	neurons	truncated MeCP2 promoter (546 bp)	AT-GTX-502, scAAV9.P546.CLN3	NCT03770572, phase 1/2	active, not recruiting
	MeP426	Rett Syndrome		MeCP2 core promoter + regulatory elements (RE)	TSHA-102	NCT05606614, phase 1/2	phase 1/2 recruiting

TTR—transthyretin; SV40—simian virus 40; ApoE—apolipoprotein E; AAT—human alpha-1-antitrypsin (hAAT—human AAT); LP1—liver promoter 1; HCR—hepatic control region; HLP—hybrid liver promoter; MPS I—mucopolysaccharidosis I; MPS II—mucopolysaccharidosis II; TBG—human thyroxine binding globulin; MPS VI—mucopolysaccharidosis type VI; OTC—ornithine transcarbamylase; LSP—liver-specific promoter; HCB—hepatic combinatorial bundle; MCK—muscle creatine kinase; DMD—Duchenne muscular dystrophy; LGMD—limb girdle muscular dystrophy; alpha-MHC—alpha-myosin heavy chain; CMT—Charcot–Marie–Tooth neuropathy; Des—desmin; hCK—human creatine kinase; OPMD—oculopharyngeal muscular dystrophy; hRS1—human retinoschisin 1; XLRS—X-linked retinoschisis; PRs—photoreceptors; hRPE65—retinal pigment epithelium-specific protein; LCA—Leber congenital amaurosis; RPEs—retinal pigment epithelial cells; GRK1—G-protein coupled receptor protein kinase 1 (hGRK1—human GRK1); RP—retinitis pigmentosa; hRK—human rhodopsin kinase; LCR—locus control region; hSyn1—human synapsin I; MTLE—mesial temporal lobe epilepsy; FTD—frontotemporal dementia; mPGK—murine phosphoglycerate kinase promoter; MeCP—methyl CpG binding protein 2.

However, the use of natural ubiquitous promoters or their optimized alternatives is often limited due to their lack of specificity for cell types. Development of tissue-specific promoters is especially important in gene therapies for eye and CNS disorders, given the numerous subtypes of cells present in target organs [141–144]. In recent years, several strong synthetic tissue- and cell-specific promoters have been developed using strategies of (1) random combination of various transcriptional CREs or (2) prediction of CREs with bioinformatic methods. Some of such promoters obtained using the former method are already in clinical trials, though they remain relatively rare [110,115]. In the last decade, the great potential of computational methods for the design of promoters and enhancers was demonstrated, but they have yet to make their way into clinical settings [121,145–148]. Still, many of the proposed promoters and enhancers have proven to be effective in mice and primates [121,149–151].

4.1. Liver-Specific Promoters

For liver-specific gene delivery, promoters derived from *alpha-1 antitrypsin* (*AAT*), *transferrin* (*TTR*), and *human thyroxine-binding globulin* (*TBG*) genes are frequently used (Table 1). Early work on the development of liver-specific promoters focused on the use of natural promoters of proteins highly expressed in the liver, such as human alpha-1 antitrypsin (hAAT) [152]. Later, a number of chimeric promoters were developed based on the *hAAT* gene, providing specific delivery of the transgene to hepatocytes. One of the first was the HCR/hAAT promoter, a hybrid of the hAAT core promoter and the hepatic control region (HCR) of the apolipoprotein E/C-I (*ApoE*) gene locus [105]. Efforts to create more compact constructs resulted in the 448 bp LP1, an improved version of the original 754 bp HCR/hAAT promoter, which provided higher transgene expression [107]. The HLP (hybrid liver promoter) promoter, only 252 bp, later emerged as an even more compact yet equally efficient option (Figure 2B) [108].

The ApoE/hAAT promoter is currently being used in clinical trials in a therapy for Fabry disease (NCT05039866), as well as in the FDA approved hemophilia B drug Beqvez [153]. Additionally, the LP1 promoter is being tested in several therapies for hemophilia B (NCT00979238, NCT03569891). HLP is being utilized in clinical trials for hemophilia A (NCT03370913, NCT03466463).

A recent study [110] applied a combinatorial method to engineer promoters by combining fragments from promoter and enhancer regions into cassettes. The resulting short HCB promoter (146 bp) demonstrated a 14-fold increase in transgene expression over HLP in vivo. This promoter is used in the ASC618 gene therapy product for the treatment of hemophilia A (NCT04676048).

4.2. Muscle-Specific Promoters

Several synthetic promoters based on muscle creatine kinase (MCK) and desmin genes are currently widely used in clinics for the delivery to muscles. Among the first constructs based on the *MCK* gene was CK6 (571 bp), which consists of a proximal MCK promoter and the 2RS5 enhancer. Delivered in an adenoviral vector, CK6 exhibited 12% of the activity seen with a CMV promoter [154]. Later, the same research group developed the MHCK7 construct, which was based on the CK6 cassette and underwent several modifications, including the addition of a 188 bp enhancer from the mouse alpha-myosin gene (α-MHC). Systemic delivery of the MHCK7 promoter in an AAV6 vector showed transgene expression similar to that observed for CMV and RSF promoters in mouse skeletal and cardiac muscles [102]. A further advancement, the CK8 cassette, had the α-MHC enhancer replaced with a modified MCK enhancer, resulting in a smaller yet more effective promoter [155]. MHCK7 is used in Elevidys for DMD [4] and in a number of therapies currently undergoing clinical trials (NCT02710500, NCT06246513, and NCT05881408). CK8 is used in a DMD gene therapy currently undergoing clinical trials (NCT03368742, NCT06138639).

Other chimeric MCK-derived promoters, such as dMCK (509 bp) and tMCK (720 bp), contain two or three copies of the 2RS5 enhancer combined with a truncated basal promoter. These promoters showed high specificity for skeletal muscle, with tMCK outperforming the CMV promoter [156]. tMCK is currently used in the development of therapies for DMD (NCT04281485), LGMD2D (NCT01976091, NCT00494195), and CMT (NCT03520751).

Another group of promoters used for muscle delivery are desmin-based promoters. These are often truncated versions of the natural desmin promoter (Des), which can be altered by introducing deletions or substitutions [157]. Multiple drugs utilizing different desmin promoter variants are currently in clinical trials (NCT01344798, NCT03199469, and NCT02240407).

To create a highly active synthetic muscle-specific promoter, Li et al. randomly combined several myogenic regulatory elements and added the resulting shuffled cassettes to the chicken α-skeletal actin core promoter, thus obtaining a library of synthetic promoters [115]. The most active promoter was SPc5-12, providing a 6–8-fold higher expression of the target gene in vivo compared to the CMV promoter. There are ongoing clinical trials using the SPc5-12 promoter for the treatment of DMD (NCT05693142, NCT06185673).

The development of bioinformatics tools gave rise to several other strong synthetic promoters. To create a hybrid MH promoter, four functional modules were selected based on an in silico analysis of various murine muscle-specific genes [158]. The MH promoter turned out to be more active in skeletal muscles compared to the CMV promoter when delivered to mice packed in an AAV vector. Another research group developed an algorithm to predict transcriptional cis-regulatory modules (Sk-CRMs) providing the highest muscle-specific expression of the transgene [95]. These modules, composed of TF binding site combinations, were cloned upstream of the desmin promoter in vectors expressing luciferase. In mice, the Sk-CRM4 module led to a dramatic increase in desmin promoter activity, enhancing it by up to 400 times in skeletal muscle.

4.3. Eye-Specific Promoters

The range of promoters used in therapies for eye disorders is the most extensive, as presented in Table 1. This is because the retina is composed of various cell types, demanding delivery vectors with cell-specific promoters. Several promoters targeting various cell types have been used in clinical trials, including NA65p (RPE cells), PR1 (cone cells), hGRK1 (rod and cone cells), and hRLBP1 (Müller glial and RPE cells) (Table 1).

The use of truncated versions of natural promoters is among the most popular strategies of creating synthetic ones. Thus, the NA65p promoter was obtained by optimizing the natural hRPE65 promoter, removing inhibitory elements, and shortening the sequence. The new version was more active than the natural promoter, providing specific transgene expression in RPE cells [159]. It is now part of clinical trials for a drug targeting Leber congenital amaurosis (LCA) (NCT02946879). Another example is hGRK1, which is used to target cones and rods. Shortened versions of this promoter are used in EDIT-101 (NCT03872479) and AGTC-501 (NCT06275620) drugs.

The PR1.7 promoter, which includes the human L-opsin enhancer and core promoter, was developed to specifically target cone photoreceptor cells [160]. When delivered to mice using an AAV vector, it demonstrated superior cone specificity compared to the mouse cone arrestin (mCAR) promoter [161]. PR1.7 is used in studies on the treatment of achromatopsia, a disease associated with the loss of cone function (NCT02599922, NCT02935517).

Bipolar cells are considered one of the best targets for optogenetic gene therapies for vision restoration [162]. Several research groups have made efforts to target gene expression to specific retinal bipolar cell types [163–166]. Thus, to target ON-bipolar cells (OBCs), Hulliger et al. designed a synthetic promoter 770En_454P(hGRM6) based on the *hGrm6* gene [166]. Using bioinformatic methods, authors predicted potential proximal promoter and enhancer regions. Enhancer–proxymal promoter combinations were packed in a vector encoding a reporter gene and tested transducing the post-mortem human retina. The selected 1.2 kbp variant 770En_454P(hGRM6) displayed strong specificity for OBCs and resulted in vision restoration in an rd1 mouse of late retinal degeneration.

With the development of bioinformatic tools for the discovery of CREs, more novel synthetic promoters are designed using computational biology methods. In this way, in the work [122], the researchers used Drop-seq data to identify novel minimal promoter elements driving restricted expression to retinal ganglion cells. Bioinformatic design was informed by a wide range of genomics datasets and resulted in seven constructs that were subsequently compared for strength and specificity in mice. The best candidate, Pre345 (NEFL), was further characterized in nonhuman primate retina, where it showed specific and robust expression in the RGCs. Later, the same group developed more MiniPromoters driving expression to various ocular cell types: ON bipolar, cone, corneal, endothelial, Müller glial, and PAX6 cells [121].

4.4. CNS-Specific Promoters

Constitutive promoters, such as CMV and CAG, remain the most used in gene therapy of CNS disorders [30]. Promoters mPGK, hSYN, and NSE are known for their specificity to neurons. For instance, it was shown that mPGK and hSYN provide stronger expression in the corticospinal tract (CST), compared to the widely used CMV and CAG [167]. Although their transition into clinical settings has been slow [30], these promoters have been used in recent trials (NCT06063850, NCT06064890, NCT03300453). Recently, truncated versions of the MeCP2 promoter have also entered clinical settings (NCT03770572, NCT06152237).

The strategy of creating chimeric constructs is also used in the development of synthetic neuron-specific promoters. As an example, a chimeric rSynI(1.0)-minCMV promoter was obtained by fusing a 1 kb fragment of the hSyn1 promoter with a minimal CMV promoter [168]. The resulting hybrid provided robust neuron-specific transgene expression.

Modern methods of creating cell-specific promoters are focused on identifying unique CREs active in certain types of neuronal cells. Previously described methods, such as open chromatin analysis, make it possible to find new enhancers that direct expression to specific cell subtypes [78,79,169]. Thus, enhancer sequences were identified by collecting information on chromatin accessibility for different cell types [79]. When delivered to the brain as AAV vectors, several enhancers were found to driveexpression to specific cell subtypes.

The effectiveness of some enhancers for cell-specific delivery of AAV vectors has been confirmed in primates. For instance, Mich et al. demonstrated the specificity of a set of enhancers targeting parvalbumin (PVALB) interneurons [149]. Vormstein-Schneider et al. identified a few enhancers specific to parvalbumin and vasoactive intestinal peptide-expressing interneurons and confirmed that their selectivity is conserved across vertebrate species [151].

4.5. Tumor-Specific Promoters

An important factor in the development of cancer gene therapies is to ensure targeted delivery of the therapeutic agent to cancer cells, while limiting the damage to normal cells. In this regard, the use of cancer-specific and tumor-specific promoters is especially relevant. Cancer-specific promoters drive gene expression in cancer cells regardless of the tumor type, while tumor-specific promoters are active in cancer cells of a certain type. Various versions of such promoters are discussed in detail by other authors [170,171].

Among them, there are optimized variants based on natural promoters, such as a2bm. This is a promoter specific to hepatocellular carcinoma (HCC), which is a modified version of the AFP promoter. It was obtained by fusing two copies of enhancer A and one copy of enhancer B with AFP, which resulted in an increase in promoter activity. Later, the Ha2bm promoter was created by adding several hypoxia-responsive elements (HREs) to a2bm, which provided higher activity compared to a2bm under hypoxic conditions. Thus, this modification allowed for a better targeting of HCC considering the hypoxic tumor environment.

Approaches aimed at predicting CRE are also used for the development of cancer- and tumor-specific promoters. For example, the human insulin promoter was used as a basis for the development of synthetic human insulin super-promoter (SHIP1) for pancreatic cancer gene therapy. The analysis of key transcriptional regulatory elements of the insulin promoter led to four SHIP variants. SHIP1 was the most active and outperformed RIP (rat insulin promoter II fragment) and CMV in vivo [172].

5. Conclusions and Future Prospects

Currently, there are various approaches to the creation of synthetic promoters. For instance, the problem of low activity and tissue specificity is generally solved by using combinations of cis-regulatory regions of different natural promoters; long promoters can be shortened by the deletion of low-conserved parts of sequence. The use of computational approaches makes it possible to significantly accelerate the process of developing new synthetic promoters, as well as reduces the time and resources needed for experimental validation. Methods of identification of CREs to design new promoters and enhancer sequences using bioinformatic methods are being actively developed. Moreover, generative algorithms are now used to de novo engineer fully synthetic CREs, that may even excel naturally occurring ones in specificity [127,173]. Further development of bioinformatic tools and generative models will enable a more accurate design of cell-specific regulatory elements, facilitating the development of targeted synthetic promoters.

In addition, other approaches are being actively explored in the field of synthetic promoter design. The creation of inducible promoters controlled by small molecules or physical actions is a good opportunity to make expression more adjustable, which is highly relevant in clinical practice. The most used tetracycline-dependent system of regulating gene expression has already been successfully applied in AAV vectors [174–176]. Along with inducible promoters, synthetic transcription factors (synTFs) based on bacterial or mammalian sequences are being developed. In the presence of low-molecular-weight exogenous or endogenous compounds, synTF can bind to regulatory elements of the inducible promoter that results in transgene expression [177,178]. Vectors that contain elements of gene expression control can be used to regulate both the level and timing of expression, allowing for the improvement of safety and efficacy of future gene therapies.

The first European Commission-approved gene therapy, Alipogene Tiparvovec (Glybera, uniQure, Lexington, KY, USA) [179], utilized a constitutive ubiquitous CMV promoter. To this day, natural promoters prevail in research and clinical trials; however, the development of new synthetic promoters may be crucial to the success of gene therapy. The use of natural promoters often requires high titers of viral vectors, which can cause serious toxicity. For example, Biogen announced the discontinuation of the development of BIIB089, citing the same concerns [139]. In contrast to natural promoters, synthetic promoters achieve higher specificity and gene expression levels that allow for an increase in transgene delivery efficiency and that reduce the risks associated with vector introduction. Currently, there are already FDA-approved in vivo gene therapies, such as the drugs Elevidys (Sarepta Therapeutics, Inc, Cambridge, UK) [4], Hemgenix (CSL Behring LLC, King of Prussia, PA, USA) [6], Beqvez (Pfizer, Inc., New York, NY, USA) [180], and Roctavian (BioMarin Pharmaceutical Inc., San Rafael, CA, USA) [5], which utilize tissue-specific synthetic promoters consisting of elements of natural promoters. There are also known examples of combining parts of enhancers and core promoter regions from strong ubiquitous promoters, as in the case of Luxturna (Spark Therapeutics, Inc.) [2] and Zolgensma (Novartis Gene Therapies, Inc., Philadelphia, PA, USA) [3], which use a CBA promoter with a CMV enhancer. Although synthetic promoters have already found application in therapeutic practice, there is a necessity for the development of new variants, both to create new gene therapies and to improve the efficacy and safety of existing ones. In the future, synthetic promoters may take a central place in clinical practice, providing safer and more targeted expression of therapeutic genes for the treatment of a wide range of diseases, such as hereditary pathologies, cancers, and neurodegenerative disorders.

Author Contributions: Conceptualization, V.A. and S.G.F.; supervision, O.M. and P.Y.V.; visualization, V.A., A.G., A.I.P. and A.A.D.; writing—original draft, V.A., A.G., A.I.P., A.D., S.G.F., A.A.D., O.M. and P.Y.V.; writing—review and editing, V.A., A.G., A.I.P., A.D., S.G.F., A.A.D., O.M. and P.Y.V. All authors have read and agreed to the published version of the manuscript.

Funding: The research was supported by the Ministry of Science and Higher Education of the Russian Federation (agreement # № 075-03-2022-107/10) and by the Russian Science Foundation (grant № 23-64-00002).

Data Availability Statement: No new data were created or analyzed in this study.

Acknowledgments: We would like to thank Anna A Maznina for her help in proofreading and adapting the text into English.

Conflicts of Interest: The authors declare no conflict of interest.

References

1. Baird, P.A.; Anderson, T.W.; Newcombe, H.B.; Lowry, R.B. Genetic Disorders in Children and Young Adults: A Population Study. *Am. J. Hum. Genet.* **1988**, *42*, 677–693. [PubMed]
2. Spark Therapeutics Inc. LUXTURNA (Voretigene Neparvovec-Rzyl). Available online: https://www.fda.gov/vaccines-blood-biologics/cellular-gene-therapy-products/luxturna (accessed on 21 October 2024).
3. Novartis Gene Therapies Inc. ZOLGENSMA (Onasemnogene Abeparvovec-Xioi). Available online: https://www.fda.gov/vaccines-blood-biologics/zolgensma (accessed on 21 October 2024).
4. Sarepta Therapeutics Inc. ELEVIDYS (Delandistrogene Moxeparvovec-Rokl). Available online: https://www.fda.gov/vaccines-blood-biologics/tissue-tissue-products/elevidys (accessed on 21 October 2024).
5. BioMarin Pharmaceutical Inc. ROCTAVIAN (Valoctocogene Roxaparvovec-Rvox). Available online: https://www.fda.gov/vaccines-blood-biologics/roctavian (accessed on 21 October 2024).
6. CSL Behring LLC. HEMGENIX (Etranacogene Dezaparvovec-Drlb). Available online: https://www.fda.gov/vaccines-blood-biologics/vaccines/hemgenix (accessed on 21 October 2024).
7. Wang, D.; Tai, P.W.L.; Gao, G. Adeno-Associated Virus Vector as a Platform for Gene Therapy Delivery. *Nat. Rev. Drug Discov.* **2019**, *18*, 358–378. [CrossRef] [PubMed]
8. Chira, S.; Jackson, C.S.; Oprea, I.; Ozturk, F.; Pepper, M.S.; Diaconu, I.; Braicu, C.; Raduly, L.Z.; Calin, G.A.; Berindan-Neagoe, I. Progresses towards Safe and Efficient Gene Therapy Vectors. *Oncotarget* **2015**, *6*, 30675–30703. [CrossRef] [PubMed]
9. Watanabe, M.; Nasu, Y.; Kusumi, N.; Nagai, A.; Kumon, H.; Kashiwakura, Y. 238: Adeno-Associated Virus 2-Mediated Intratumoral Prostate Cancer Gene Therapy: Long-Term Maspin Expression Efficiently Suppresses Tumor Growth. *J. Urol.* **2005**, *173*, 65. [CrossRef]
10. Ng, S.S.M.; Gao, Y.; Chau, D.H.W.; Li, G.H.Y.; Lai, L.H.; Huang, P.T.; Huang, C.F.; Huang, J.J.; Chen, Y.C.; Kung, H.F.; et al. A Novel Glioblastoma Cancer Gene Therapy Using AAV-Mediated Long-Term Expression of Human TERT C-Terminal Polypeptide. *Cancer Gene Ther.* **2007**, *14*, 561–572. [CrossRef]
11. Hacker, U.T.; Bentler, M.; Kaniowska, D.; Morgan, M.; Büning, H. Towards Clinical Implementation of Adeno-Associated Virus (Aav) Vectors for Cancer Gene Therapy: Current Status and Future Perspectives. *Cancers* **2020**, *12*, 1889. [CrossRef]
12. He, L.F.; Wang, Y.G.; Xiao, T.; Zhang, K.J.; Li, G.C.; Gu, J.F.; Chu, L.; Tang, W.H.; Tan, W.S.; Liu, X.Y. Suppression of Cancer Growth in Mice by Adeno-Associated Virus Vector-Mediated IFN-β Expression Driven by HTERT Promoter. *Cancer Lett.* **2009**, *286*, 196–205. [CrossRef]
13. Münch, R.C.; Muth, A.; Muik, A.; Friedel, T.; Schmatz, J.; Dreier, B.; Trkola, A.; Plückthun, A.; Büning, H.; Buchholz, C.J. Off-Target-Free Gene Delivery by Affinity-Purified Receptor-Targeted Viral Vectors. *Nat. Commun.* **2015**, *6*, 6246. [CrossRef]
14. Duan, D. Lethal Immunotoxicity in High-Dose Systemic AAV Therapy. *Mol. Ther.* **2023**, *31*, 3123–3126. [CrossRef]
15. Korneyenkov, M.A.; Zamyatnin, A.A. Next Step in Gene Delivery: Modern Approaches and Further Perspectives of Aav Tropism Modification. *Pharmaceutics* **2021**, *13*, 750. [CrossRef]
16. Lee, E.J.; Guenther, C.M.; Suh, J. Adeno-Associated Virus (AAV) Vectors: Rational Design Strategies for Capsid Engineering. *Curr. Opin. Biomed. Eng.* **2018**, *7*, 58–63. [CrossRef] [PubMed]
17. Ghauri, M.S.; Ou, L. AAV Engineering for Improving Tropism to the Central Nervous System. *Biology* **2023**, *12*, 186. [CrossRef] [PubMed]
18. Paulk, N.K.; Pekrun, K.; Zhu, E.; Nygaard, S.; Li, B.; Xu, J.; Chu, K.; Leborgne, C.; Dane, A.P.; Haft, A.; et al. Bioengineered AAV Capsids with Combined High Human Liver Transduction In Vivo and Unique Humoral Seroreactivity. *Mol. Ther.* **2018**, *26*, 289–303. [CrossRef]
19. Cabanes-Creus, M.; Hallwirth, C.V.; Westhaus, A.; Ng, B.H.; Liao, S.H.Y.; Zhu, E.; Gale, R.; Baltazar, G.; Drouyer, M.; Scott, S.; et al. Restoring the Natural Tropism of AAV2 Vectors for Human Liver. *Sci. Transl. Med.* **2020**, *12*, eaba3312. [CrossRef] [PubMed]
20. Pavlou, M.; Schön, C.; Occelli, L.M.; Rossi, A.; Meumann, N.; Boyd, R.F.; Bartoe, J.T.; Siedlecki, J.; Gerhardt, M.J.; Babutzka, S.; et al. Novel AAV Capsids for Intravitreal Gene Therapy of Photoreceptor Disorders. *EMBO Mol. Med.* **2021**, *13*, e13392. [CrossRef]
21. Yin, L.; Greenberg, K.; Hunter, J.J.; Dalkara, D.; Kolstad, K.D.; Masella, B.D.; Wolfe, R.; Visel, M.; Stone, D.; Libby, R.T.; et al. Intravitreal Injection of AAV2 Transduces Macaque Inner Retina. *Investig. Ophthalmol. Vis. Sci.* **2011**, *52*, 2775–2783. [CrossRef]
22. Fraldi, A.; Hemsley, K.; Crawley, A.; Lombardi, A.; Lau, A.; Sutherland, L.; Auricchio, A.; Ballabio, A.; Hopwood, J.J. Functional Correction of CNS Lesions in an MPS-IIIA Mouse Model by Intracerebral AAV-Mediated Delivery of Sulfamidase and *SUMF1* Genes. *Hum. Mol. Genet.* **2007**, *16*, 2693–2702. [CrossRef]

23. Bosch, A.; Perret, E.; Desmaris, N.; Heard, J.M. Long-Term and Significant Correction of Brain Lesions in Adult Mucopolysaccharidosis Type VII Mice Using Recombinant AAV Vectors. *Mol. Ther.* **2000**, *1*, 63–70. [CrossRef]
24. Zhao, L.; Gottesdiener, A.J.; Parmar, M.; Li, M.; Kaminsky, S.M.; Chiuchiolo, M.J.; Sondhi, D.; Sullivan, P.M.; Holtzman, D.M.; Crystal, R.G.; et al. Intracerebral Adeno-Associated Virus Gene Delivery of Apolipoprotein E2 Markedly Reduces Brain Amyloid Pathology in Alzheimer's Disease Mouse Models. *Neurobiol. Aging* **2016**, *44*, 159–172. [CrossRef]
25. Fraites, T.J.; Schleissing, M.R.; Shanely, R.A.; Walter, G.A.; Cloutier, D.A.; Zolotukhin, I.; Pauly, D.F.; Raben, N.; Plotz, P.H.; Powers, S.K.; et al. Correction of the Enzymatic and Functional Deficits in a Model of Pompe Disease Using Adeno-Associated Virus Vectors. *Mol. Ther.* **2002**, *5*, 571–578. [CrossRef]
26. Massaro, G.; Geard, A.F.; Nelvagal, H.R.; Gore, K.; Clemo, N.K.; Waddington, S.N.; Rahim, A.A. Comparison of Different Promoters to Improve AAV Vector-Mediated Gene Therapy for Neuronopathic Gaucher Disease. *Hum. Mol. Genet.* **2024**, *33*, 1467–1480. [CrossRef] [PubMed]
27. Salabarria, S.M.; Nair, J.; Clement, N.; Smith, B.K.; Raben, N.; Fuller, D.D.; Byrne, B.J.; Corti, M. Advancements in AAV-Mediated Gene Therapy for Pompe Disease. *J. Neuromuscul. Dis.* **2020**, *7*, 15–31. [CrossRef] [PubMed]
28. Kulkarni, A.; Chen, T.; Sidransky, E.; Han, T.U. Advancements in Viral Gene Therapy for Gaucher Disease. *Genes* **2024**, *15*, 364. [CrossRef]
29. Milenkovic, I.; Blumenreich, S.; Hochfelder, A.; Azulay, A.; Biton, I.E.; Zerbib, M.; Oren, R.; Tsoory, M.; Joseph, T.; Fleishman, S.J.; et al. Efficacy of an AAV Vector Encoding a Thermostable Form of Glucocerebrosidase in Alleviating Symptoms in a Gaucher Disease Mouse Model. *Gene Ther.* **2024**, *31*, 439–444. [CrossRef] [PubMed]
30. Au, H.K.E.; Isalan, M.; Mielcarek, M. Gene Therapy Advances: A Meta-Analysis of AAV Usage in Clinical Settings. *Front. Med.* **2022**, *8*, 809118. [CrossRef]
31. Brooks, A.R.; Harkins, R.N.; Wang, P.; Qian, H.S.; Liu, P.; Rubanyi, G.M. Transcriptional Silencing Is Associated with Extensive Methylation of the CMV Promoter Following Adenoviral Gene Delivery to Muscle. *J. Gene Med.* **2004**, *6*, 395–404. [CrossRef]
32. Shirley, J.L.; de Jong, Y.P.; Terhorst, C.; Herzog, R.W. Immune Responses to Viral Gene Therapy Vectors. *Mol. Ther.* **2020**, *28*, 709–722. [CrossRef]
33. Xiong, W.; Wu, D.M.; Xue, Y.; Wang, S.K.; Chung, M.J.; Ji, X.; Rana, P.; Zhao, S.R.; Mai, S.; Cepko, C.L. AAV Cis-Regulatory Sequences Are Correlated with Ocular Toxicity. *Proc. Natl. Acad. Sci. USA* **2019**, *116*, 5785–5794. [CrossRef]
34. Novartis Pharmaceuticals Corporation KYMRIAH (Tisagenlecleucel). Available online: https://www.fda.gov/vaccines-blood-biologics/cellular-gene-therapy-products/kymriah (accessed on 21 October 2024).
35. Labbé, R.P.; Vessillier, S.; Rafiq, Q.A. Lentiviral Vectors for t Cell Engineering: Clinical Applications, Bioprocessing and Future Perspectives. *Viruses* **2021**, *13*, 1528. [CrossRef]
36. Booth, C.; Gaspar, H.B.; Thrasher, A.J. Treating Immunodeficiency through HSC Gene Therapy. *Trends Mol. Med.* **2016**, *22*, 317–327. [CrossRef]
37. Rintz, E.; Higuchi, T.; Kobayashi, H.; Galileo, D.S.; Wegrzyn, G.; Tomatsu, S. Promoter Considerations in the Design of Lentiviral Vectors for Use in Treating Lysosomal Storage Diseases. *Mol. Ther. Methods Clin. Dev.* **2022**, *24*, 71–87. [CrossRef] [PubMed]
38. Parr-Brownlie, L.C.; Bosch-Bouju, C.; Schoderboeck, L.; Sizemore, R.J.; Abraham, W.C.; Hughes, S.M. Lentiviral Vectors as Tools to Understand Central Nervous System Biology in Mammalian Model Organisms. *Front. Mol. Neurosci.* **2015**, *8*, 14. [CrossRef] [PubMed]
39. Michels, K.R.; Sheih, A.; Hernandez, S.A.; Brandes, A.H.; Parrilla, D.; Irwin, B.; Perez, A.M.; Ting, H.A.; Nicolai, C.J.; Gervascio, T.; et al. Preclinical Proof of Concept for VivoVec, a Lentiviral-Based Platform for in Vivo CAR T-Cell Engineering. *J. Immunother. Cancer* **2023**, *11*, e006292. [CrossRef] [PubMed]
40. Santilli, G.; Almarza, E.; Brendel, C.; Choi, U.; Beilin, C.; Blundell, M.P.; Haria, S.; Parsley, K.L.; Kinnon, C.; Malech, H.L.; et al. Biochemical Correction of *X-CGD* by a Novel Chimeric Promoter Regulating High Levels of Transgene Expression in Myeloid Cells. *Mol. Ther.* **2011**, *19*, 122–132. [CrossRef]
41. Garcia-Perez, L.; van Eggermond, M.; van Roon, L.; Vloemans, S.A.; Cordes, M.; Schambach, A.; Rothe, M.; Berghuis, D.; Lagresle-Peyrou, C.; Cavazzana, M.; et al. Successful Preclinical Development of Gene Therapy for Recombinase-Activating Gene-1-Deficient SCID. *Mol. Ther. Methods Clin. Dev.* **2020**, *17*, 666–682. [CrossRef]
42. Levin, M.C.; Lidberg, U.; Jirholt, P.; Adiels, M.; Wramstedt, A.; Gustafsson, K.; Greaves, D.R.; Li, S.; Fazio, S.; Linton, M.F.; et al. Evaluation of Macrophage-Specific Promoters Using Lentiviral Delivery in Mice. *Gene Ther.* **2012**, *19*, 1041–1047. [CrossRef] [PubMed]
43. Latorre-Rey, L.J.; Wintterle, S.; Dütting, S.; Kohlscheen, S.; Abel, T.; Schenk, F.; Wingert, S.; Rieger, M.A.; Nieswandt, B.; Heinz, N.; et al. Targeting Expression to Megakaryocytes and Platelets by Lineage-Specific Lentiviral Vectors. *J. Thromb. Haemost.* **2017**, *15*, 341–355. [CrossRef]
44. Kerns, H.M.; Ryu, B.Y.; Stirling, B.V.; Sather, B.D.; Astrakhan, A.; Humblet-Baron, S.; Liggitt, D.; Rawlings, D.J. B Cell-Specific Lentiviral Gene Therapy Leads to Sustained B-Cell Functional Recovery in a Murine Model of X-Linked Agammaglobulinemia. *Blood* **2010**, *115*, 2146–2155. [CrossRef]
45. Pucci, F.; Rickelt, S.; Newton, A.P.; Garris, C.; Nunes, E.; Evavold, C.; Pfirschke, C.; Engblom, C.; Mino-Kenudson, M.; Hynes, R.O.; et al. PF4 Promotes Platelet Production and Lung Cancer Growth. *Cell Rep.* **2016**, *17*, 1764–1772. [CrossRef]

46. Montiel-Equihua, C.A.; Zhang, L.; Knight, S.; Saadeh, H.; Scholz, S.; Carmo, M.; Alonso-Ferrero, M.E.; Blundell, M.P.; Monkeviciute, A.; Schulz, R.; et al. The B-Globin Locus Control Region in Combination with the EF1α Short Promoter Allows Enhanced Lentiviral Vector-Mediated Erythroid Gene Expression with Conserved Multilineage Activity. *Mol. Ther.* **2012**, *20*, 1400–1409. [CrossRef]
47. Moreau-Gaudry, F.; Xia, P.; Jiang, G.; Perelman, N.P.; Bauer, G.; Ellis, J.; Surinya, K.H.; Mavilio, F.; Shen, C.K.; Malik, P. High-Level Erythroid-Specific Gene Expression in Primary Human and Murine Hematopoietic Cells with Self-Inactivating Lentiviral Vectors. *Blood* **2001**, *98*, 2664–2672. [CrossRef] [PubMed]
48. Milone, M.C.; O'Doherty, U. Clinical Use of Lentiviral Vectors. *Leukemia* **2018**, *32*, 1529–1541. [CrossRef]
49. Masiuk, K.E.; Laborada, J.; Roncarolo, M.G.; Hollis, R.P.; Kohn, D.B. Lentiviral Gene Therapy in HSCs Restores Lineage-Specific Foxp3 Expression and Suppresses Autoimmunity in a Mouse Model of IPEX Syndrome. *Cell Stem Cell* **2019**, *24*, 309–317. [CrossRef] [PubMed]
50. Nicolas, C.T.; VanLith, C.J.; Hickey, R.D.; Du, Z.; Hillin, L.G.; Guthman, R.M.; Cao, W.J.; Haugo, B.; Lillegard, A.; Roy, D.; et al. In Vivo Lentiviral Vector Gene Therapy to Cure Hereditary Tyrosinemia Type 1 and Prevent Development of Precancerous and Cancerous Lesions. *Nat. Commun.* **2022**, *13*, 1–15. [CrossRef]
51. Houghton, B.C.; Booth, C. Gene Therapy for Primary Immunodeficiency. *HemaSphere* **2021**, *5*, e509. [CrossRef] [PubMed]
52. Fassler, J.S.; Gussin, G.N. Promoters and Basal Transcription Machinery in Eubacteria and Eukaryotes: Concepts, Definitions, and Analogies. *Methods Enzymol.* **1996**, *273*, 3–29. [CrossRef]
53. Cooper, S.J.; Trinklein, N.D.; Anton, E.D.; Nguyen, L.; Myers, R.M. Comprehensive Analysis of Transcriptional Promoter Structure and Function in 1% of the Human Genome. *Genome Res.* **2006**, *16*, 1–10. [CrossRef]
54. Carninci, P.; Sandelin, A.; Lenhard, B.; Katayama, S.; Shimokawa, K.; Ponjavic, J.; Semple, C.A.M.; Taylor, M.S.; Engström, P.G.; Frith, M.C.; et al. Genome-Wide Analysis of Mammalian Promoter Architecture and Evolution. *Nat. Genet.* **2006**, *38*, 626–635. [CrossRef] [PubMed]
55. Juven-Gershon, T.; Kadonaga, J.T. Regulation of Gene Expression via the Core Promoter and the Basal Transcriptional Machinery. *Dev. Biol.* **2010**, *339*, 225–229. [CrossRef]
56. Emami, K.H.; Jain, A.; Smale, S.T. Mechanism of Synergy between TATA and Initiator: Synergistic Binding of TFIID Following a Putative TFIIA-Induced Isomerization. *Genes Dev.* **1997**, *11*, 3007–3019. [CrossRef]
57. Zabidi, M.A.; Arnold, C.D.; Schernhuber, K.; Pagani, M.; Rath, M.; Frank, O.; Stark, A. Enhancer-Core-Promoter Specificity Separates Developmental and Housekeeping Gene Regulation. *Nature* **2015**, *518*, 556–559. [CrossRef] [PubMed]
58. Butler, J.E.F.; Kadonaga, J.T. Enhancer-Promoter Specificity Mediated by DPE or TATA Core Promoter Motifs. *Genes Dev.* **2001**, *15*, 2515–2519. [CrossRef] [PubMed]
59. Gershenzon, N.I.; Trifonov, E.N.; Ioshikhes, I.P. The Features of Drosophila Core Promoters Revealed by Statistical Analysis. *BMC Genom.* **2006**, *7*, 161. [CrossRef] [PubMed]
60. Thomas, M.C.; Chiang, C.M. The General Transcription Machinery and General Cofactors. *Crit. Rev. Biochem. Mol. Biol.* **2006**, *41*, 105–178. [CrossRef]
61. Kadonaga, J.T. Perspectives on the RNA Polymerase II Core Promoter. *Wiley Interdiscip. Rev. Dev. Biol.* **2012**, *1*, 40–51. [CrossRef]
62. Spitz, F.; Furlong, E.E.M. Transcription Factors: From Enhancer Binding to Developmental Control. *Nat. Rev. Genet.* **2012**, *13*, 613–626. [CrossRef]
63. Hernandez-Garcia, C.M.; Finer, J.J. Identification and Validation of Promoters and Cis-Acting Regulatory Elements. *Plant Sci.* **2014**, *217*, 109–119. [CrossRef]
64. Powell, S.K.; Rivera-Soto, R.; Gray, S.J. Viral Expression Cassette Elements to Enhance Transgene Target Specificity and Expression in Gene Therapy. *Discov. Med.* **2015**, *19*, 49.
65. Calo, E.; Wysocka, J. Modification of Enhancer Chromatin: What, How, and Why? *Mol. Cell* **2013**, *49*, 825–837. [CrossRef]
66. Huminiecki, Ł.; Horbańczuk, J. Can We Predict Gene Expression by Understanding Proximal Promoter Architecture? *Trends Biotechnol.* **2017**, *35*, 530–546. [CrossRef]
67. Haberle, V.; Stark, A. Eukaryotic Core Promoters and the Functional Basis of Transcription Initiation. *Nat. Rev. Mol. Cell Biol.* **2018**, *19*, 621–637. [CrossRef] [PubMed]
68. Shlyueva, D.; Stampfel, G.; Stark, A. Transcriptional Enhancers: From Properties to Genome-Wide Predictions. *Nat. Rev. Genet.* **2014**, *15*, 272–286. [CrossRef] [PubMed]
69. Spitz, F. Gene Regulation at a Distance: From Remote Enhancers to 3D Regulatory Ensembles. *Semin. Cell Dev. Biol.* **2016**, *57*, 57–67. [CrossRef]
70. Beagrie, R.A.; Scialdone, A.; Schueler, M.; Kraemer, D.C.A.; Chotalia, M.; Xie, S.Q.; Barbieri, M.; De Santiago, I.; Lavitas, L.M.; Branco, M.R.; et al. Complex Multi-Enhancer Contacts Captured by Genome Architecture Mapping. *Nature* **2017**, *543*, 519–524. [CrossRef] [PubMed]
71. Ho, S.C.L.; Yang, Y. Identifying and Engineering Promoters for High Level and Sustainable Therapeutic Recombinant Protein Production in Cultured Mammalian Cells. *Biotechnol. Lett.* **2014**, *36*, 1569–1579. [CrossRef]
72. Song, E.S.; Lee, Y.H.; So, M.K.; Kuk, M.U.; Park, J.H.; Yoon, J.H.; Lee, Y.J.; Kim, D.; So, B.; Byun, Y.; et al. Establishment of a New Promoter Trapping Vector Using 2A Peptide. *Biotechnol. Bioprocess Eng.* **2024**, *29*, 520–528. [CrossRef]
73. Nakato, R.; Sakata, T. Methods for ChIP-Seq Analysis: A Practical Workflow and Advanced Applications. *Methods* **2021**, *187*, 44–53. [CrossRef]

74. Zhou, P.; VanDusen, N.J.; Zhang, Y.; Cao, Y.; Sethi, I.; Hu, R.; Zhang, S.; Wang, G.; Ye, L.; Mazumdar, N.; et al. Dynamic Changes in P300 Enhancers and Enhancer-Promoter Contacts Control Mouse Cardiomyocyte Maturation. *Dev. Cell* **2023**, *58*, 898–914. [CrossRef]
75. He, H.H.; Meyer, C.A.; Hu, S.S.; Chen, M.W.; Zang, C.; Liu, Y.; Rao, P.K.; Fei, T.; Xu, H.; Long, H.; et al. Refined DNase-Seq Protocol and Data Analysis Reveals Intrinsic Bias in Transcription Factor Footprint Identification. *Nat. Methods* **2014**, *11*, 73–78. [CrossRef]
76. Giresi, P.G.; Kim, J.; McDaniell, R.M.; Iyer, V.R.; Lieb, J.D. FAIRE (Formaldehyde-Assisted Isolation of Regulatory Elements) Isolates Active Regulatory Elements from Human Chromatin. *Genome Res.* **2007**, *17*, 877–885. [CrossRef]
77. Starks, R.R.; Biswas, A.; Jain, A.; Tuteja, G. Combined Analysis of Dissimilar Promoter Accessibility and Gene Expression Profiles Identifies Tissue-Specific Genes and Actively Repressed Networks. *Epigenetics Chromatin* **2019**, *12*, 16. [CrossRef] [PubMed]
78. Nair, R.R.; Blankvoort, S.; Lagartos, M.J.; Kentros, C. Enhancer-Driven Gene Expression (EDGE) Enables the Generation of Viral Vectors Specific to Neuronal Subtypes. *iScience* **2020**, *23*, 100888. [CrossRef] [PubMed]
79. Graybuck, L.T.; Daigle, T.L.; Sedeño-Cortés, A.E.; Walker, M.; Kalmbach, B.; Lenz, G.H.; Morin, E.; Nguyen, T.N.; Garren, E.; Bendrick, J.L.; et al. Enhancer Viruses for Combinatorial Cell-Subclass-Specific Labeling. *Neuron* **2021**, *109*, 1449–1464. [CrossRef] [PubMed]
80. Schlaeger, T.M.; Daheron, L.; Brickler, T.R.; Entwisle, S.; Chan, K.; Cianci, A.; DeVine, A.; Ettenger, A.; Fitzgerald, K.; Godfrey, M.; et al. A Comparison of Non-Integrating Reprogramming Methods. *Nat. Biotechnol.* **2015**, *33*, 58–63. [CrossRef]
81. Zhang, M.; Jia, C.; Li, F.; Li, C.; Zhu, Y.; Akutsu, T.; Webb, G.I.; Zou, Q.; Coin, L.J.M.; Song, J. Critical Assessment of Computational Tools for Prokaryotic and Eukaryotic Promoter Prediction. *Brief. Bioinform.* **2022**, *23*, bbab551. [CrossRef]
82. Le, N.Q.K.; Yapp, E.K.Y.; Nagasundaram, N.; Yeh, H.Y. Classifying Promoters by Interpreting the Hidden Information of DNA Sequences via Deep Learning and Combination of Continuous FastText N-Grams. *Front. Bioeng. Biotechnol.* **2019**, *7*, 305. [CrossRef]
83. Xiao, X.; Xu, Z.C.; Qiu, W.R.; Wang, P.; Ge, H.T.; Chou, K.C. IPSW(2L)-PseKNC: A Two-Layer Predictor for Identifying Promoters and Their Strength by Hybrid Features via Pseudo K-Tuple Nucleotide Composition. *Genomics* **2019**, *111*, 1785–1793. [CrossRef]
84. Wang, Y.; Tai, S.; Zhang, S.; Sheng, N.; Xie, X. PromGER: Promoter Prediction Based on Graph Embedding and Ensemble Learning for Eukaryotic Sequence. *Genes* **2023**, *14*, 1441. [CrossRef]
85. Huang, G.; Wu, L.; Ma, X.; Zhang, W.; Fan, J.; Yu, X.; Zeng, W.; Zhou, H. Evaluation of CatBoost Method for Prediction of Reference Evapotranspiration in Humid Regions. *J. Hydrol.* **2019**, *574*, 1029–1041. [CrossRef]
86. Kari, H.; Bandi, S.M.S.; Kumar, A.; Yella, V.R. DeePromClass: Delineator for Eukaryotic Core Promoters Employing Deep Neural Networks. *IEEE/ACM Trans. Comput. Biol. Bioinform.* **2023**, *20*, 802–807. [CrossRef]
87. Meylan, P.; Dreos, R.; Ambrosini, G.; Groux, R.; Bucher, P. EPD in 2020: Enhanced Data Visualization and Extension to NcRNA Promoters. *Nucleic Acids Res.* **2020**, *48*, D65–D69. [CrossRef] [PubMed]
88. Kolchanov, N.A.; Podkolodnaya, O.A.; Ananko, E.A.; Ignatieva, E.V.; Stepanenko, I.L.; Kel-Margoulis, O.V.; Kel, A.E.; Merkulova, T.I.; Goryachkovskaya, T.N.; Busygina, T.V.; et al. Transcription Regulatory Regions Database (TRRD): Its Status in 2000. *Nucleic Acids Res.* **2000**, *28*, 298–301. [CrossRef] [PubMed]
89. Yamashita, R.; Sugano, S.; Suzuki, Y.; Nakai, K. DBTSS: DataBase of Transcriptional Start Sites Progress Report in 2012. *Nucleic Acids Res.* **2012**, *40*, D150–D154. [CrossRef]
90. Forrest, A.R.R.; Kawaji, H.; Rehli, M.; Baillie, J.K.; De Hoon, M.J.L.; Haberle, V.; Lassmann, T.; Kulakovskiy, I.V.; Lizio, M.; Itoh, M.; et al. A Promoter-Level Mammalian Expression Atlas. *Nature* **2014**, *507*, 462–470. [CrossRef]
91. Salgado, H.; Gama-Castro, S.; Lara, P.; Mejia-Almonte, C.; Alarcón-Carranza, G.; López-Almazo, A.G.; Betancourt-Figueroa, F.; Peña-Loredo, P.; Alquicira-Hernández, S.; Ledezma-Tejeida, D.; et al. RegulonDB V12.0:Ã Comprehensive Resource of Transcriptional Regulation in E. Coli K-12. *Nucleic Acids Res.* **2024**, *52*, D255–D264. [CrossRef] [PubMed]
92. Sierro, N.; Makita, Y.; De Hoon, M.; Nakai, K. DBTBS: A Database of Transcriptional Regulation in Bacillus Subtilis Containing Upstream Intergenic Conservation Information. *Nucleic Acids Res.* **2008**, *36*, D93–D96. [CrossRef]
93. Roberts, M.L.; Katsoupi, P.; Tseveleki, V.; Taoufik, E. Bioinformatically Informed Design of Synthetic Mammalian Promoters. *Methods Mol. Biol.* **2017**, *1651*, 93–112.
94. Roberts, M.L. The Use of Functional Genomics in Synthetic Promoter Design. In *Computational Biology and Applied Bioinformatics*; IntechOpen: London, UK, 2011.
95. Sarcar, S.; Tulalamba, W.; Rincon, M.Y.; Tipanee, J.; Pham, H.Q.; Evens, H.; Boon, D.; Samara-Kuko, E.; Keyaerts, M.; Loperfido, M.; et al. Next-Generation Muscle-Directed Gene Therapy by in Silico Vector Design. *Nat. Commun.* **2019**, *10*, 1–16. [CrossRef]
96. Mogno, I.; Vallania, F.; Mitra, R.D.; Cohen, B.A. TATA Is a Modular Component of Synthetic Promoters. *Genome Res.* **2010**, *20*, 1391–1397. [CrossRef]
97. Domenger, C.; Grimm, D. Next-Generation AAV Vectors-Do Not Judge a Virus (Only) by Its Cover. *Hum. Mol. Genet.* **2019**, *28*, R3–R14. [CrossRef]
98. Nguyen, T.A.; Jones, R.D.; Snavely, A.R.; Pfenning, A.R.; Kirchner, R.; Hemberg, M.; Gray, J.M. High-Throughput Functional Comparison of Promoter and Enhancer Activities. *Genome Res.* **2016**, *26*, 1023–1033. [CrossRef] [PubMed]
99. Rushton, P.J. What Have We Learned about Synthetic Promoter Construction? *Methods Mol. Biol.* **2016**, *1482*, 1–13. [PubMed]
100. Aysha, J.; Noman, M.; Wang, F.; Liu, W.; Zhou, Y.; Li, H.; Li, X. Synthetic Promoters: Designing the Cis Regulatory Modules for Controlled Gene Expression. *Mol. Biotechnol.* **2018**, *60*, 608–620. [CrossRef] [PubMed]

101. Cazier, A.P.; Blazeck, J. Advances in Promoter Engineering: Novel Applications and Predefined Transcriptional Control. *Biotechnol. J.* **2021**, *16*, 2100239. [CrossRef] [PubMed]
102. Salva, M.Z.; Himeda, C.L.; Tai, P.W.L.; Nishiuchi, E.; Gregorevic, P.; Allen, J.M.; Finn, E.E.; Nguyen, Q.G.; Blankinship, M.J.; Meuse, L.; et al. Design of Tissue-Specific Regulatory Cassettes for High-Level RAAV-Mediated Expression in Skeletal and Cardiac Muscle. *Mol. Ther.* **2007**, *15*, 320–329. [CrossRef]
103. Lee, Y.; Messing, A.; Su, M.; Brenner, M. GFAP Promoter Elements Required for Region-Specific and Astrocyte-Specific Expression. *Glia* **2008**, *56*, 481–493. [CrossRef]
104. Yasmeen, E.; Wang, J.; Riaz, M.; Zhang, L.; Zuo, K. Designing Artificial Synthetic Promoters for Accurate, Smart, and Versatile Gene Expression in Plants. *Plant Commun.* **2023**, *4*, 100558. [CrossRef]
105. Miao, C.H.; Ohashi, K.; Patijn, G.A.; Meuse, L.; Ye, X.; Thompson, A.R.; Kay, M.A. Inclusion of the Hepatic Locus Control Region, an Intron, and Untranslated Region Increases and Stabilizes Hepatic Factor IX Gene Expression in Vivo but Not in Vitro. *Mol. Ther.* **2000**, *1*, 522–532. [CrossRef]
106. Davidoff, A.M.; Gray, J.T.; Ng, C.Y.C.; Zhang, Y.; Zhou, J.; Spence, Y.; Bakar, Y.; Nathwani, A.C. Comparison of the Ability of Adeno-Associated Viral Vectors Pseudotyped with Serotype 2, 5, and 8 Capsid Proteins to Mediate Efficient Transduction of the Liver in Murine and Nonhuman Primate Models. *Mol. Ther.* **2005**, *11*, 875–888. [CrossRef]
107. Nathwani, A.C.; Gray, J.T.; Ng, C.Y.C.; Zhou, J.; Spence, Y.; Waddington, S.N.; Tuddenham, E.G.D.; Kemball-Cook, G.; McIntosh, J.; Boon-Spijker, M.; et al. Self-Complementary Adeno-Associated Virus Vectors Containing a Novel Liver-Specific Human Factor IX Expression Cassette Enable Highly Efficient Transduction of Murine and Nonhuman Primate Liver. *Blood* **2006**, *107*, 2653–2661. [CrossRef]
108. McIntosh, J.; Lenting, P.J.; Rosales, C.; Lee, D.; Rabbanian, S.; Raj, D.; Patel, N.; Tuddenham, E.G.D.; Christophe, O.D.; McVey, J.H.; et al. Therapeutic Levels of FVIII Following a Single Peripheral Vein Administration of RAAV Vector Encoding a Novel Human Factor VIII Variant. *Blood* **2013**, *121*, 3335–3344. [CrossRef] [PubMed]
109. Ede, C.; Chen, X.; Lin, M.Y.; Chen, Y.Y. Quantitative Analyses of Core Promoters Enable Precise Engineering of Regulated Gene Expression in Mammalian Cells. *ACS Synth. Biol.* **2016**, *5*, 395–404. [CrossRef] [PubMed]
110. Brown, H.C.; Zakas, P.M.; George, S.N.; Parker, E.T.; Spencer, H.T.; Doering, C.B. Target-Cell-Directed Bioengineering Approaches for Gene Therapy of Hemophilia A. *Mol. Ther. Methods Clin. Dev.* **2018**, *9*, 57–69. [CrossRef]
111. Müller, O.J.; Leuchs, B.; Pleger, S.T.; Grimm, D.; Franz, W.M.; Katus, H.A.; Kleinschmidt, J.A. Improved Cardiac Gene Transfer by Transcriptional and Transductional Targeting of Adeno-Associated Viral Vectors. *Cardiovasc. Res.* **2006**, *70*, 70–78. [CrossRef] [PubMed]
112. Yew, N.S.; Wysokenski, D.M.; Wang, K.X.; Ziegler, R.J.; Marshall, J.; Mcneilly, D.; Cherry, M.; Osburn, W.; Cheng, S.H. Optimization of Plasmid Vectors for High-Level Expression in Lung Epithelial Cells. *Hum. Gene Ther.* **1997**, *8*, 575–584. [CrossRef]
113. Liu, Y.L.; Mingozzi, F.; Rodriguéz-Colôn, S.M.; Joseph, S.; Dobrzynski, E.; Suzuki, T.; High, K.A.; Herzog, R.W. Therapeutic Levels of Factor IX Expression Using a Muscle-Specific Promoter and Adeno-Associated Virus Serotype 1 Vector. *Hum. Gene Ther.* **2004**, *15*, 783–792. [CrossRef]
114. Hitoshi, N.; Ken-ichi, Y.; Jun-ichi, M. Efficient Selection for High-Expression Transfectants with a Novel Eukaryotic Vector. *Gene* **1991**, *108*, 193–199. [CrossRef]
115. Li, X.; Eastman, E.M.; Schwartz, R.J.; Draghia-Akli, R. Synthetic Muscle Promoters: Activities Exceeding Naturally Occurring Regulatory Sequences. *Nat. Biotechnol.* **1999**, *17*, 241–245. [CrossRef] [PubMed]
116. Wu, M.R.; Nissim, L.; Stupp, D.; Pery, E.; Binder-Nissim, A.; Weisinger, K.; Enghuus, C.; Palacios, S.R.; Humphrey, M.; Zhang, Z.; et al. A High-Throughput Screening and Computation Platform for Identifying Synthetic Promoters with Enhanced Cell-State Specificity (SPECS). *Nat. Commun.* **2019**, *10*, 1–10. [CrossRef]
117. Shibata, T.; Giaccia, A.J.; Brown, J.M. Development of a Hypoxia-Responsive Vector for Tumor-Specific Gene Therapy. *Gene Ther.* **2000**, *7*, 493–498. [CrossRef]
118. Xia, J.B.; Wu, H.Y.; Lai, B.L.; Zheng, L.; Zhou, D.C.; Chang, Z.S.; Mao, C.Z.; Liu, G.H.; Park, K.S.; Zhao, H.; et al. Gene Delivery of Hypoxia-Inducible VEGF Targeting Collagen Effectively Improves Cardiac Function after Myocardial Infarction. *Sci. Rep.* **2017**, *7*, 1–13. [CrossRef]
119. Lemken, M.L.; Wybranietz, W.A.; Schmidt, U.; Graepler, F.; Armeanu, S.; Bitzer, M.; Lauer, U.M. Expression Liver-Directed Genes by Employing Synthetic Transcriptional Control Units. *World J. Gastroenterol.* **2005**, *11*, 5295–5302. [CrossRef]
120. Maturana, C.J. Engineered Compact Pan-Neuronal Promoter from Alphaherpesvirus LAP2 Enhances Target Gene Expression in the Mouse Brain and Reduces Tropism in the Liver. *Gene Ther.* **2024**, *31*, 335–344. [CrossRef]
121. Korecki, A.J.; Cueva-Vargas, J.L.; Fornes, O.; Agostinone, J.; Farkas, R.A.; Hickmott, J.W.; Lam, S.L.; Mathelier, A.; Zhou, M.; Wasserman, W.W.; et al. Human MiniPromoters for Ocular-RAAV Expression in ON Bipolar, Cone, Corneal, Endothelial, Müller Glial, and PAX6 Cells. *Gene Ther.* **2021**, *28*, 351–372. [CrossRef]
122. Simpson, E.M.; Korecki, A.J.; Fornes, O.; McGill, T.J.; Cueva-Vargas, J.L.; Agostinone, J.; Farkas, R.A.; Hickmott, J.W.; Lam, S.L.; Mathelier, A.; et al. New MiniPromoter Ple345 (NEFL) Drives Strong and Specific Expression in Retinal Ganglion Cells of Mouse and Primate Retina. *Hum. Gene Ther.* **2019**, *30*, 257–272. [CrossRef]
123. Wang, Y.; Wang, H.; Wei, L.; Li, S.; Liu, L.; Wang, X. Synthetic Promoter Design in *Escherichia Coli* Based on a Deep Generative Network. *Nucleic Acids Res.* **2020**, *48*, 6403–6412. [CrossRef]

124. Qiao, H.; Zhang, S.; Xue, T.; Wang, J.; Wang, B. IPro-GAN: A Novel Model Based on Generative Adversarial Learning for Identifying Promoters and Their Strength. *Comput. Methods Programs Biomed.* **2022**, *215*, 106625. [CrossRef]
125. Zhang, P.; Wang, H.; Xu, H.; Wei, L.; Liu, L.; Hu, Z.; Wang, X. Deep Flanking Sequence Engineering for Efficient Promoter Design Using DeepSEED. *Nat. Commun.* **2023**, *14*, 1–14. [CrossRef]
126. Seo, E.; Choi, Y.N.; Shin, Y.R.; Kim, D.; Lee, J.W. Design of Synthetic Promoters for *Cyanobacteria* with Generative Deep-Learning Model. *Nucleic Acids Res.* **2023**, *51*, 7071–7082. [CrossRef]
127. Gosai, S.J.; Castro, R.I.; Fuentes, N.; Butts, J.C.; Mouri, K.; Alasoadura, M.; Kales, S.; Nguyen, T.T.L.; Noche, R.R.; Rao, A.S.; et al. Machine-Guided Design of Cell-Type-Targeting Cis-Regulatory Elements. *Nature* **2024**, *634*, 1211–1220. [CrossRef]
128. Linder, J.; Seelig, G. Fast Activation Maximization for Molecular Sequence Design. *BMC Bioinform.* **2021**, *22*, 1–20. [CrossRef] [PubMed]
129. Wang, H.; Du, Q.; Wang, Y.; Xu, H.; Wei, Z.; Wang, X. GPro: Generative AI-Empowered Toolkit for Promoter Design. *Bioinformatics* **2024**, *40*, btae123. [CrossRef]
130. Belokopytova, P.S.; Nuriddinov, M.A.; Mozheiko, E.A.; Fishman, D.; Fishman, V. Quantitative Prediction of Enhancer–Promoter Interactions. *Genome Res.* **2020**, *30*, 72–84. [CrossRef]
131. Zheng, L.; Liu, L.; Zhu, W.; Ding, Y.; Wu, F. Predicting Enhancer-Promoter Interaction Based on Epigenomic Signals. *Front. Genet.* **2023**, *14*, 1133775. [CrossRef]
132. Noman, M.Z.; Desantis, G.; Janji, B.; Hasmim, M.; Karray, S.; Dessen, P.; Bronte, V.; Chouaib, S. PD-L1 Is a Novel Direct Target of HIF-1α, and Its Blockade under Hypoxia Enhanced: MDSC-Mediated T Cell Activation. *J. Exp. Med.* **2014**, *211*, 781–790. [CrossRef]
133. de Wet, J.R.; Wood, K.V.; DeLuca, M.; Helinski, D.R.; Subramani, S. *Firefly luciferase* Gene: Structure and Expression in Mammalian Cells. *Mol. Cell. Biol.* **1987**, *7*, 725–737. [CrossRef]
134. Cayrou, C.; Bayliss, C.D. Assessment of FHbp Expression Level by Reverse Transcriptase Quantitative PCR and Promoter Sequence Analysis. *Methods Mol. Biol.* **2019**, *1969*, 237–249.
135. Rizzuto, R.; Brini, M.; Pizzo, P.; Murgia, M.; Pozzan, T. Chimeric Green Fluorescent Protein as a Tool for Visualizing Subcellular Organelles in Living Cells. *Curr. Biol.* **1995**, *5*, 635–642. [CrossRef]
136. Kang, J.; Huang, L.; Zheng, W.; Luo, J.; Zhang, X.; Song, Y.; Liu, A. Promoter CAG Is More Efficient than Hepatocyte-Targeting TBG for Transgene Expression via RAAV8 in Liver Tissues. *Mol. Med. Rep.* **2022**, *25*, 1–9. [CrossRef]
137. Mogno, I.; Kwasnieski, J.C.; Cohen, B.A. Massively Parallel Synthetic Promoter Assays Reveal the in Vivo Effects of Binding Site Variants. *Genome Res.* **2013**, *23*, 1908–1915. [CrossRef]
138. Shen, S.Q.; Myers, C.A.; Hughes, A.E.O.; Byrne, L.C.; Flannery, J.G.; Corbo, J.C. Massively Parallel Cis-Regulatory Analysis in the Mammalian Central Nervous System. *Genome Res.* **2016**, *26*, 238–255. [CrossRef]
139. Kuzmin, D.A.; Shutova, M.V.; Johnston, N.R.; Smith, O.P.; Fedorin, V.V.; Kukushkin, Y.S.; van der Loo, J.C.M.; Johnstone, E.C. The Clinical Landscape for AAV Gene Therapies. *Nat. Rev. Drug Discov.* **2021**, *20*, 173–174. [CrossRef]
140. Burdett, T.; Nuseibeh, S. Changing Trends in the Development of AAV-Based Gene Therapies: A Meta-Analysis of Past and Present Therapies. *Gene Ther.* **2023**, *30*, 323–335. [CrossRef] [PubMed]
141. Rheaume, B.A.; Jereen, A.; Bolisetty, M.; Sajid, M.S.; Yang, Y.; Renna, K.; Sun, L.; Robson, P.; Trakhtenberg, E.F. Single Cell Transcriptome Profiling of Retinal Ganglion Cells Identifies Cellular Subtypes. *Nat. Commun.* **2018**, *9*, 2759. [CrossRef]
142. Macosko, E.Z.; Basu, A.; Satija, R.; Nemesh, J.; Shekhar, K.; Goldman, M.; Tirosh, I.; Bialas, A.R.; Kamitaki, N.; Martersteck, E.M.; et al. Highly Parallel Genome-Wide Expression Profiling of Individual Cells Using Nanoliter Droplets. *Cell* **2015**, *161*, 1202–1214. [CrossRef] [PubMed]
143. Darmanis, S.; Sloan, S.A.; Zhang, Y.; Enge, M.; Caneda, C.; Shuer, L.M.; Gephart, M.G.H.; Barres, B.A.; Quake, S.R. A Survey of Human Brain Transcriptome Diversity at the Single Cell Level. *Proc. Natl. Acad. Sci. USA* **2015**, *112*, 7285–7290. [CrossRef]
144. Raj, B.; Wagner, D.E.; McKenna, A.; Pandey, S.; Klein, A.M.; Shendure, J.; Gagnon, J.A.; Schier, A.F. Simultaneous Single-Cell Profiling of Lineages and Cell Types in the Vertebrate Brain. *Nat. Biotechnol.* **2018**, *36*, 442–450. [CrossRef] [PubMed]
145. Visel, A.; Taher, L.; Girgis, H.; May, D.; Golonzhka, O.; Hoch, R.V.; McKinsey, G.L.; Pattabiraman, K.; Silberberg, S.N.; Blow, M.J.; et al. A High-Resolution Enhancer Atlas of the Developing Telencephalon. *Cell* **2013**, *152*, 895–908. [CrossRef]
146. Silberberg, S.N.; Taher, L.; Lindtner, S.; Sandberg, M.; Nord, A.S.; Vogt, D.; Mckinsey, G.L.; Hoch, R.; Pattabiraman, K.; Zhang, D.; et al. Subpallial Enhancer Transgenic Lines: A Data and Tool Resource to Study Transcriptional Regulation of GABAergic Cell Fate. *Neuron* **2016**, *92*, 59–74. [CrossRef]
147. Jüttner, J.; Szabo, A.; Gross-Scherf, B.; Morikawa, R.K.; Rompani, S.B.; Hantz, P.; Szikra, T.; Esposti, F.; Cowan, C.S.; Bharioke, A.; et al. Targeting Neuronal and Glial Cell Types with Synthetic Promoter AAVs in Mice, Non-Human Primates and Humans. *Nat. Neurosci.* **2019**, *22*, 1345–1356. [CrossRef]
148. Markenscoff-Papadimitriou, E.; Whalen, S.; Przytycki, P.; Thomas, R.; Binyameen, F.; Nowakowski, T.J.; Kriegstein, A.R.; Sanders, S.J.; State, M.W.; Pollard, K.S.; et al. A Chromatin Accessibility Atlas of the Developing Human Telencephalon. *Cell* **2020**, *182*, 754–769. [CrossRef] [PubMed]
149. Mich, J.K.; Graybuck, L.T.; Hess, E.E.; Mahoney, J.T.; Kojima, Y.; Ding, Y.; Somasundaram, S.; Miller, J.A.; Kalmbach, B.E.; Radaelli, C.; et al. Functional Enhancer Elements Drive Subclass-Selective Expression from Mouse to Primate Neocortex. *Cell Rep.* **2021**, *34*, 108754. [CrossRef] [PubMed]

150. Goldberg, E.M. Getting a Foot IN the Door: GABAergic INterneuron-Specific Enhancers. *Epilepsy Curr.* **2021**, *21*, 114–116. [CrossRef]
151. Vormstein-Schneider, D.; Lin, J.D.; Pelkey, K.A.; Chittajallu, R.; Guo, B.; Arias-Garcia, M.A.; Allaway, K.; Sakopoulos, S.; Schneider, G.; Stevenson, O.; et al. Viral Manipulation of Functionally Distinct Interneurons in Mice, Non-Human Primates and Humans. *Nat. Neurosci.* **2020**, *23*, 1629–1636. [CrossRef]
152. Schiedner, G.; Morral, N.; Parks, R.J.; Wu, Y.; Koopmans, S.C.; Langston, C.; Graham, F.L.; Beaudet, A.L.; Kochanek, S. Genomic DNA Transfer with a High-Capacity Adenovirus Vector Results in Improved in Vivo Gene Expression and Decreased Toxicity. *Nat. Genet.* **1998**, *18*, 180–183. [CrossRef]
153. FDA. Available online: https://www.fda.gov/ (accessed on 21 October 2024).
154. Hauser, M.A.; Robinson, A.; Hartigan-O'Connor, D.; Williams-Gregory, D.A.; Buskin, J.N.; Apone, S.; Kirk, C.J.; Hardy, S.; Hauschka, S.D.; Chamberlain, J.S. Analysis of Muscle Creatine Kinase Regulatory Elements in Recombinant Adenoviral Vectors. *Mol. Ther.* **2000**, *2*, 16–25. [CrossRef]
155. Martari, M.; Sagazio, A.; Mohamadi, A.; Nguyen, Q.; Hauschka, S.D.; Kim, E.; Salvatori, R. Partial Rescue of Growth Failure in Growth Hormone (GH)-Deficient Mice by a Single Injection of a Double-Stranded Adeno-Associated Viral Vector Expressing the *GH* Gene Driven by a Muscle-Specific Regulatory Cassette. *Hum. Gene Ther.* **2009**, *20*, 759–766. [CrossRef]
156. Wang, B.; Li, J.; Fu, F.H.; Chen, C.; Zhu, X.; Zhou, L.; Jiang, X.; Xiao, X. Construction and Analysis of Compact Muscle-Specific Promoters for AAV Vectors. *Gene Ther.* **2008**, *15*, 1489–1499. [CrossRef] [PubMed]
157. Skopenkova, V.V.; Egorova, T.V.; Bardina, M.V. Muscle-Specific Promoters for Gene Therapy. *Acta Naturae* **2021**, *13*, 47–58. [CrossRef]
158. Piekarowicz, K.; Bertrand, A.T.; Azibani, F.; Beuvin, M.; Julien, L.; Machowska, M.; Bonne, G.; Rzepecki, R. A Muscle Hybrid Promoter as a Novel Tool for Gene Therapy. *Mol. Ther. Methods Clin. Dev.* **2019**, *15*, 157–169. [CrossRef]
159. Georgiadis, A.; Duran, Y.; Ribeiro, J.; Abelleira-Hervas, L.; Robbie, S.J.; Sünkel-Laing, B.; Fourali, S.; Gonzalez-Cordero, A.; Cristante, E.; Michaelides, M.; et al. Development of an Optimized AAV2/5 Gene Therapy Vector for Leber Congenital Amaurosis Owing to Defects in RPE65. *Gene Ther.* **2016**, *23*, 857–862. [CrossRef] [PubMed]
160. Ye, G.J.; Budzynski, E.; Sonnentag, P.; Nork, T.M.; Sheibani, N.; Gurel, Z.; Boye, S.L.; Peterson, J.J.; Boye, S.E.; Hauswirth, W.W.; et al. Cone-Specific Promoters for Gene Therapy of Achromatopsia and Other Retinal Diseases. *Hum. Gene Ther.* **2016**, *27*, 72–82. [CrossRef]
161. Khabou, H.; Garita-Hernandez, M.; Chaffiol, A.; Reichman, S.; Jaillard, C.; Brazhnikova, E.; Bertin, S.; Forster, V.; Desrosiers, M.; Winckler, C.; et al. Noninvasive Gene Delivery to Foveal Cones for Vision Restoration. *JCI Insight* **2018**, *3*, e96029. [CrossRef]
162. Kralik, J.; van Wyk, M.; Stocker, N.; Kleinlogel, S. Bipolar Cell Targeted Optogenetic Gene Therapy Restores Parallel Retinal Signaling and High-Level Vision in the Degenerated Retina. *Commun. Biol.* **2022**, *5*, 1–15. [CrossRef] [PubMed]
163. Bi, A.; Cui, J.; Ma, Y.P.; Olshevskaya, E.; Pu, M.; Dizhoor, A.M.; Pan, Z.H. Ectopic Expression of a Microbial-Type Rhodopsin Restores Visual Responses in Mice with Photoreceptor Degeneration. *Neuron* **2006**, *50*, 23–33. [CrossRef]
164. Lagali, P.S.; Balya, D.; Awatramani, G.B.; Münch, T.A.; Kim, D.S.; Busskamp, V.; Cepko, C.L.; Roska, B. Light-Activated Channels Targeted to ON Bipolar Cells Restore Visual Function in Retinal Degeneration. *Nat. Neurosci.* **2008**, *11*, 667–675. [CrossRef]
165. Busskamp, V.; Picaud, S.; Sahel, J.A.; Roska, B. Optogenetic Therapy for Retinitis Pigmentosa. *Gene Ther.* **2012**, *19*, 169–175. [CrossRef] [PubMed]
166. Hulliger, E.C.; Hostettler, S.M.; Kleinlogel, S. Empowering Retinal Gene Therapy with a Specific Promoter for Human Rod and Cone ON-Bipolar Cells. *Mol. Ther. Methods Clin. Dev.* **2020**, *17*, 505–519. [CrossRef]
167. Nieuwenhuis, B.; Haenzi, B.; Hilton, S.; Carnicer-Lombarte, A.; Hobo, B.; Verhaagen, J.; Fawcett, J.W. Optimization of Adeno-Associated Viral Vector-Mediated Transduction of the Corticospinal Tract: Comparison of Four Promoters. *Gene Ther.* **2021**, *28*, 56–74. [CrossRef]
168. Matsuzaki, Y.; Oue, M.; Hirai, H. Generation of a Neurodegenerative Disease Mouse Model Using Lentiviral Vectors Carrying an Enhanced Synapsin I Promoter. *J. Neurosci. Methods* **2014**, *223*, 133–143. [CrossRef]
169. Rubin, A.N.; Malik, R.; Cho, K.K.A.; Lim, K.J.; Lindtner, S.; Schwartz, S.E.R.; Vogt, D.; Sohal, V.S.; Rubenstein, J.L.R. Regulatory Elements Inserted into Aavs Confer Preferential Activity in Cortical Interneurons. *eNeuro* **2020**, *7*, ENEURO.0211-20.2020. [CrossRef] [PubMed]
170. Chen, C.; Yue, D.; Lei, L.; Wang, H.; Lu, J.; Zhou, Y.; Liu, S.; Ding, T.; Guo, M.; Xu, L. Promoter-Operating Targeted Expression of Gene Therapy in Cancer: Current Stage and Prospect. *Mol. Ther. Nucleic Acids* **2018**, *11*, 508–514. [CrossRef] [PubMed]
171. Montaño-Samaniego, M.; Bravo-Estupiñan, D.M.; Méndez-Guerrero, O.; Alarcón-Hernández, E.; Ibáñez-Hernández, M. Strategies for Targeting Gene Therapy in Cancer Cells With Tumor-Specific Promoters. *Front. Oncol.* **2020**, *10*, 605380. [CrossRef]
172. Liu, S.H.; Yu, J.; Sanchez, R.; Liu, X.; Heidt, D.; Willey, J.; Nemunaitis, J.; Brunicardi, F.C. A Novel Synthetic Human Insulin Super Promoter for Targeting PDX-1-Expressing Pancreatic Cancer. *Cancer Lett.* **2018**, *418*, 75–83. [CrossRef]
173. Reddy, A.J.; Geng, X.; Herschl, M.H.; Kolli, S.; Kumar, A.; Hsu, P.D.; Levine, S.; Ioannidis, N.M. Designing Cell-Type-Specific Promoter Sequences Using Conservative Model-Based Optimization. *bioRxiv* **2024**. [CrossRef]
174. Le Guiner, C.; Stieger, K.; Toromanoff, A.; Guilbaud, M.; Mendes-Madeira, A.; Devaux, M.; Guigand, L.; Cherel, Y.; Moullier, P.; Rolling, F.; et al. Transgene Regulation Using the Tetracycline-Inducible TetR-KRAB System after AAV-Mediated Gene Transfer in Rodents and Nonhuman Primates. *PLoS ONE* **2014**, *9*, e102538. [CrossRef] [PubMed]

175. Sohn, J.; Takahashi, M.; Okamoto, S.; Ishida, Y.; Furuta, T.; Hioki, H. A Single Vector Platform for High-Level Gene Transduction of Central Neurons: Adeno-Associated Virus Vector Equipped with the Tet-off System. *PLoS ONE* **2017**, *12*, e0169611. [CrossRef]
176. O'Callaghan, J.; Crosbie, D.E.; Cassidy, P.S.; Sherwood, J.M.; Flügel-Koch, C.; Lütjen-Drecoll, E.; Humphries, M.M.; Reina-Torres, E.; Wallace, D.; Kiang, A.S.; et al. Therapeutic Potential of AAV-Mediated MMP-3 Secretion from Corneal Endothelium in Treating Glaucoma. *Hum. Mol. Genet.* **2017**, *26*, 1230–1246. [CrossRef] [PubMed]
177. Bhatt, B.; García-Díaz, P.; Foight, G.W. Synthetic Transcription Factor Engineering for Cell and Gene Therapy. *Trends Biotechnol.* **2024**, *42*, 449–463. [CrossRef]
178. Galvan, S.; Madderson, O.; Xue, S.; Teixeira, A.P.; Fussenegger, M. Regulation of Transgene Expression by the Natural Sweetener Xylose. *Adv. Sci.* **2022**, *9*, e2203193. [CrossRef]
179. uniQure. Glybera (Alipogene Tiparvovec). Available online: https://www.ema.europa.eu/en/medicines/human/EPAR/glybera (accessed on 21 October 2024).
180. Pfizer Inc. BEQVEZ (Fidanacogene Elaparvovec-Dzkt). Available online: https://www.fda.gov/vaccines-blood-biologics/cellular-gene-therapy-products/beqvez (accessed on 21 October 2024).

cells

MDPI

Review

Prime Editing: Mechanistic Insights and DNA Repair Modulation

Astrid Mentani *, Marcello Maresca and Anna Shiriaeva *

Genome Engineering, Discovery Science, BioPharmaceuticals R&D, AstraZeneca, 43183 Mölndal, Sweden; marcello.maresca@astrazeneca.com
* Correspondence: astrid.mentani@astrazeneca.com (A.M.); anna.shiriaeva@astrazeneca.com (A.S.)

Abstract: Prime editing is a genome editing technique that allows precise modifications of cellular DNA without relying on donor DNA templates. Recently, several different prime editor proteins have been published in the literature, relying on single- or double-strand breaks. When prime editing occurs, the DNA undergoes one of several DNA repair pathways, and these processes can be modulated with the use of inhibitors. Firstly, this review provides an overview of several DNA repair mechanisms and their modulation by known inhibitors. In addition, we summarize different published prime editors and provide a comprehensive overview of associated DNA repair mechanisms. Finally, we discuss the delivery and safety aspects of prime editing.

Keywords: Cas9; prime editing; DNA repair mechanisms; DNA repair modulation; homologous recombination; non-homologous end joining; microhomology-mediated end joining; single-strand annealing

1. Introduction

For many decades, scientists have aspired to introduce targeted, precise changes into the human genome. With the development of zinc finger nucleases (ZFNs) and transcription activator-like effector nucleases (TALENs), it became evident that this dream could become a reality [1,2]. However, the breakthrough occurred only after 2012, when the mechanism of CRISPR-Cas-mediated immunity was discovered in bacteria *Streptococcus pyogenes* and *Streptococcus thermophilus* [3,4]. The CRISPR-Cas system in these bacteria (later classified as type II-A) is an example of the diverse prokaryotic CRISPR-Cas systems, which include at least six types and thirty-three subtypes [5]. Cas9 is central to the mechanism of the type II CRISPR-Cas immunity. This protein binds to a short CRISPR RNA (crRNA) molecule, which contains a 20 nt sequence originating from a mobile genetic element. Guided by this sequence, Cas9 finds a complementary target in the DNA of an invading bacteriophage or plasmid and introduces a double-strand break (DSB) [3,4]. In bacteria, crRNA functions in tandem with another RNA molecule called tracrRNA. Jinek et al. demonstrated that crRNA and tracrRNA can be fused into a single guide RNA (sgRNA), providing researchers with a simple, programmable, two-component Cas9-sgRNA tool for the targeted introduction of DSBs into a genome of virtually any organism [3].

Knocking out genes using Cas9-sgRNA has already proven to be an efficient therapeutic strategy. Following the successful results of ongoing clinical trials [6–8], the first CRISPR-Cas9 knockout-based therapy, Casgevy (exagamglogene autotemcel), received regulatory approval for the treatment of transfusion-dependent β-thalassemia and sickle cell disease in 2023 [9]. Although Cas9-mediated knockouts can be relatively easily established, reverting pathogenic variants to a wild-type sequence is much more challenging due to several competing double-strand break repair (DSBR) pathways present in human cells. Numerous studies have identified key proteins involved in DNA repair pathways. Using this knowledge, it is possible to modulate the cellular DNA damage response in favor of a desired editing outcome. In the first part of this review, we provide an overview of DSBR in human cells and possible ways to promote the desired editing outcome. In the second part, we shift our focus to prime editing (PE)—an alternative gene editing strategy relying on a Cas9 nickase instead of a nuclease. We then discuss how the method has evolved in recent years and review the DNA repair factors involved with a focus on mismatch repair proteins.

Finally, we discuss the use of Cas9 nuclease in PE, the manipulation of DSBR to promote precise Cas9 nuclease-mediated PE, various delivery methods and the safety of this approach.

2. DSBR Pathways and Genome Editing Mediated by Homology-Directed Repair

2.1. DSBR in Human Cells

A DSB in human cells is repaired via one of four DSBR pathways, called homologous recombination (HR), canonical or classical non-homologous end joining (c-NHEJ), microhomology-mediated end joining (MMEJ, also called alt-EJ) and single-strand annealing (SSA) (Figure 1) [10–14]. HR can be further divided into three subpathways: the double Holliday junction pathway (dHJ), break-induced replication (BIR) and synthesis-dependent strand annealing (SDSA). In addition to HR, a related term, homology-directed repair (HDR), is often used in the literature. While some authors use this term interchangeably with HR, others define HR as a broader term relative to HDR or vice versa. Throughout this review, we will use 'HR' as a term referring to the mechanisms of DSBR and 'HDR' as a term describing any known or unknown homology-directed pathway for gene editing with exogeneous templates.

2.2. Homologous Recombination (HR)

HR is a DSBR pathway in which a homologous sequence is used as a template to extend one or both 3′ ends of a broken chromosome such that they gain sufficient homology between the two ends (Figure 1A). This allows for the two halves to reanneal and complete the repair without the loss of the sequence at the site of the break.

HR starts with the resection of 5′ ends by the MRN complex (MRE11, RAD50 and NBS1) [15,16], assisted by several nucleases and regulated by multiple factors, including CtIP, BRCA1 and BARD1 [17–25]. CtIP activates MRE11 endonuclease activity [26]. MRE11 incises the 5′-terminated strand at up to several hundred nucleotides from the break and degrades the cleaved strand through its 3′-5′ exonuclease activity in a process called short-range resection [26,27]. The short-range resection is followed by long-range resection, which removes up to several kilobases of DNA due to the coordinated action of exonucleases (EXO1, DNA2) and helicases (BLM, WRN) [21,28–30]. The process is directly stimulated by BRCA1-BARD1 [17,18]. Another important function of the BRCA1-BARD1 complex is to prevent the loading of the c-NHEJ factor 53BP1 (see below), and thus commit the cell to HR [31,32]. The resection step generates a nucleoprotein filament consisting of 3′-tailed ssDNA coated with RAD51 recombinase [33]. Since naked ssDNA is rapidly bound by the single-strand binding protein RPA, mediator proteins are required to replace RPA with RAD51 [34]. BRCA2 is the key mediator protein in human cells promoting nucleoprotein filament assembly [35]. It is recruited to a DSB via its partner PALB2, which is, in turn, recruited to BRCA1 [36–38]. Next, the nucleoprotein filament catalyzes the homology search and strand invasion, resulting in a displacement loop (D-loop) in a process stimulated by RAD54, PALB2, BRCA1-BARD1 and RAD51AP1-UAF1 [39–46]. Although the two ends of the DSB are processed in a similar manner, only one forms the D-loop [47]. The D-loop is then extended by a DNA polymerase capable of performing displacement synthesis within the D-loop, likely POL δ, with possible contributions from other DNA polymerases [48]. Depending on the subsequent steps, HR can be divided into three subpathways: the double Holliday junction pathway (dHJ), break-induced replication (BIR) and synthesis-dependent strand annealing (SDSA) [10]. SDSA is the predominant mechanism in mitotic cells. In this pathway, the D-loop disintegrates quickly and the extended 3′ end anneals to the non-extended 3′ end on the other side of the break [49,50]. The dHJ pathway occurs when both 3′ ends pair with a homologous sequence, prime DNA synthesis and are ligated to form classic dHJ structures that can be resolved into crossover or non-crossover products [51]. BIR happens when one of the two parts of the broken chromosome is inaccessible., and D-loop formation is followed by replisome assembly and replication through the rest of the chromosome [52,53].

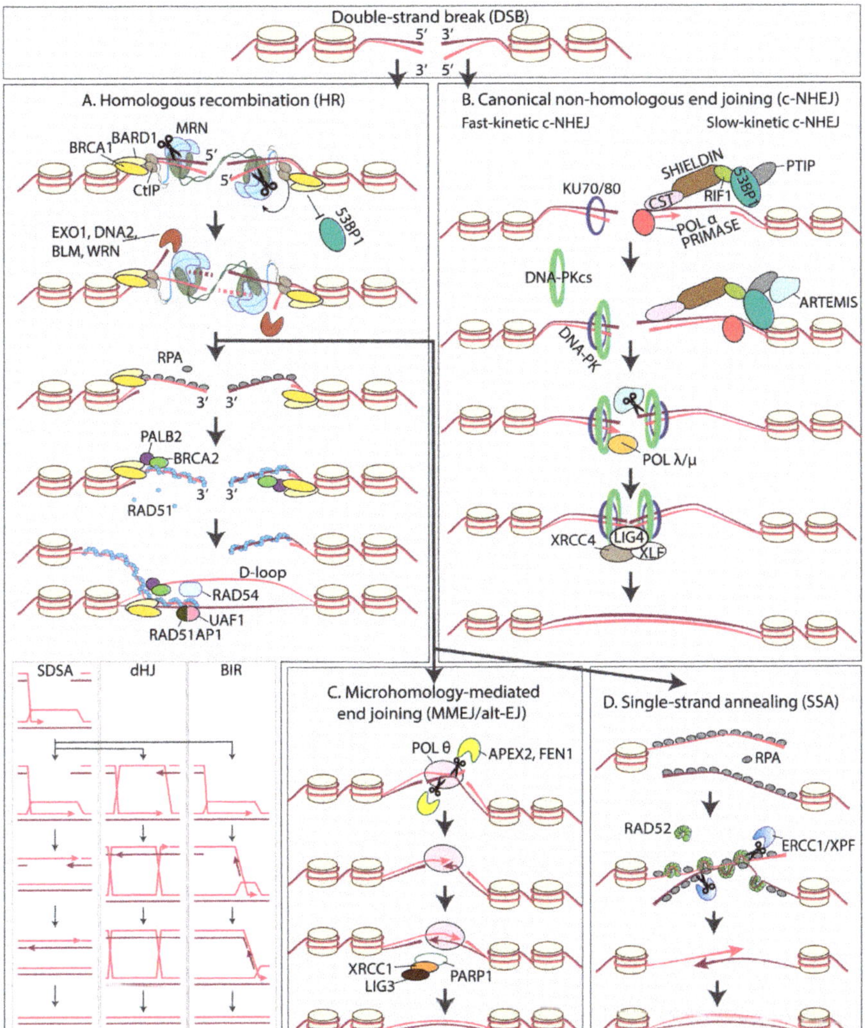

Figure 1. Double-strand break (DSB) repair pathways in human cells. (**A**) Homologous recombination (HR). A DSB is recognized by the MRN complex. Activated by CtIP, the MRN complex incises the

5′-terminated strand at some distance from the DSB and degrades the 5′-terminal part due to MRE11 3′-5′ exonuclease activity. Such 'short-range resection' is followed by 'long-range resection' performed by EXO1 and DNA2 exonucleases and BLM and WRN helicases. The BRCA1-BARD1 complex recognizes histone modifications specific to DSBs, stimulates resection, prevents classical non-homologus end joining (c-NHEJ) factor 53BP1 from loading onto the chromatin and recruits PALB2-BRCA2. RPA binds to the generated 3′ overhang. PALB2-BRCA2 stimulates the replacement of RPA with RAD51 recombinase. The RAD51 filament invades a homologous donor sequence, resulting in the formation of a displacement loop (D-loop) with the assistance of RAD54, PALB2-BRCA2, BRCA1-BARD1 and RAD51AP1-UAF1. The 3′ end is then extended by a DNA polymerase. The subsequent process is subdivided into synthesis-dependent strand annealing (SDSA), a double Holliday junction pathway (dHJ) and break-induced replication (BIR) depending on the presence of one or two ends and the interaction between the ends and the donor. (**B**) c-NHEJ. Depending on the complexity of the ends, c-NHEJ proceeds with fast or slow kinetics. Fast-kinetic c-NHEJ starts with the binding of the KU70/80 heterodimer to a blunt end or an end with a relatively short overhang. KU70/80 activates the DNA-dependent protein kinase catalytic subunit (DNA-PKcs), and, together, they form DNA-PK. In the case of slow-kinetic c-NHEJ, additional factors are necessary to counteract end resection and prepare the end for ligation. 53BP1 recognizes histone modifications specific to DSBs. This leads to the recruitment of additional factors: RIF1, SHIELDIN, CST, POL α/primase, PTIP and ARTEMIS. POL α/primase fills the 3′ overhangs. ARTEMIS nuclease removes the 3′ overhangs (or other overhangs if present). DNA POL λ or μ also contributes to the generation of compatible ends. The subsequent ligation is performed by the XRCC4-LIG4/XLF complex. (**C**) Following end resection, a DSB may be repaired via microhomology-mediated end joining (MMEJ), also called alternative end joining (alt-EJ). DNA POL θ promotes the annealing of the two 3′ ends due to microhomologies which are several nucleotides long. APEX2 and FEN1 remove 3′ flaps, preparing 3′ ends to be extended by POL θ. Single-strand breaks are ligated by LIG3/XRCC1/PARP1. (**D**) Single-strand annealing (SSA) is associated with hyper-resection of 5′ ends. The two 3′ ends are annealed through an extensive homology region mediated by RAD52 and RPA. 3′ flaps are removed by the ERCC1/XPF endonuclease. The process is finalized by gap filling and ligation, although it is not clear which DNA polymerase and DNA ligase are involved.

In most cases, a sister chromatid is used as a donor in HR and the pathway is considered error-free compared to other DSBR pathways [54]. Nevertheless, HR can also be mutagenic to some extent due to possible genome rearrangements and small-scale mutations. Genome rearrangements arise from erroneous donor choice, while the increased rate of polymerase errors and the intrinsic chemical instability of ssDNA may lead to small-scale mutations [48,55].

2.3. *Canonical Non-Homologous End Joining (c-NHEJ)*

c-NHEJ is a predominant DSBR pathway in human cells (Figure 1B). Fundamentally, it centers around the re-ligation of broken DNA. Ligatable DNA ends can be directly processed by core c-NHEJ factors KU70/80, DNA-PKcs, XRCC4, XLF and DNA ligase 4 (LIG4) in a process starting within seconds after DSB formation [56–58]. The KU70/80 heterodimer is a DSB sensor that tightly encircles DNA and serves as a recruitment point for downstream c-NHEJ effectors [59]. Among these effectors is DNA-PKcs, a constituent of the phosphoinositide 3-kinase-related kinase family. DNA-PKcs, together with DNA-Ku complexes, forms the DNA-PK enzyme [60]. The kinase activity of DNA-PKcs is stimulated upon DNA binding, promoting the phosphorylation of various DNA repair proteins and the autophosphorylation of DNA-PKcs [61].

The ligation step requires precise alignment of DNA ends within a synaptic complex, comprised of two DNA ends, two KU70/80 heterodimers and two DNA-PKcs molecules. The synaptic complex adopts long-range and short-range conformations [62]. The long-range synaptic complex brings the two DNA ends into proximity, while, in the short-range synaptic complex, the DNA ends are aligned for ligation [63]. The kinase activity of DNA-PKcs, along with XRCC4-LIG4 and XLF factors, is required for the transition from the long-range to the short-range complex [62]. A single LIG4 molecule, stimulated by XRCC4,binds to both DNA termini and catalyzes the ligation of compatible ends immediately after the formation of the short-range complex,. This minimizes error-prone processing by other factors [64,65]. Interestingly, several studies show that LIG4 can tolerate certain terminal mismatches and damaged bases, a unique feature among vertebrate ligases [66,67].

DSBs that cannot be directly ligated may still be processed through a slow-kinetic subpathway of c-NHEJ, involving the activation of complex signaling pathways and the recruitment of multiple proteins [56,68,69]. The key player of slow-kinetic cNHEJ is 53BP1, also known as TP53BP1 [70]. 53BP1 recruits additional effectors, such as RIF1, SHIELDIN, CST and POL α/primase [31,71–77]. These proteins prevent the loading of HR factors and counteract the 5′ end resection, which is crucial for HR [31,71–83]. While 53BP1/RIF1/SHIELDIN/CST/POL α/primase largely counteract 5′ end resection, limited end trimming may still be necessary for c-NHEJ to generate ligatable ends. 53BP1 recruits PTIP [78], which, in turn, recruits ARTEMIS [84], a nuclease with both exo- and endonuclease activities [85]. The ARTEMIS nuclease activity requires the presence of KU70/80 and autophosphorylated DNA-PKcs [86]. ARTEMIS can remove both 3′ and 5′ overhangs, although only the processing of 5′ overhangs produces perfectly blunt ends [87]. DNA POL λ or POL μ carries out additional template-dependent or template-independent synthesis at 3′ termini to promote DNA-end synapsis and ligation [88].

2.4. Microhomology-Mediated End Joining (MMEJ or alt-EJ)

Although c-NHEJ is active in most cell types, early studies of c-NHEJ-deficient cells identified an alternative error-prone mechanism, often referred to as alt-EJ or microhomology-mediated end joining (MMEJ) (Figure 1C) [89,90]. Similarly to HR, MMEJ starts with the resection of 5′-terminated strands on both sides of the break, mediated by the MRN complex and CtIP, which is crucial for exposing microhomology-containing single-stranded regions [15,16,26]. Next, the two 3′ ends are aligned through the annealing of the exposed microhomologies, which can be as short as a few nucleotides (nt) in mammals [91]. If microhomologies are located at some distance from the break site, end bridging generates 3′ flaps that must be removed to complete the repair, followed by gap filling and ligation [89,90]. As a result, MMEJ is associated with deletions flanking the original DSB. The deleted region includes one of the two microhomologies and the region between them. Due to its mutagenic characteristics, MMEJ contributes to the plasticity of the genome but can also lead to chromosomal translocations, telomere fusions and carcinogenesis [16,89,90,92].

Poly ADP-ribose polymerase 1 (PARP1) and DNA polymerase theta (POL θ) are the key MMEJ factor [93–95]. PARP1 competes with KU for DNA end binding [96], promotes MRN recruitment [97] and facilitates end synapsis [98]. In addition, a recent study demonstrated the direct poly-(ADP)-ribosylation (PARylation) of POL θ by PARP1, which facilitates POL θ recruitment to DSBs [99].

POL θ is a multifunctional enzyme composed of an A-family DNA polymerase domain and an SF2 helicase-like domain separated by a large, unstructured central domain [100]. The helicase domain of POL θ promotes annealing [101], unwinds DNA with 3′-5′ polarity and facilitates strand displacement synthesis by the polymerase domain [102]. The low-fidelity polymerase domain fills in the gaps after the alignment of the 3′ ends and can accommodate various DNA structures including mismatched termini and ssDNA [95,103–106]. The 3′ flaps produced during MMEJ are removed by APEX2 and FEN1 [95,105,106]. The final ligation steps conclude the MMEJ repair pathway. In contrast to c-NHEJ, which relies on LIG4, LIG3 is the major contributor in MMEJ [98]. LIG3 forms a stable complex with the scaffold protein XRCC1, and both proteins interact with PARP1 [107,108].

2.5. Single-Strand Annealing (SSA)

Single-strand annealing (SSA) is the fourth DSBR pathway, discovered over 40 years ago but not studied as extensively as HR or c-NHEJ (Figure 1C) [109]. Mechanistically, SSA resembles MMEJ but relies on long direct repeats instead of microhomologies. It is usually associated with a higher degree of end resection and requires a different set of proteins [110,111]. SSA depends on resection factors MRE11, CtIP, EXO1 and DNA2 [112]. A characteristic feature of SSA is the annealing of 3′ ends by RAD52 in the presence of RPA [113,114]. In line with other studies, a recent cryo-EM structure [115] demonstrated that RAD52 forms a ring composed of ~10 subunits that binds ssDNA on the outer surface [113,116–120]. The structure also revealed a single RPA protein at the site of ring opening [115]. Based on observations that RPA stimulates SSA, excess RAD52 inhibits the reaction and there is no apparent interaction between RAD52 rings, the authors proposed a model of SSA in which RAD52 sporadically binds to RPA-coated ssDNA, replacing some of the RPA molecules on each of the two ends. Subsequently, the annealing of the two ends is promoted by the interaction between the loaded RAD52 rings on one strand and the remaining RPA on the other strand [115].

The annealing is followed by the removal of 3′ flaps by the ERCC1/XPF endonuclease [121–123]. The process concludes with gap filling and ligation, although it remains unclear which DNA polymerase and DNA ligase are involved [48].

2.6. DBSR Pathway Choice

c-NHEJ and HR are considered the major DSBR pathways. Their activity is regulated in many ways but is primarily controlled by the cell cycle [10,124]. In G1, c-NHEJ is a predominant pathway, whereas HR is largely considered inactive [125–127]. An exception to this rule is the demonstration of HR in G1 within highly repetitive ribosomal genes [128] and centromeric regions [129].

In S phase, both HR and c-NHEJ are active and contribute to the repair of two-ended DSBs [126,127]. At the same time, one-ended DSBs caused by replication fork blockage are repaired by HR but not by c-NHEJ [126,130]. In G2 phase, both c-NHEJ and HR are active [125]. In general, HR is considered to contribute less than c-NHEJ in G2. However, the exact proportion depends on the chromatin state and the complexity of the DNA ends induced by various damaging agents [131].

Repair kinetics studies suggest that c-NHEJ can be categorized into two types: fast-kinetic c-NHEJ, which is completed within 2–4 h after the break occurs, and slow-kinetic c-NHEJ, which operates over a timescale of approximately 24 h [125,131]. Fast-kinetic c-NHEJ is a predominant pathway in the G1 and G2 phases and accounts for the repair of ~70–80% of DSBs. The remaining DSBs are believed to be repaired by slow-kinetic c-NHEJ in G1 and by HR in G2, respectively [131,132]. Fast-kinetic c-NHEJ requires only the core c-NHEJ proteins, while slow-kinetic c-NHEJ is characterized by the recruitment of additional proteins, such as 53BP1 and ARTEMIS [133]. HR can be solely described by slow kinetics, operating on a timescale of ~24 h [125,134].

In the context of gene editing using SpCas9, repair half-life times ranging from 1.4 to 10.7 h have been demonstrated for different targets in K562 cells [135]. Interestingly, experimental data did not support a hypothesis of multiple rounds of SpCas9 cleavage and error-free repair until insertions or deletions (indels) are eventually generated. Instead, evidence supporting slow, error-prone repair was obtained [135]. This was an unexpected conclusion, given that SpCas9 produces blunt or nearly blunt ligatable ends and HR is active in K562 cells. A possible explanation for this is that SpCas9 remains bound to its target after cleavage [136,137]. This, in theory, may preclude fast-kinetic c-NHEJ and direct repair toward slow-kinetic c-NHEJ with the recruitment of additional factors that promote SpCas9 eviction and concomitant indels.

While c-NHEJ and HR can potentially result in error-free DSBR, SSA and MMEJ are inherently error-prone and must be tightly regulated. SSA is predominantly active during the S/G2 phase [122], consistent with its reliance on BRCA1, CtIP and the extensive end resection performed by HR proteins [110,112,138,139].

Mutations in BRCA1 that prevent its binding to PALB2 promote SSA [140]. SSA is also increased when RAD51 or BRCA2 is impaired [138]. Altogether, these findings suggest that SSA regulation throughout the cell cycle is similar to HR, but SSA is typically suppressed by the formation of the RAD51 filament.

MMEJ typically requires less extensive resection compared to SSA and HR, but it still relies on MRN and CtIP [110,141]. However, MMEJ is independent of BRCA1 and long-range resection factors BLM and EXO1 [25,141], meaning that unlike SSA, the decision between HR and MMEJ occurs during the resection stage. Experiments with synchronized cells have shown that MMEJ activity is lowest in G0/G1, gradually increases throughout the cell cycle and peaks in cells arrested in early mitosis [141]. In fact, MMEJ becomes the dominant DSBR pathway during mitosis when c-NHEJ and HR are repressed [142,143]. It is believed that MMEJ evolved to repair condensed chromosomes before cell division, thereby preventing genome instability at the cost of small deletions.

The network of DSBR pathways is highly complex, involving complicated signaling and chromatin remodeling that falls beyond the scope of this review. Additionally, there are multiple connections between different pathways, such as c-NHEJ proteins contributing to HR [125] and, vice versa, HR proteins contributing to c-NHEJ [133]. Numerous studies explore how to manipulate cells into selecting a specific DSBR pathway, a topic that will be discussed in the next chapter.

3. Inhibitors of DNA Repair Mechanisms in Gene Editing

3.1. HDR and Inhibitors of c-NHEJ and MMEJ in Gene Editing

While Cas9 and sgRNA alone are enough for disrupting genes, an exogenous double-stranded or single-stranded donor DNA molecule is required for targeted insertions (Figure 2). Double-stranded DNA or single-stranded oligodeoxyribonucleotides (ssODNs) can be used, but the latter approach is more popular due to its higher efficiency [137,144,145]. HDR with dsDNA templates uses the RAD51-dependent HR pathway discussed in Chapter 2, while HDR with ssDNA is RAD51-independent and involves proteins from the Fanconi Anemia pathway through an unknown mechanism [137]. Both approaches can be compromised by c-NHEJ and MMEJ competing with HDR for the substrate and diluting the intended edit with undesired indels. Hence, many groups have been working on inhibiting c-NHEJ/MMEJ for efficient gene editing (Figure 2).

3.1.1. Inhibitors of c-NHEJ

Given that c-NHEJ has the fastest kinetics in terms of repairing DNA in human cells [146], inhibiting c-NHEJ is of particular interest. c-NHEJ is initiated by the binding of the KU70/KU80 heterodimer to DNA. In 2016, Weterings et al. developed STL127705, the first compound to inhibit the interaction between KU70/KU80 and DNA, both in vitro and in vivo [147]. While STL127705 has been applied in cancer therapy [148], no studies to date have shown its effect on improving the efficiency of precise genome editing. Moreover, treatment with STL127685, a 4-fluorophenyl analog of STL127705, showed no improvement in Cas9 editing with an ssDNA donor [148].

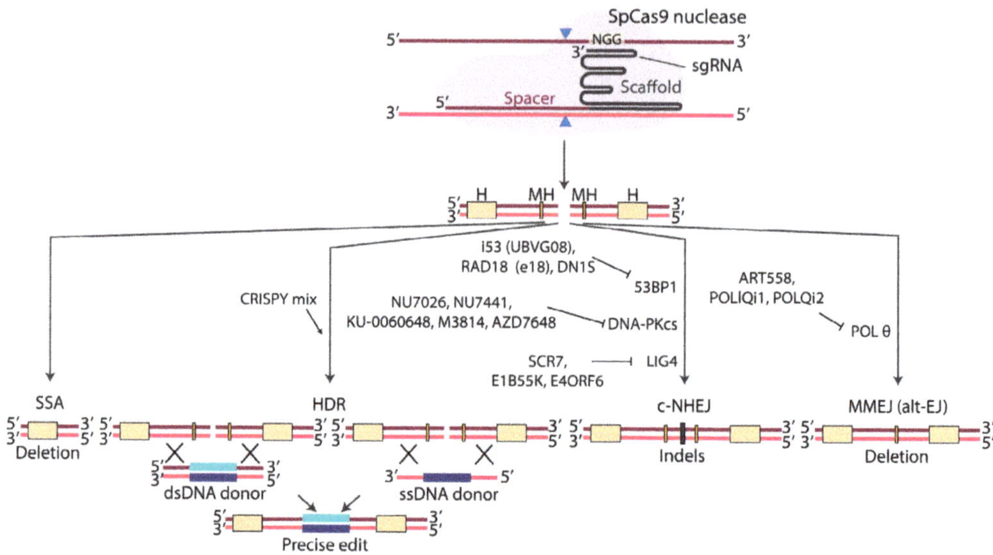

Figure 2. SpCas9-mediated targeted knock-in is installed via homology-directed repair (HDR) and is counteracted by competing repair pathways. SpCas9 paired with a single guide RNA (sgRNA) recognizes a DNA target and introduces a DSB. sgRNA includes a spacer and a scaffold. The target must be complementary to the spacer and must contain an adjacent Protospacer Adjacent Motif (PAM) sequence (NGG in the case of SpCas9). The resulting DSB can be repaired using an exogenous double-stranded or single-stranded donor via HDR to install the desired edit. Alternatively, c-NHEJ may result in generation of indels, MMEJ may lead to deletions between microhomologies (MH) and SSA may lead to deletions between larger homology sequences (H). Small-molecule or protein inhibitors of c-NHEJ or MMEJ proteins enhancing HDR are shown. The CRISPY mix consists of several components, including compounds with poorly characterized mechanisms of action. Thus, we present the CRISPY mix as promoting HDR without specifying its targets.

In 2012, Srivastava et al. identified a compound called SCR7, which inhibited DNA LIG4 binding to DNA, leading to the accumulation of DNA breaks and the activation of intrinsic apoptotic pathways [149]. Maruyama et al. have successfully used SCR7 to promote HDR using dsDNA in epithelial cells (A549) and melanoma cells (MelJuSo) and ssODN in mouse embryos [150]. At least two other studies confirmed the positive effect of SCR7 on HDR [151,152]. However, other studies did not detect significant improvements in gene editing upon treatment with SCR7, questioning its potency, selectivity and utility as an HDR booster [153–157]. An alternative approach resulted in up to a seven-fold enhancement in HDR through the co-expression of adenovirus 4 E1B55K and E4ORF6 proteins, mediating the ubiquitination and proteasomal degradation of DNA LIG4 [151].

DNA-PKcs is probably the most promising c-NHEJ target, with multiple small-molecule inhibitors developed and tested in gene editing experiments, such as NU7026, NU7441, KU-0060648, M3814 and AZD7648 [157–163]. In 2013, Maresca et al. showed that a DNA-PK inhibitor, NU7026, increases the efficiency of HDR with zinc finger nucleases [164]. Other groups later demonstrated a 1.2–2.5-fold HDR improvement with NU7026 and Cas9 [146,153,165].

In 2015, Robert et al. demonstrated that NU7441 and KU-0060648 decrease c-NHEJ events with a concomitant ~2-fold increase in HDR in HEK293T cells edited with Cas9 [166]. In line with this, up to a 1.4-fold improvement in HDR upon NU7441 treatment was reported for iPSCs [146,167].

Another potent DNA-PKcs inhibitor is M3814 (also known as MSC2490484A, nedisertib or peposertib) [163,168]. Riesenberg et al. demonstrated that M3814 increases ssDNA-mediated HDR in hiPSCs, hESCs and K562 cells by ~2–10-fold depending on the cell line and the target [162,169]. In another study, M3814 caused a 3-fold increase in HDR with Cas9 and an AAV6 DNA donor [146].

In 2019, Fok et al. described AZD7648 as a potent and selective DNA-PKcs inhibitor that promotes tumor regression in combination with other agents that target the DNA damage response [161]. Our group demonstrated that AZD7648 led to a 2.9-fold improvement in HDR with an ssODN in HEK293T cells [157]. Recently, Matthew Porteus's lab compared several DNA-PKcs inhibitors, including AZD7648 and M3814, and concluded that AZD7648 was more potent [170].

53BP1 is another target for c-NHEJ inhibition since it prevents end resection and BRCA1 recruitment [31,171]. To our knowledge, no small-molecule inhibitors of 53BP1 have been tested thus far while other approaches have been successfully applied to inhibit 53BP1. Since 53BP1 recognizes ubiquitylated histones at DSB sites, Canny et al. suggested that ubiquitin variants can be used as 53BP1 inhibitors. Indeed, several ubiquitin variants binding to 53BP1 were found. One of them, UBVG08 (also called i53), selectively binds 53BP1 in cells, inhibiting its accumulation at DSBs while promoting BRCA1 recruitment and HDR with dsDNA and ssODN donors [172]. The expression of an engineered variant of RAD18 (e18) is another way to improve HDR, since this protein competes with 53BP1 for binding ubiquitylated histones [173]. Fusing Cas9 to a dominant negative variant of 53BP1 (DN1S) increases HDR by reducing recruitment of 53BP1 specifically to Cas9-induced breaks without globally affecting the c-NHEJ pathway [174]

3.1.2. Inhibitors of MMEJ

The repertoire of developed inhibitors of MMEJ is notably smaller compared to that of c-NHEJ, with DNA polymerase θ (encoded by the *POLQ* gene) being the primary target. Several studies have indicated that knockout or knockdown of *POLQ* partially decreases MMEJ-associated deletions and reduces the unwanted on-target effects of Cas9, such as translocations or large deletions [91,94]. However, only a limited number of small-molecule inhibitors targeting POL θ have been documented to date.

In 2021, Lord et al. presented ART558, targeting the polymerase function of POL θ. In their study, they demonstrated its efficacy in inhibiting the principal POL-θ-mediated DNA repair pathway, theta end joining [175]. Furthermore, in 2022, Heald et al. validated the effects of ART558, showing that it replicates the phenotype of POL θ loss [176]. Specifically in the context of gene editing, a 2023 study reported that this inhibitor prevents the formation of large deletions and facilitates HDR at a *GFP* gene stably integrated into the genome of mESCs [177]. However, no significant HDR enhancement was revealed for another target in mESCs and several human cell lines [177]. Similarly, the results from our lab demonstrate that a single *POLQ* knockout does not influence HDR efficiency in HEK293T cells [157].

3.1.3. Combination of Inhibitors

While the inactivation of POL θ alone only has a marginal effect or no effect on HDR, dual inhibition of c-NHEJ with NU7441 and MMEJ with ART558 leads to a consistent improvement across multiple cell lines [177]. Our research group tested POLlQi1 (WO2021/028643) and POLQi2 (WO2022/0243459) inhibitors in combination with the DNA-PKcs inhibitor AZD7648 [157]. POLQi1 targets the polymerase domain of POL θ and POlQi2 inhibits its helicase activity. AZD7648 in combination with POLQi1 or POLQi2 results in a consistent increase in HDR rate across various cell lines and targets, outperforming treatment with AZD7648 alone. The combined treatment, termed 2iHDR, not only enhances templated insertion efficiency but also greatly decreases indels and Cas9 off-target effects [157]. Riesenberg et al. came to a similar conclusion using the DNA-PKcs inhibitor M3814 and *POLQ* siRNAs, a combination called "HDRobust" [169].

Simultaneous inhibition of c-NHEJ and MMEJ is not the only available strategy. Riesenberg and Maricic tested various small-molecule inhibitors and found a combination, termed the "CRISPY" mix, which improves targeted insertions with ssODNs [153]. In addition to NU7026, discussed above, the mix includes trichostatin A, MLN4924 and NSC 15520. Trichostatin A is a histone deacetylase inhibitor that activates an ATM-dependent DNA damage signaling pathway [178]. MLN4924 promotes HDR by inhibiting the neddylation of unknown CtIP-interacting proteins and consequently promoting DNA end resection [179]. NSC15520 (fumaropimaric acid) blocks the binding of RPA to p53 and RAD9, possibly increasing the abundance of available RPA [180,181]. The "CRISPY" mix promoted HDR in hiPSCs and hESCs. However, in non-pluripotent cell types, some of the "CRISPY" components had an opposite effect, highlighting the difference in DNA repair in different cell lines [153].

4. Evolution of Prime Editing

4.1. At the Origin of Double-Strand-Break-Free Editing Methods

DSBs impose a risk of inversions, translocations, chromotripsis or long deletions spanning several kilobases [182–184]. To avoid DSBs, David Liu's group developed a base editing approach. Base editing relies on a catalytically-inactive Cas9 nuclease (dCas9) fused to a cytidine or adenosine deaminase [185,186]. The binding of dCas9-sgRNA to the target results in an R-loop, in which ssDNA is accessible to the deaminase, eventually leading to C→T or A→G substitutions. Although this technology is a safer alternative to a classic HDR-based approach initiated by a DSB, the application of base editing is limited only to transition mutations; no insertions or deletions can be installed using this method.

In pursuit of a technique to install various types of edits without DSBs, the same group developed the revolutionary PE technology, which can be used to install any types of base substitutions, small deletions or insertions, greatly expanding the genome editing toolbox (Figure 3) [187]. PE relies on a prime editing guide RNA (pegRNA) and a Cas9 nickase fused to a reverse transcriptase (RT) (Figure 3A). Similarly to a regular guide RNA, the pegRNA is comprised of a spacer on the 5′ end and a scaffold sequence but also contains a unique 3′-terminal extension. This extension includes the sequence complementary to the region upstream of the cleavage site (primer-binding site, PBS), the desired edit, and the sequence homologous to the region downstream of the cleavage site (the homology arm, HA). Once Cas9 introduces a nick to a DNA strand displaced after the spacer annealing, the PBS hybridizes to the PAM-distal part of the nicked strand. The RT reverse transcribes the desired edit and HA into the genome using the 3′ end of the nicked strand as a primer (Figure 3B). This results in a single-stranded 3′ flap (Figure 3C), which is able to hybridize to the PAM-proximal side of the nick, generating a 5′ flap that does not contain the edit (Figure 3D). The 5′ flap removal, nick sealing and the processing of the resulting heteroduplex by DNA repair proteins lead to a stably incorporated genomic edit (Figure 3E–L).

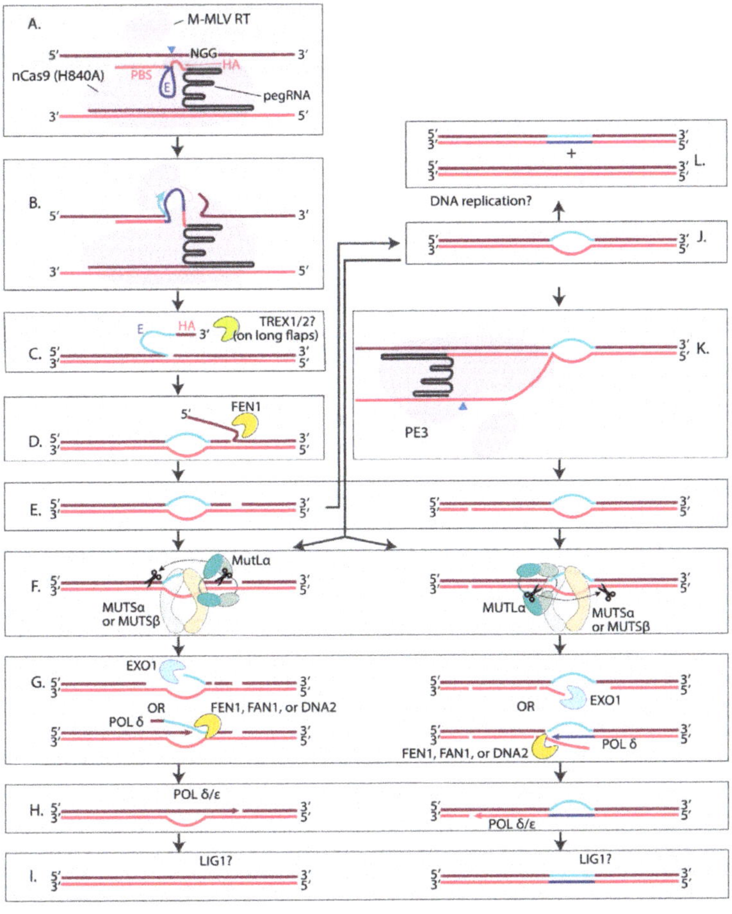

Figure 3. Prime editing (PE) and its interaction with mismatch repair (MMR) and other factors. (**A**) A typical prime editor consisting of an SpCas9 H840A nickase (nCas9) and a Moloney Murine Leukemia Virus (M-MLV) reverse transcriptase (RT) cleaves the nontarget strand of DNA (the strand that is not complementary to the spacer as opposed to the target strand annealed to the spacer). The **PE** guide RNA (pegRNA) includes a spacer, scaffold, homology arm (HA), the desired edit (E) and a primer-binding site (PBS). (**B**) The PBS binds to the cleaved nontarget strand. RT extends the DNA 3′ end using the edit and homology arm RNA sequence as a template. (**C**) The generated 3′ flap is likely subjected to degradation by cellular nucleases. TREX1/2 is a possible candidate that may act on long insertions. (**D**) The 3′ flap containing the edit hybridizes to the PAM-proximal side of the nick, replacing the initial sequence. The displaced 5′ flap containing the initial sequence is removed by the FEN1 endonuclease. (**E**) Following the removal of the 5′ flap, an intermediate is formed which contains a mismatch and single-strand break. (**F**) Mismatches of up to 13 nt introduced by the prime editor are recognized by either the MUTSα or MUTSβ complex, which recruit MUTLα. MUTLα cuts the strand containing a pre-existing break. In the context of **PE**, there is a higher probability of cleaving the edited strand due to the nick introduced by nCas9 (**F, left**). MUTLα cuts 5′ and 3′ to the mismatch. (**G**) The part of the incised strand which contains the mismatch is removed due to EXO1 5′-3′ exonuclease activity. Alternatively, POL δ synthesizes DNA while simultaneously displacing the incised strand without EXO1 involvement. In this case, the generated 5′ flap is removed by FEN1, FAN1 or DNA2. (**H**) The remaining gap is filled by a DNA polymerase, likely POL δ or POL ε. (**I**) The nicks are ligated, presumably by LIG1. As a result, either the initial sequence (**I, left**) or the desired edit is installed (**I, right**). (**J**) In the case where the nCas9-introduced nick is ligated prior to MMR, the

subsequent MUTLα cleavage may happen either in the edit-containing strand (**F, left**) or the strand without the edit (**F, right**), depending on which strand contains pre-existing single-strand breaks generated independently from **PE**. (**K**) When the PE3 system is used, a second guide RNA without a pegRNA extension directs nCas9-RT to cleave the unedited DNA strand, thus promoting MMR to act on the unedited strand (**F, right**). This eventually leads to the installation of the desired edit (**I, right**). The nicking guide can direct Cas9-RT to cleave 5′ or 3′ of the mismatch (here, only cleavage of 3′ of the mismatch is depicted). The precise order of events has not been studied; PE3 cleavage may also occur **earlier.** (**L**) Given that **PE** is observed even when MMR is inhibited and since MMR is inactive on long contiguous mismatches, there should be an alternative way of resolving **PE** intermediates. Our hypothesis is that DNA replication may serve as this resolution mechanism in dividing cells, although other mechanisms may exist.

In the years following the first publication on PE, many research groups have been working on optimizing different components of the system (including the guide RNA, Cas9 and RT), developing new approaches inspired by the original PE concept, or manipulating cellular factors to improve the efficiency and favor the desired editing outcome. In the next part of the review, we summarize major achievements in the field with a focus on the mechanisms of DNA repair and possible ways to control this process.

4.2. From PE1 to PE7 and Beyond

Seven generations of SpCas9-based prime editors have been developed (summarized in Table 1). The first-generation prime editor (PE1) was composed of the wild-type M-MLV RT fused to the C-terminus of SpCas9 (H840A) nickase [187]. This approach enabled 0.7–5.5% editing efficiency for transversion mutations, ~4% for a small deletion and ~10–17% for short insertions. The second generation (PE2) introduced five amino acid substitutions into M-MLV RT, improving its in vitro substrate binding, processivity and thermostability [188,189]. This led to a ~5-fold increase in editing efficiency compared to PE1 [187]. PE3, PE4 and PE5 further enhanced PE by reducing the inhibitory effect of mismatch repair through the addition of a nicking guide or a dominant negative variant of the MLH1 protein (see more details below) [187,190]. Protein optimization using phage-assisted continuous (PACE) or non-continuous (PANCE) evolution combined with rational engineering led to the development of PE6, further subdivided into PE6a-g [191]. RT mutants PE6a-c were optimized for a smaller editor size, while PE6c-d demonstrated an increased activity on long, highly-structured templates. In addition, several SpCas9 variants (PE6e-g) with enhanced activity on a subset of targets appeared in the course of the evolution (PE6e-g [192]. Finally, PE7 was developed, which is a prime editor fused to the RNA-binding domain of the LA protein [191]. The LA protein binds to U tracts of the 3′ ends of RNA polymerase III transcripts and stabilizes 3′ ends of polyuridylated pegRNAs in the context of PE [191].

Table 1. Summary of seven generations of prime editors. The characteristic feature of each editor differentiating it from previous generations is underlined.

Prime Editor	Cas9 Variant	Reverse Transcriptase	Guide RNA	Helper Protein
PE1 [187]	<u>nSpCas9(H840A) nickase</u>	<u>wild-type M-MLV RT</u>	<u>pegRNA</u>	-
PE2 [187]	nSpCas9(H840A) nickase as in PE1	<u>M-MLV RT (D200N/L603W/ T330P/T306K/W313F)</u>	pegRNA	-
PE3 [187]	nSpCas9(H840A) nickase as in PE1	M-MLV RT (D200N/L603W/ T330P/T306K/W313F) as in PE2	<u>pegRNA + sgRNA complementary to the edited strand up- or downstream of the edit</u>	-

Table 1. *Cont.*

Prime Editor	Cas9 Variant	Reverse Transcriptase	Guide RNA	Helper Protein
PE3b [187]	nSpCas9(H840A) nickase as in PE1	M-MLV RT(D200N/L603W/ T330P/T306K/W313F) as in PE2	pegRNA + sgRNA complementary to the edit established by the prime editor	-
PE4 [190]	nSpCas9(H840A) nickase as in PE1	M-MLV RT(D200N/L603W/ T330P/T306K/W313F) as in PE2	pegRNA as in PE2	MLH1dn
PE5 [190]	nSpCas9(H840A) nickase as in PE1	M-MLV RT(D200N/L603W/ T330P/T306K/W313F) as in PE2	pegRNA + nicking sgRNA as in PE3	MLH1dn
PE6a [192]	PEmax: nSpCas9(H840A) nickase with additional R221K and N394K substitutions	evo-Ec48: *Escherichia coli* Ec48 retron-derived RT with substitutions E60K/K87E/E165D/ D243N/R267I/E279K/K318E/ K343N evolved using PANCE	pegRNA as in PE2 or pegRNA + nicking sgRNA as in PE3	Optional: MLH1dn
PE6b [192]	nSpCas9(H840A) nickase with additional R221K and N394K substitutions as in PEmax	Evo-Tf1: *Schizosaccharomyces pombe* Tf1 retrotransposon-derived RT with substitutions P70T/G72V/ S87G/M102I/K106R/K118R/ I128V/L158Q/F269L/A363V/ K413E/ S492N evolved using PANCE.	pegRNA as in PE2 or pegRNA + nicking sgRNA as in PE3	Optional: MLH1dn
PE6c [192]	nSpCas9(H840A) nickase with additional R221K and N394K substitutions as in PEmax	PE6b with additional rationally designed substitutions (Tf1 RT P70T/ G72V/ S87G/M102I/ K106R/K118R/I128V/L158Q/ F269L/A363V/K413E/S492N/ K118R/S188K/I260L/S297Q/ R288Q)	pegRNA as in PE2 or pegRNA + nicking sgRNA as in PE3	Optional: MLH1dn
PE6d [192]	nSpCas9(H840A) nickase with additional R221K and N394K substitutions as in PEmax	M-MLV RT with the truncated RNaseH domain and additional mutations T128N/ V223Y/ D200C	pegRNA as in PE2 or pegRNA + nicking sgRNA as in PE3	Optional: MLH1dn
PE6e [192]	PEmax with additional substitutions K918A/ K775R	M-MLV RT (D200N/ L603W/T330P/T306K/W313F) as in PE2 but ΔRNaseH	pegRNA as in PE2 or pegRNA + nicking sgRNA as in PE3	Optional: MLH1dn
PE6f [192]	PEmax with additional substitutions E471K/ H99R/I632V/ H721Y/D645N/ K918A	M-MLV RT (D200N/L603W/ T330P/T306K/W313F) as in PE2 but ΔRNaseH	pegRNA as in PE2 or pegRNA + nicking sgRNA as in PE3	Optional: MLH1dn
PE6g [192]	PEmax with additional substitutions E471K/ H99R/I632V/ H721Y/R654C/ D645N	M-MLV RT (D200N/L603W/ T330P/T306K/W313F) as in PE2 but ΔRNaseH	pegRNA as in PE2 or pegRNA + nicking sgRNA as in PE3	Optional: MLH1dn
PE7 [191]	PEmax	M-MLV RT (D200N/L603W/ T330P/T306K/W313F) as in PEmax	pegRNA as in PE2 or pegRNA + nicking sgRNA as in PE3	The N-terminal part of a small RNA-binding protein La(1–194) fused to the C terminus of PEmax. Optional: MLH1dn

It is worth noting that there are two major drawbacks of the current PE naming system. The first is a lack of a standard nomenclature for PE systems composed of several elements from different PE generations. For example, any PE6a-d RT variants can be combined with any PE6e-f Cas9 variants and may include a nicking guide (as in PE3 systems) and MLH1dn (as in PE4/5), but there is no consensus on how the resulting editor should be named. Second, many reported modifications have been made to PE systems to address challenges and increase editing efficiency, but these have not been formally classified as new PE generations.

One such challenge is the guide stability. Since the 3′ extension of the pegRNA is not protected by Cas9, it is prone to degradation by RNA exonucleases. The PE7 approach likely solves this problem by stabilizing the polyuridylated 3′ ends of pegRNAs [191]. But other approaches have been suggested as well, such as adding structured RNA motifs to 3′ ends [193–195], using untethered circular RT templates [196,197] or using exonuclease-resistant synthetic guides [198].

Recruiting helper proteins to the editor through aptamers, the Suntag system, or covalently fused linkers (as in PE7) is a common approach used to influence editing efficiency. The pioneer transcription factor P65 [199], chromatin-modulating peptides [200], RAD51 DNA-binding domain [201] and T5 5′ exonuclease [202] are examples of helper proteins integrated into PE systems to promote chromatin accessibility, stabilize ssDNA and remove 5′ flaps, respectively. Two helper peptides (NFATC2IPP1 and IGF1PM1) that boost the translation efficiency of the PE2 editor were discovered because of extensive screening [203].

Several groups have focused on modifying the SpCas9-RT sequence to improve the protein architecture, optimize Nuclear Localization Signals (NLSs) and RT codon usage, remove potential splice sites and introduce mutations to increase nuclease activity. These efforts resulted in more efficient prime editors, such as PE* [204], PEmax [190] and iPE-C/iPE-N [202]. Among them, PEmax is a widely used variant (Table 1), which includes SpCas9 with additional R221K and N394K substitutions that improve nuclease activity, a 34-aa linker containing a bipartite SV40 NLS, human codon-optimized RT and an additional C-terminal c-Myc NLS [190].

The proximity of a PAM to the target is a huge challenge for CRISPR-based editing (Figure 2). The PAM of the most widely used SpCas9, NGG, occurs only once in every 16 genomic loci. To broaden the set of editable targets, a near-PAM-less SpRY SpCas9 variant and SpCas9 variants recognizing NGA, NGCC and NG were engineered and tested in combination with PE [205–210]. In addition, orthologous Cas9 variants from Staphylococcus aureus (SaCas9, NNGRRT PAM, SaCas9KKH, NNNRRT PAM) and Francisella novicida (FnCas9, NGG PAM, RHA-FnCas9, YG PAM) have been tested. The FnCas9 variants cleave 6–8 nt upstream of the PAM, in contrast to SpCas9, which cleaves 3 nt upstream of the PAM. On average, they demonstrated lower editing efficiencies compared to SpCas9 [204,210,211]. Recently, Cas12a has been developed, which is the first prime editor based on the type V CRISPR-Cas effector protein. Cas12a demonstrated up to 40% editing efficiency in vitro [212].

Several groups have tried to substitute M-MLV RT with RTs from other organisms, however, so far, none of the natural RT proteins has surpassed the efficiency of M-MLV RT. The most successful attempts so far include the evolution of *Escherichia coli* Ec48 retron-derived RT and *Schizosaccharomyces pombe* Tf1 retrotransposon-derived RT, which have led to the development of PE6a, PE6b and PE6c (Table 1) [209].

Another promising approach is to replace RT with a DNA-dependent DNA polymerase. A fundamental difference from classic PE in this case is the requirement of a DNA template. Three research groups have tested this concept in different configurations. Liu et al. previously developed a split PE approach with untethered SpCas9 nickase and RT, but, in their recent work, they substituted RT with a replicative polymerase from *Bacillus subtilis* phage phi29 [196,213]. In this system, the phi29 DNA polymerase is fused to the MS2 coat protein (MCP) and tethered via the MS2 stem–loop to the template, which is separated from the guide RNA. Ferreira da Silva et al. developed an approach called click editing (CE) [214]. The CE editor is a fusion of an SpCas9 nickase, a 3′-5′ exonuclease-deficient Klenow fragment from *E. coli* DNA polymerase I, and an HUH endonuclease (HUHe) from porcine circovirus 2 (PCV2). The PCV2 protein forms a covalent phosphotyrosine adduct with the HUHe recognition sequence in the click DNA (clkDNA) template. Similarly to the approach of Liu et al., the DNA polymerase template is not fused to the guide in this case. Finally, the recent pre-print by Nguyen et al. describes a two-component chimeric oligonucleotide-directed editing (CODE) system with an architecture similar to the original PE design [215]. The CODE system includes a thermophilic Bst DNA polymerase from *Geobacillus stearothermophilus* fused to the SpCas9 nickase and a long chimeric cpegRNA with a DNA extension on the 3′ end composed of a PBS and the DNA polymerase template. These DNA-polymerase-based technologies will likely expand the toolbox of editors with new capabilities in the future.

While rational design and phage-assisted protein evolution have accelerated progress in PE optimization, AI-assisted in silico protein optimization is undoubtedly the next step. In a recent pioneering study, Jiang et al. combined protein language models with a top-layer regression model to perform several rounds of PE2 evolution [216]. In the first round, random mutants were selected and tested experimentally. In subsequent cycles, the model actively learnt from the data and predicted variants with improved efficiency for experimental validation. In seven rounds, with just 12 mutants tested in each, the model predicted several variants, with a ~1.5-fold improvement in editing relative to PE2. Surprisingly, most of the substitutions were in the C-terminal RNase H domain and were not uncovered by previous in vitro optimization methods, demonstrating that there is still room for improvement even for the relatively well-optimized PE2 editor.

4.3. Bidirectional Prime Editing Systems (Bi-PE)

PE successfully installs insertions of up to ~40 bp and deletions of up to ~80 bp [187]. To enable larger modifications, several groups independently tested TwinPE [217], GRAND [218], PRIME-Del [219], HOPE [220] and Bi directional PE (Bi-PE) systems [221] in human cell lines (Figure 4A–C). The five approaches share the idea of using two pegRNAs that target opposite DNA strands, encoding complementary or partially-complementary 3′ flaps that hybridize to each other. Deletions of ~1 kbp were generated using these methods with up to 30–80% efficiencies reported by different groups [217,221]. Choi et al. were able to obtain 10 kbp deletions at the *HPRT1* locus using PRIME-Del with ~0.8% efficiency measured by ddPCR [219]. Remarkably, sequencing amplicons with the deletion revealed that only ~3% of them contained additional indels [219]. Similarly, TwinPE demonstrated precise deletions of up to 780 nucleotides, including exon deletions with efficiencies up to 28% and only 5.1% indels [217]. These results demonstrate the high precision of deletions introduced by Bi-PE systems, as shown by several independent groups [217,221].

In addition to targeted deletions, dual-pegRNA strategies can be used to replace the target locus with a heterologous sequence [217,219–221]. Although such strategies inherit the inability of PE to integrate very long sequences, ~150 bp fragments were inserted with up to 63% efficiency due to splitting the sequence between the two partially overlapping flaps [218].

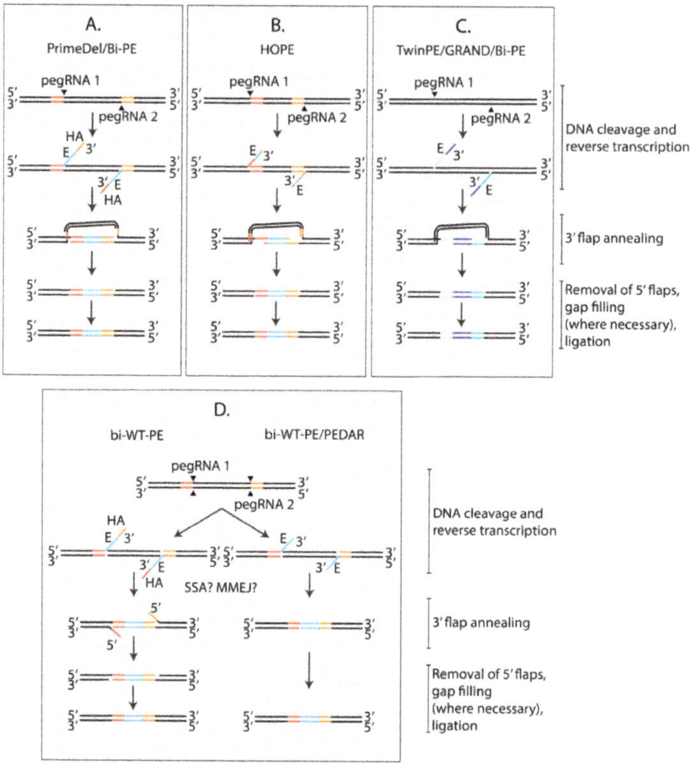

Figure 4. Bidirectional prime editing (Bi-PE) systems tested in human cells. (**A–C**) Nickase-based **Bi-PE** systems. Several methods are presented, all sharing the same principle: a nickase-based prime editor targets opposite strands of the same chromosome using a pair of pegRNAs. The pegRNA extensions are designed so that the resulting 3′ flaps are at least partially complementary to each other. The exact mechanism following the annealing of the 3′ flaps is unknown. The suggested process includes removal of 5′ flaps, gap filling, and ligation. Based on the composition of the 3′ flaps, **Bi-PE** approaches can be roughly divided into three categories: (**A**) In PrimeDel and a variation of Bi-PE, the 3′ flaps contain an edit (E) and a homology arm (HA). The HA of a flap connected to cleavage site 1 (pegRNA 1) is complementary to the region next to cleavage site 2 (pegRNA 2) and vice versa. Thus, in this case, the 3′ terminal part of each flap (the homology arm, HA) invades the homologous duplex while the remaining portions of both flaps hybridize to each other. This approach leads to the installation of edits accompanied by a deletion of the entire region between the two cleavage sites. Note that the edit may be removed from the flaps, resulting in clean deletions. (**B**) In the HOPE system, the 3′ flaps also share homology with the chromosome, but the homologous sequence directly follows the same nick site (black arrowheads). With this approach, a part of the region between the two nicks may be replaced with another sequence. (**C**) In TwinPE, GRAND and another variation of the Bi-PE approach, the 3′ flaps do not contain endogenous sequences. This approach allows one to replace the entire region between the two nicks with an exogenous sequence. (**D**) Nuclease-based **Bi-PE** systems. The method called PEDAR (Precise and Specific Deletion and Repair) or Bidirectional Wild-Type Prime Editing (Bi-WT-PE) involves a nuclease-based prime editor together with a pair of pegRNAs targeting opposite strands of the same chromosome. **The** cleavage (black arrowheads) splits the chromosome into three parts: the central part enclosed between the two cleavage sites (lost during subsequent repair) and two parts with 3′ flaps generated by prime editing. The two pegRNAs are designed with complementary edits. In addition, a HA may be present in the bi-WT-PE approach, similarly to the approach presented in A. The annealing of the two 3′ flaps generated by the nuclease-based systems may be mediated by SSA or MMEJ, but the precise mechanism is unknown.

An alternative strategy for targeted insertions called template-jumping PE (TJ-PE) was suggested by Zheng et al., inspired by retrotransposon replication [222]. This method starts as regular PE, but the reverse-transcribed sequence includes a region at its 3′ end that serves as a PBS for a second reverse transcription reaction. A second guide RNA is used to introduce a nick in the opposite DNA strand at some distance from the initial nick. The nicked DNA 3′ end hybridizes to the initial 3′ flap, and RT initiates the second reverse transcription reaction. Using TJ-PE, 200 bp sequences were accurately integrated with ~34% efficiency and ~2% integration of 800 bp *EGFP* fragments was observed.

Overall, these results demonstrate the great potential of Bi-PE systems for skipping or rewriting exons, inserting minigenes and correcting complex genetic rearrangements.

4.4. Recombinase-Based Prime Editing Systems

Several PE proteins capable of making precise insertions and deletions in human cells have been described. However, these proteins currently cannot mediate insertions or deletions of sizes typical of exons or entire gene coding sequences. Such large DNA changes require extended pegRNA reverse transcription templates and long-range DNA polymerization, which significantly reduce editing efficiency. To address this limitation, several research groups have focused on recombinase-based prime editors, which use site-specific recombinases (SSRs). SSRs are enzymes that can excise, invert and integrate large DNA sequences in mammalian cells by catalyzing recombination between attachment sites *attB* and *attP* [223]. For example, one of the most used recombinases is Bxb1, a 500-amino-acid protein that binds *attP* and *attB* recognition sites that are 39 and 35 bp, respectively [224]. When PE is used to install one of the two attachment sites (e.g., *attB*) at a genomic location, a larger cargo can be subsequently integrated inside the *attB* sequence by Bxb1 if a donor DNA containing the *attP* site is provided.

This Prime-Editing-Assisted Site-Specific Integrase Gene Editing technology (PASSIGE) was introduced together with TwinPE, mentioned above [217,225]. At the time of initial publication, PASSIGE enabled highly efficient integration of the attachment site into the genome with more than 50% efficiency. When a clonal *attB*-containing cell line was isolated, subsequent transfection with Bxb1 and a donor plasmid led to up to 17% integration efficiency, demonstrating the huge potential of this technology. Yet a low efficiency when all reagents were delivered in a single transfection step (up to ~6.8%) remained a limitation [217]. In 2024, PASSIGE was significantly improved through the evolution of Bxb1 using PACE and PANCE. The evolved and engineered variants, evoBxb1 and eeBxb1 respectively, achieved 2.7-fold and 4.2-fold average improvements in the efficiency of targeted DNA integration, respectively.

These improvements resulted in evolved PASSIGE (evoPASSIGE) and engineered PASSIGE (eePASSIGE). These systems achieved integration efficiencies ranging from 20% to 46% for multi-kilobase gene-sized cargo at commonly used and therapeutic loci after a single transfection [225].

In 2023, Gootenberg and colleagues introduced PASTE (Programmable Addition via Site-Specific Targeting Elements). In contrast to PASSIGE, which relies on TwinPE, PASTE relies on a classic PE approach with a single pegRNA. In addition, a serine integrase is directly fused to PE, rather than provided in trans. PASTE achieved the integration of DNA cargo of up to 36 kb in a single delivery reaction, with efficiencies of 50–60% in cell lines and 4–5% in primary human hepatocytes and T cells [226].

Interestingly, recombinase-based methods can also be used to correct genomic rearrangements. For example, in a model of Hunter Syndrome, multiplex TwinPE insertion of *attB* and *attP* into the genome, combined with Bxb1 recombinase, facilitated a 40 kb inversion [217]. Meanwhile, two other studies reported dual-pegRNAPE systems capable of inducing large deletions in human cells and plants [219,227].

Future advances in PE technologies together with the discovery and directed evolution of new recombinases are expected to enhance the capabilities and applications of recombinase-based technologies in PE [228].

4.5. The Role of DNA Repair Genes in Prime Editing

4.5.1. Mismatch Repair (MMR)

In human cells, short mismatches or insertion/deletion loops (IDLs) of up to 13 nt are recognized by either the MUTSα complex (MSH2/MSH6 heterodimer) or the MUTSβ complex (MSH2/MSH3 heterodimer), which have partially-overlapping binding specificities (Figure 3F) [229–234]. MUTSα predominantly binds to base mispairs and short insertion/deletion loops (IDLs) that are up to 3 nt long [230,232,233]. In comparison, MUTSβ is essential for recognizing longer IDLs (~10 nt). It binds some shorter IDLs, but does not bind mispairs [229,233].

MUTSα/MUTSβ bound to a lesion recruits the MUTL complex, the main form of which, MUTLα, is a heterodimer of MLH1 and PMS2 (Figure 3F [235,236]. In the presence of MUTSα/MUTSβ, RFC and PCNA, MUTLα cuts the strand containing a pre-existing break near the mismatch [237]. MUTLα makes incisions 5′ and 3′ to the mispair [237]. The cuts made to the 5′ of the mismatch are of particular importance, because they serve as an entry point for the exonuclease EXO1, which is activated by MUTSα and removes a mismatch-containing patch in the 5′-3′ direction (Figure 3G) [238,239]. The generated gap is filled by DNA polymerase δ or ε holoenzyme (Figure 3H) [240–242]. Alternatively, POL δ synthesizes DNA while simultaneously displacing the mismatch-containing strand without EXO1 involvement [238]. The generated 5′ flap is removed by a 5′ flap endonuclease, FEN1, FAN1 or DNA2 (Figure 3G [238,243,244]. The final ligation step is likely performed by DNA LIG1, based on its interaction with PCNA, involvement in replication and ability to complete MMR in vitro (Figure 3I) [245,246].

The requirement of a pre-existing break for the cleavage by MUTLα is fundamental in eukaryotic MMR. Since the main function of MMR is the correction of replication errors, discontinuities in the newly synthesized DNA direct the repair machinery towards the daughter strand to preserve the initial sequence. This mechanism poses a challenge in PE, where the strand with an RT-introduced edit also contains a Cas9-generated nick. This problem was identified by Anzalone et al., who developed PE3 in an attempt to trick MMR by introducing a second nick in the nonedited strand (Figure 3K) [187]. The second cut is generated by the same SpCas9(H840A)-RT protein coupled to an sgRNA targeting the region 5′ or 3′ from the edit. This strategy increased editing efficiency ~1.5–4.2-fold compared to PE2 at four out of five tested targets. The efficiency of PE varied between different secondary cut sites with no apparent correlation between the percentage of intended edits and the location of the secondary cut relative to the initial cut site. In addition, the percentage of indels varied greatly depending on the location of the secondary cut; some sgRNAs did not change the indel level relative to PE2, while the others led to more than a 20-fold increase [187]. Habib et al. compared the frequency of indels between cells treated with PE3 and a similar system where the pegRNA was substituted with an sgRNA targeting the same site [247]. The PE3 system led to ~10% indels at two tested sites, while the same system with two nicks but lacking a 3′ flap resulted in almost no indels. These observations suggest that indels in PE3 are not directly caused by two nicks but rather arise because of the processing of flaps in the presence of a second nick in the opposite strand [247].

To decrease the probability of DSB formation, a strategy called PE3b was proposed in the same study [187]. The sgRNA in PE3b is designed to target the established edit, but not the initial sequence. To eliminate the risk of DSBs, the 5′ flap must be removed, and the edited strand must be ligated for the sgRNA-mediated nick to occur. The PE3b approach indeed demonstrated decreased levels of indels compared to PE3, although, in some cases, the percentage of prime edits was higher with the PE3 approach [187]. These findings highlight the importance of screening various sgRNAs to achieve the optimal edit/indel ratio when using PE3.

Though it was expected that MMR counteracted the establishment of the edit at the time when the concept of PE was introduced by David Liu [187], it was not proven genetically until two years later when the same group published the results of a CRISPRi screen [190]. In this study, Chen et al. tested how knocking down each of 476 genes involved in DNA repair or associated processes affected the efficiency of PE2 or PE3. *MLH1*, *PMS2*, *MSH2* and *MSH6* knockdowns led to the highest increase in the intended base substitutions, corroborating the inhibitory effect of MMR on PE. The increase in the intended edit was higher for PE2 (up to a 5.8-fold increase) than for PE3 (up to a 2.5-fold increase), but, in the case of PE3, the inhibition of the MMR genes also had a positive effect on the purity of editing outcomes by reducing indels. Deletions outside of the region between the pegRNA and sgRNA cut sites were especially responsive to MMR inactivation, suggesting that a part of them is caused by MMR. *EXO1* inactivation improved editing efficiency to a lesser extent compared to MUTSα or MUTLα, which is in line with the existence of EXO1-dependent and EXO1-independent pathways of MMR (Figure 3G) [237–239].

Another study by da Silva et al. explored the effect of 32 gene knockouts on PE and confirmed the inhibitory effect of MMR [248]. The only marked differences were the absence of the *MSH6* effect, observed by Chen et al., and an increase in PE in the Δ*MSH3* background. This discrepancy can be explained by the different substrates used in the two studies. Chen et al. used a single-nucleotide mismatch, the substrate of MUTSα (MSH2/MSH6 heterodimer), while da Silva et al. used a 5 bp deletion loop recognized by MUTSβ (MSH2/MSH3 heterodimer) [190,248]. A recent study by Park et al. highlights these differences, demonstrating that base mispairs and 1 nt IDLs are enhanced in Δ*MSH6* knockout cells; 5–10 nt IDLs are enhanced in Δ*MSH3* mutants; 3 nt IDLs are stimulated by either of the two deletions; and 15–34 nt IDLs are not affected by *MSH2/MSH3/MSH6* [249]. In addition, an shRNA screen performed by Li et al. revealed an inhibitory effect of MMR components *PMS2* and *MLH1* on 6 nt insertions, though no consistent results were observed for other MMR factors [250]. Overall, these genetic studies unequivocally demonstrate the inhibitory effect of MMR on PE. These findings are further supported by microscopy, revealing colocalization of a prime editor with MLH1 or MSH2 [248].

To our knowledge, there is no available small-molecule inhibitor of MMR. However, several other approaches have been suggested to leverage the effect of MMR inactivation on PE. Chen et al. suggested that contiguous mismatches are not well recognized by MMR, and therefore increasing the size of the edit by introducing silent mutations should promote PE [190]. Indeed, PE2 works ~2.7-fold better on 3-5-base contiguous substitutions compared to shorter 1-2-base mutations [190].

If the intended outcome is an insertion or deletion, the inhibitory effect of MMR should decrease with the increase in the IDL length. IDLs of up to 12 nt are still repaired by MMR in vitro when treated with extracts of HeLa cells, although at a lower efficiency compared to shorter IDls, and IDLs longer than 16 nt are unlikely to be recognized by MMR [251]. In line with this, Koeppel et al. demonstrated that the fold change difference in PE efficiency between *wt* and Δ*MLH1* cells decreases exponentially with increasing insertion length. According to their model, a ~23–28-fold difference is expected for 1 nt insertions, which drops by 40–48% for every additional nucleotide, approaching 1 for insertions longer than 13 nt. However, increasing the indel size is not always possible when it comes to editing a therapeutically-relevant target. In addition, if the intended outcome is a long insertion, other factors, such as RT processivity or the presence of secondary structures in the reverse transcription template, may lower efficiency [252].

Transient inactivation of MMR components using siRNA is another strategy to enhance PE, but it requires pre-treatment with siRNA for 2–3 days [190]. Transient degradation of the MLH1 protein tagged with dTAG is another alternative [248]. However, this method is even more time-consuming since it requires introduction of the tag into the genome.

Chen et al. suggested co-expressing the dominant negative variants of *MMR* genes together with a prime editor to achieve transient MMR inactivation [190]. Among several tested engineered MLH1, PMS2, MSH2 and MSH6 proteins, the MLH1 Δ754–756 variant (referred to as MLH1dn) added in trans demonstrated the highest results. The combination of PE2 with MLH1dn was designated as PE4, and the combination of PE3 with MLH1dn was designated as PE5 (Table 1). Alternatively, an MLH1NTD–NLSSV40 variant, which is less efficient but smaller in size (355 aa vs. 753 aa for MLH1dn), can be used. Overall, the addition of MLH1dn improved the installation of all types of substitutions, although the effect was smaller for G-C to C-G edits. The G-C to C-G mutation is formed due to C-C mismatches that are not efficiently repaired by MMR and therefore have a higher basal editing level [190,253]. The efficiency of short 1 or 3 bp indel installation was also greatly improved by MLH1dn. However, the effect decreased with the size of the indel, and almost no difference between PE2 and PE4 was observed for indels $\geq$ 15 nt in length. The effect of MLH1dn on PE also depends on the cell line. The improvement is smaller in cells like HEK293T, in which MMR is already partially-inactivated, compared to MMR-proficient cell lines such as HeLa, K562 and U2OS [190]. These results demonstrate the great potential of PE4 and PE5 technologies for cell engineering and therapeutic application. However, there is also a possibility for a further improvement since the use of MLH1dn does not completely inhibit MMR, as evidenced by the higher activity of PE in Δ*MLH1* knockout cells.

4.5.2. FEN1

Following the reverse transcription step, an intermediate with a 3′ flap, which can be converted into a 5′ flap, is formed (Figure 3C,D). The removal of the 5′ flap is a favorable process for PE, while the cleavage of the 3′ flap may lead to unintended editing outcomes. FEN1, a 5′ flap endonuclease that removes RNA primers of Okazaki fragments, is an obvious candidate for the role of the enzyme that removes 5′ flaps in PE [254]. Indeed, the results of the CRISPRi screen by Chen et al. revealed a decrease in PE2 and PE3 efficiency upon *FEN1* knockdown [190]. Li et al. also reported a significant decrease in PE in cells expressing *FEN1* shRNA [250]. The observed decrease was less than 2-fold in both studies, suggesting either functional redundancy with other proteins or knockdown that was not efficient.

4.5.3. HLTF

DNA damage that hasve not been repaired prior to replication can stall the progression of a replication fork. To achieve the completion of replication despite the presence of damage, two DNA damage tolerance (DDT) pathways have evolved: translesion synthesis (TLS) and template switching (TS) [255,256]. The TLS pathway is initiated by the monoubiquitination of PCNA at K164 and relies on low-fidelity DNA polymerases replicating DNA across damaged bases [257]. The TS pathway is initiated by the polyubiquitination of PCNA at K164 [258–260]. It operates by switching the template from the damaged strand to the nascent daughter strand on the sister chromatid to bypass the damage [256,261].

The HLTF (Helicase-Like Transcription Factor) protein was initially characterized as a transcription factor binding to promoters and enhancers of various genes and hence its name he high degree of sequence similarity between HLTF and its closest ortholog, RAD5, which is central to the TS pathway in *Saccharomyces cerevisiae*, prompted studies of HLTF function in the DNA damage response and led to the discovery of its role in TS. Similarly to RAD5, HLTF is an E3 ubiquitin ligase (E3) and catalyzes PCNA polyubiquitination [260]. The DNA-binding domain of HTLF recognizes 3′ ends of stalled replication forks and the helicase domain promotes fork reversal [262,263]. Interestingly, a recent study demonstrated a direct interaction between HLTF and MSH2 in human cells [264]. However, the functional significance of this interaction for MMR or TS remains unknown.

Chen et al. demonstrated a weak stimulating effect of *HLTF* on PE2 and an inhibiting effect on PE3 in two different cell lines with a single-nucleotide substitution as the desired edit [190]. In contrast, Li et al. observed an upregulation of 6 nt insertions installed via PE2 upon knocking down *HLTF* [250]. This study also revealed that the inhibitory effect of *HLTF* on PE2 varies across genomic targets, with actively transcribed genes being more responsive to *HLTF* knockdown compared to non-transcribed regions. At the same time, only minimal changes in gene expression and chromatin accessibility were revealed in cells with *HLTF* knockdown, suggesting that its effect on PE is not related to its role as a transcription factor. It is yet to be determined whether HLTF influences PE through its function in TS, the interaction with MSH2, or another uncharacterized mechanism. The reason for the opposite effects of *HLTF* on PE2 vs. PE3, as well as different types of edits, also remains unknown.

4.5.4. TREX1 and TREX2

TREX1 (DNase III) and TREX2 are 3′-5′ exonucleases that are active on ssDNA and dsDNA [265–267]. The preferable substrate for both enzymes is a partial DNA duplex with mispaired 3′ termini like the 3′ flap generated during PE [265]. TREX1's main role is to degrade cytosolic DNA and prevent inappropriate immune responses through a cGAS-STING pathway of DNA sensing [268,269]. Accordingly, TREX1 is predominantly localized in the perinuclear space or endoplasmic reticulum [270,271]. However, several studies detected TREX1 in nuclei under certain circumstances. For example, when cytolytic T cells and NK cells release Granzyme A (GZMA) into target cells through an immunological synapse, TREX1 moves to the nucleus to enhance DNA degradation during the caspase-independent cell death pathway [272]. More importantly, in the context of gene editing, TREX1 translocates to the nucleus upon UV, γ irradiation or hydroxyurea treatment in growing mouse cells, suggesting that TREX1 may play a role in DNA repair or damage tolerance, although the mechanism remains unknown [273,274]. ~3-fold decrease in unintended deletions with the PE3 approach was detected upon knocking down *TREX1* [190]. Therefore, it is likely that TREX1 is present in the nucleus and ready to trim DNA ends when PE occurs.

TREX2 is a nuclear protein that participates in DDT, processing stalled replication forks and promoting mutations through its 3′-5′ exonuclease activity and ability to ubiquitinate PCNA at K164 [252,275,276]. TREX2's impact on genome stability depends on genetic backgrounds, but, at least in some cases, TREX2 degrades unprotected 3′ ends. For example, in mouse embryonic stem cells expressing human mutant RAD51 K133A defective in filament assembly, TREX2 contributes to the nascent strand degradation after treatment with hydroxyurea [276].

Although both TREX1 and TREX2 may theoretically antagonize PE by degrading 3′ flaps, their role in PE is not clear. Koeppel et al. tested the effect of TREX1 or TREX2 overexpression on the efficiency of insertions installed by PE2 and found that both enzymes interfere with prime insertions in a length-dependent manner [252]. While only up to 3-fold differences were observed for 1 nt insertions, *TREX1/2* overexpression led to a 20-180-fold decrease in editing for longer 30 nt insertions. However, the effect of *TREX1/2*'s loss of function on short vs. long PE insertions has not been studied and therefore it is not clear whether the observed differences reflect what happens in cells with a normal expression level. No effect of *TREX1* knockdown on base substitutions installed by PE2 or PE3 was detected in the CRISPRi screen by Chen et al. [190]. Similarly, the shRNA screen by Li et al. did not detect statistically significant differences for 6 bp prime insertions between cells with or without *TREX1* shRNA [250] It is unknown whether TREX1/2 only interferes with long insertions, and further experiments are required to support or refute this.

4.5.5. LIG1

The 3′ flap must be ligated for the successful completion of PE. If the resulting heteroduplex is resolved through MMR, the MMR outcomes should also be ligated to restore DNA integrity. Chen et al. demonstrated a decrease in PE2 and PE3 upon *LIG1* knockdown in HeLa and K562 cells [190]. In contrast, Li et al. did not observe an inhibitory effect of *LIG1* knockdown on PE2 in K562 cells [250]. Therefore, further experiments are required to elucidate the impact of DNA ligases on PE.

4.6. Prime Editing Nuclease (PEn) and Associated DNA Repair Pathways

The efficiency of PE varies between cell lines. Adikusuma et al. replaced SpCas9 nickase with SpCas9 nuclease, as they hypothesized that the low efficiency of PE in HeLa and K562 cells might be due to inefficient 5′ flap resection and removal of the nonedited strand (Figure 5A). SpCas9 nuclease boosted the overall PE efficiency from 22% to 71% in K562 cells and from 6.7% to 30% in HeLa cells [277]. Motivated by the high efficiency in vitro, Adikusuma et al. next applied PEn to generate mice through zygote microinjection. Though the efficiency varied depending on the target and the edit type, remarkable results were achieved. For example, in some cases, up to 87.5% of the mice contained the intended edit with 3 nt insertions at the *CHD2* and *COL12A1* sites while 100% of the mice had the intended edit at the *TYR* site [277].

A few months after the first paper on PEn was released, three other research groups, including ours, published results on PE using SpCas9 nuclease [278–280]. Two of these studies demonstrated that, similarly to the TwinPE/Bi-PE/PRIME-Del/HOPE/GRAND technologies, PEn combined with a pair of pegRNAs targeting complementary strands can be used to delete large pieces of DNA and insert a short sequence encoded in the 3′ flaps (Figure 4D [279,280]. This method, called PE-Cas9-based deletion and repair (PEDAR)by Jiang et al. or bi-WT-PE by Tao et al., achieved ~3% deletion of a 16.8-megabase region in cell cultures, which is much larger than was previously reported for PRIME-Del [219,280]. In vivo, PEDAR enabled precise correction of a 1.38-kilobase pathogenic insertion disrupting the *FAH* gene in a mouse model of tyrosinemia [279]. In the presence of a tyrosine catabolic pathway inhibitor, ~1% of hepatocytes in PEDAR-treated mice expressed *FAH*. In the absence of the inhibitor, the corrected hepatocytes gained a growth advantage and repopulated the liver, leading to ~78% corrected alleles, demonstrating the potential therapeutic relevance of this method.

Tao et al. showed that this technology can also be used for installing inter-chromosomal translocations. Interestingly, most of the translocations were unbalanced, likely because the complementary 3′ flaps promote the joining of distant regions, while the respective PAM-proximal blunt ends remain non-ligated [280]. However, the exact mechanism is not known. The two complementary 3′ flaps are required for PEDAR, since the concomitant use of a pegRNA and an sgRNA fails to produce the correct edit [279]. Therefore, it is likely that the process is driven by MMEJ or SSA, but this awaits experimental validation.

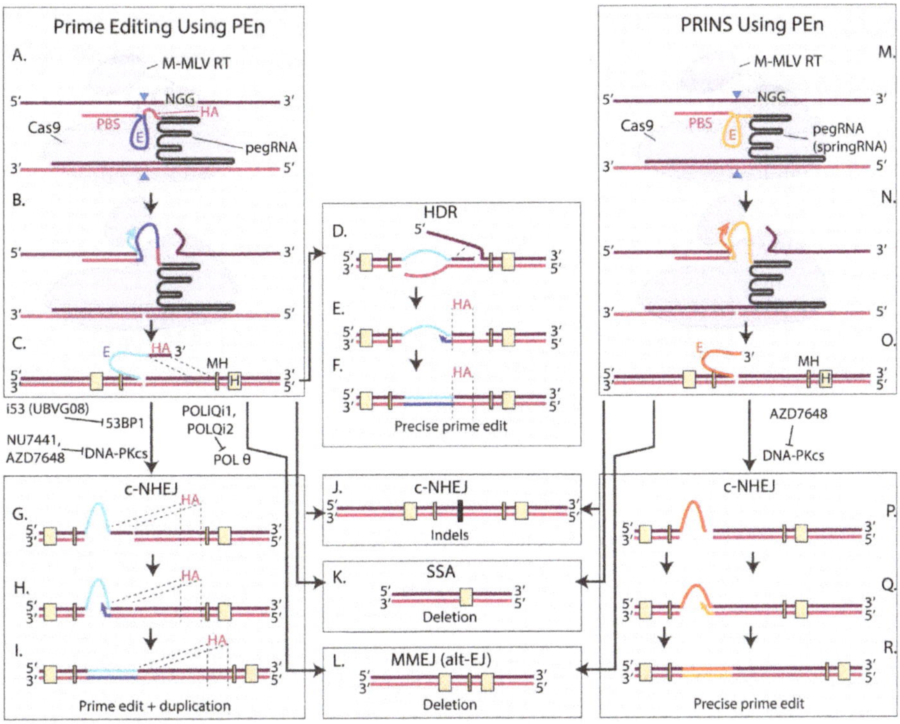

Figure 5. Nuclease-based prime editing(PE)technologies. (**A–C**) DNA cleavage and reverse transcription mediated by PEn (prime editing nuclease) and pegRNA. DNA cleavage and 3′ flap generation proceed similarly to the classic PE approach, except that a DSB is generated instead of a nick (**A**,**B**). Cleaved DNA with a 3′ flap containing a homology arm (HA, magenta) and an edit (E, light blue) is shown (**C**). Dashed lines denote the region in the chromosome that corresponds to the HA in the flap. This schematic depicts a scenario in which PEn substitutes a region downstream of the break with an edit. (**D–F**) A homology-directed mechanism leading to the correctly installed prime edit. While the exact mechanism is not known, it is governed by the HA in the flap, which needs to invade the homologous duplex at the other side of the break (**D**). In the case of a continuous substitution (depicted), the invasion will generate two flaps with the original sequence. The removal of the flaps, subsequent gap filling, and ligation (**E**) lead to successful installation of the edit (**F**). (**G–J**) Imprecise integration of the edit via c-NHEJ. In the case when strand invasion fails, c-NHEJ can align the two sides of the break, one of which contains the 3′ flap (**G**). Although the detailed mechanisms are not clear, subsequent gap filling and ligation (**H**) lead to the installation of the edit which is accompanied by the duplication of the HA (**I**). c-NHEJ may be accompanied by limited end resection, leading to indels (**J**). Inhibitors of DNA-PKcs and 53BP1 which have been shown to inhibit c-NHEJ following PEn editing are shown beside arrows that depict the sequential flow of events. (**K**) End resection may result in DSB repair via single-strand annealing (SSA), resulting in a deletion between two long homologous sequences denoted as 'H'. (**L**) End resection may result in DSB repair via MMEJ, resulting in a deletion between two microhomologies denoted as 'MH'. Inhibitors of POL θ which have been shown to inhibit MMEJ following PEn editing are shown beside arrows that depict the sequential flow of events. (**M–O**) DNA cleavage and reverse transcription mediated by PEnwith a pegRNA missing an HA (also called springRNA). The method is called PRimed INSertions (PRINS). DNA cleavage and 3′ flap generation proceed similarly to the stages depicted in A-C (**M**,**N**). Cleaved DNA with a 3′ flap containing only the edit (E, dark orange) is shown (**O**). Due to the absence of an HA, insertions are the only possible edit that can be installed. (**P–R**) Precise integration of the edit via c-NHEJ. c-NHEJ aligns the two sides of the break, one of which contains the 3′ flap (**P**), and ligates

them (**Q**), although the detailed mechanisms are not clear. Gap filling and subsequent ligation lead to the installation of the intended insertion (**R**). Similarly to PE using PEn and pegRNA, processing of the break caused by PRINS editing may result in erroneous c-NHEJ (**J**), SSA (**K**) or MMEJ (**L**). The DNA-PKcs inhibitor, which has been shown to inhibit Choi PPRINS, is shown beside an arrow that depicts the sequential flow of events.

More clarity on the DNA repair mechanism has been achieved for PEn with a single pegRNA. Adikusama et al. and later others noted that while PEn greatly improves the overall PE efficiency, many edits contain duplications of the homology arm sequence [277,278,280]. These imprecise edits might be products of DSB repair mediated by c-NHEJ when the 3′ flap is directly ligated to the PAM-proximal side of the break (Figure 5G–I), while the precise insertions could occur through a homology-dependent process (Figure 5D–F). Indeed, members of our group, demonstrated that the additional RT template integrations were abolished upon treatment with the selective inhibitor of DNA-PK, AZD7648 [161,278]. Moreover, at several loci, DNA-PK inhibition also led to an increase in total rates of correct insertions [278]. In 2023, Li et al. applied another DNA-PK inhibitor (NU7441) and demonstrated improved purity of the RT-dependent edits, reaching up to 75% precision, thus confirming the results obtained by our group [281]. They also tested the effect of inhibiting 53BP1, a protein promoting c-NHEJ by blocking the 5′ end resection required for HDR [281]. Previously it was shown that engineered ubiquitin variants prevent the recruitment of 53BP1 to DSBs [172]. Li et al. demonstrated that the UBVG08 variant, and especially its derivative G08(144A), also called i53, consistently promoted the levels of accurate edits by PEn at three tested endogenous sites [281]. This approach was called uPEn (ubiquitin-variant-assisted PEn). uPEn efficiently installed insertions (38%), deletions (43%) and substitutions (52%) in HEK293T cells [281]. Altogether, the results obtained by Peterka and Li demonstrate that inhibiting c-NHEJ is a promising approach to improve the precision of PEn.

While DNA-PK inhibition greatly reduced imprecise prime edits in experiments by Peterka et al., it did not change the percentage of unrelated indels. Strikingly, DNA-PK inhibition in *POLQ-/-* cells almost completely abolished indels without compromising precise editing [278]. At the same time, no difference in indels was observed between *wt* and *POLQ-/-* cells without DNA-PK inhibition. This important experiment demonstrates that c-NHEJ and MMEJ are redundant pathways responsible for the generation of byproducts, and a remarkable purity of prime edits can be achieved by the simultaneous inhibition of both pathways. Recently, Antoniou et al. tested this idea in *wt* cells treated with AZD7648 targeting DNA-PK and PolQi1/PolQi2 inhibiting POL θ [157,282]. Simultaneous treatment with all three compounds led to almost 100% purity in HEK293T and HeLa cells. This approach was called 2$^+$iPEn. Similar precision was achieved with the nickase-based PE5 approach, but 2$^+$iPEn's overall efficiency greatly surpassed PE5 on some of the targets (although was lower on the others) [282].

Though homology arm duplications are very frequent in PEn without inhibitors, these events also occur in nickase-based PE approaches utilizing a second nicking guide. Adikusuma et al. demonstrated that such events were present in all tested PE3 target sites and comprised 5–40% of the unintended edits [277]. Therefore, it would be interesting to test if DNA-PK inhibitors further improve the purity of PE3 and PE5.

Once it became clear that PEn-mediated imprecise edits could be inserted through a homology-independent c-NHEJ mechanism, testing a pegRNA without a HA was the next step. The method was called PRINS (Single Primed INsertion) and the pegRNA without an HA was called springRNA (Single PRimed INsertion gRNA) (Figure 5M–R) [278]. Peterka et al. showed that PRINS can install an intended insertion with up to 50% efficiency across a panel of targets in various cell lines [278]. Data from an independent study also confirmed the ability of PRINS to install an insertion at three tested targets [280]. Precise PRINS editing was completely abrogated if cells were treated with the DNA-PK inhibitor, thus confirming that NHEJ is responsible for PRINS-mediated insertion [278]. We expect that PRINS will become a useful technology for editing cell lines with inefficient homology-based pathways. The limitation of PRINS is its inability to install modifications other than insertions.

5. Delivery of Prime Editing

Delivery strategies for PEcan be categorized into three main types: physical, chemical and viral. Physical delivery methods include microinjection and electroporation. Chemical delivery strategies involve liposomes and nanoparticles, while viral delivery commonly utilizes retroviruses, adenoviruses and adeno-associated viruses (AAVs), which are the most widely used vectors for delivering PE components [283]. In this chapter, we will focus on the delivery of PE using AAVs, lipid nanoparticles (LNPs) and virus-like particles (VLP).

AAV vectors are currently the leading platform not only for PE but also for various gene therapy applications. They can transduce both dividing and non-dividing cells and rarely integrate into the host genome [284]. Additionally, they are considered safe, as they are only mildly immunogenic [285]. However, the large size of the prime editor (~6.3 kb) exceeds the packaging capacity of AAVs (~4.7 kb) [286]. To overcome this limitation, several split PE systems compatible with dual-AAV vectors were developed, such as a system with untethered Cas9 and RT, several intein-split PE systems and a system with RT and Cas9 dimerized through coiled-coil peptides [210,287–292]. To further reduce the size of the editor, several groups tested RT mutants lacking the ΔRNaseH domain, which is dispensable for PE [210,289,291].

To achieve a single AAV delivery, a 4.5 kb mini-PE editor composed of compact *Campylobacter jejuni* Cas9 (CjCas9 H559A) and truncated M-MLV RT was developed [293]. However, it demonstrated only a maximum of 10% editing efficiency in vitro and less than 1% in vivo [293]. The application of the CjCas9 editor was also limited by a long N3VRYAC PAM [294,295]. Recently, a promising evoCjCas9 variant has been derived using PANCE and PACE [296]. This variant demonstrates higher PE efficiencies and supports editing at non-canonical PAMs. Though the evoCjCas9-RT prime editor has not been tested in vivo with an AAV yet, a single AAV delivery of an evoCjCas9 base editor led to 41% editing in hepatocytes and up to 34% in mouse brain [296]. Several new approaches have since targeted models for brain, eye and liver diseases [196,204,292,297]. A study published in 2024 reported that dual-AAV delivery achieved up to 42% editing efficiency in the brain cortex, 46% in the liver and 11% in the heart [287]. Another recent study demonstrated 17.5% editing efficiency with split-AAV9 delivery at the *PCSK9* gene, which is involved in cholesterol homeostasis [298]. To further enhance therapeutic relevance, size reduction of PE components could improve the dual-AAV system by simplifying usage and increasing efficiency.

To address the size limitations of viral delivery and improve safety, alternative methods such as lipid nanoparticles (LNPs) and virus-like particles (VLPs) have been developed. LNPs typically consist of four components: ionizable lipids, cholesterol, a helper lipid and a PEG–lipid conjugate. These components form uniform spheres capable of encapsulating RNA payloads. The component ratio significantly affects LNP activity, toxicity and transfection efficiency. LNPs can deliver PE components and other CRISPR tools, such as mRNA-Cas9 LNPs, which offer efficient loading, design flexibility, and biocompatibility, making them a key player in clinical-stage gene editing for CRISPR therapies [299]. In 2023, Chen et al. successfully delivered PE mRNA via LNPs in an immunodeficient mouse model [300]. Herrera-Barrer and colleagues achieved 54% PE efficiency in a reporter cell line using enhanced LNPs (eLNPs) [301]. Despite these advancements, LNP accumulation in the liver restricts their applicability to non-hepatic tissues. Strategies for targeting specific cells or organs are under development to address this issue [299].

Virus-like particles (VLPs) are non-infectious viral capsid/envelope structures that deliver gene editing agents such as mRNA or proteins, reducing the risk of viral genome integration and off-target effects. Currently, lenti/retrovirus-based VLPs are the most used VLPs for PE delivery [302]. While most studies focus on base editing, An et al. demonstrated in 2024 that subretinal injection of v3-PE-eVLPs achieved 15% editing efficiency in a mouse model of retinal degeneration [303]. Very recently, Nanoscribes, a new type of engineered VLP, achieved up to 25% editing efficiency in myoblasts, hiPSCs and hiPSC-derived hematopoietic stem cells [304]. Given the potential of VLPs in base editing, their application in PE is expected to expand significantly in the coming years.

6. Safety

6.1. Off-Target Effects

Evaluating the safety is a pivotal aspect when considering PE as a potential therapeutic tool. PE requires not only target–guide RNA complementarity, as with other Cas9-based methods, but also target DNA-pegRNA PBS complementarity to initiate pegRNA-templated reverse transcription and target DNA-RT product complementarity for flap resolution. Hence, researchers hypothesized that these two additional DNA hybridization steps could reduce off-target PE.

In 2019, Liu and colleagues showed that PE2 and PE3 induce much lower off-target editing than Cas9 at known Cas9 off-target sites [187]. With the gradual improvement of PE systems and their on-target editing efficiencies, off-target effects have been evaluated multiple times and remained consistently low. For example, PE4, PE6 and PE7 did not substantially increase off-target PE [190–192]. Engineered pegRNA (epegRNA) did not exhibit a higher level of off-target editing compared with pegRNAs [195]. At the same time, while nuclease-based prime editors such as PEn and uPEnshowed very high on-target editing efficiency they also promoted off-targets at a level comparable to the Cas9 nuclease alone, highlighting the need for stringent peg/springRNAs and high-fidelity Cas9 enzymes.

Recently, several new methods to assess the profile of PE off-target sites have been developed. In 2023, Wolfe and colleagues published PE-tag, an approach for the genome-wide identification of prime editor activity. In the same study, they show that off-target editing rates are influenced by pegRNA design [305]. In 2022, Chen and colleagues developed a platform to profile guide-independent off-target effects in human cells. Using this approach, they demonstrated that PE3 does not cause guide-independent off-target mutations in DNA or RNA, as well as alterations in telomeres, confirming the high specificity of its reverse transcriptase moiety. In the same year, Lee and colleagues published TAgmentation of Prime Editor sequencing (TAPE-seq), another method to determine off-target candidates for PE [306].

6.2. DNA Repair Considerations

While modulating DNA repair may be an efficient strategy to improve PE, associated perturbations potentially leading to genome instability should be carefully examined. Nuclease-based PE is often accompanied by the integration of the pegRNA scaffold and duplications of the HA sequence [221,277,278]. This is something rarely observed with a nickase-based PE approach, where a 1.7% average total insertion of any number of pegRNA scaffold nucleotides has been documented [187]. The difference is due to c-NHEJ precluding correct editing in the case of the nuclease-based approaches. While c-NHEJ inhibition with mutations or small molecules like DNA-PK inhibitors AZD7648, M3814 or NU7441 are effective in decreasing small c-NHEJ-mediated indels, they are also associated with an increased rate of genomic rearrangements [157,307,308].

Since a short-range PCR is commonly used to evaluate editing outcomes, such complex events are often overlooked, while the frequency of the detected alleles is overestimated. A recent study by Cullot et al. demonstrated that AZD7648 causes frequent kilobase-scale and megabase-scale deletions, chromosome arm loss and translocations when used for genome editing [309]. The kilobase-scale deletions can be partially prevented by simultaneous treatment with AZD7648 and a POL θ inhibitor, meaning that such deletions are caused by MMEJ [157,309]. However, MMEJ inhibition did not influence megabase-scale deletions [309], suggesting that another DSBR pathway may be responsible for their generation.

Another study suggests that SSA accounts for large deletions accumulating in cells with inhibited c-NHEJ [112]. Using a multi-pathway DNA repair reporter, which is able to discriminate between unmodified alleles, HR-repaired alleles, SSA-repaired alleles and small indels, van de Kooij et al. demonstrated that several tested DNA-PK inhibitors (NU7441, M3814 and AZD7648) led to a decrease in indels accompanied by a ~1.5–2.5-fold increase in both HR and SSA. The reporter system used in this study was, by design, prone to SSA due to the presence of long homologous sequences within 3 kb of the DSB. However, the data suggest that the same principle applies when long repeats are present at endogenous sites, such as at the *HBB* and *HBD* loci or abundant targets in the human genome flanked by *Alu* repeats [112].

Given that HR and SSA share the initial resection step it may be difficult to devise a strategy to selectively inhibit SSA without any effect on HR. Indeed, depletion of BRCA1 or short-range resection factors (CtIP, MRE11) led to a decrease in both HR and SSA, while depletion of BRCA2 inhibited HR but promoted SSA, consistent with previous findings on HR and SSA regulation [112]. Surprisingly, a knockdown of long-range resection nucleases selectively inhibited SSA. siRNA against *EXO1* demonstrated the most promising results, bringing SSA after NU7441 treatment to the level observed in cells that have not been treated with siRNAs and DNA-PK inhibitors. While SSA was decreased upon NU7441 + EXO1si treatment, HR was further promoted compared to cells treated with NU7441 + nontargeting siRNA [112].

Thus, while DNA-PK inhibition alone promotes large deletions and translocations, this effect may be partially prevented by using a combination of inhibitors targeting different DSBR pathways. These studies highlight the pressing need to investigate multiple potential editing outcomes using various techniques while continuing to explore DNA repair pathways and selective inhibitors.

Author Contributions: A.M. and A.S. prepared the original draft which was reviewed and edited by M.M. All authors have read and agreed to the published version of the manuscript.

Funding: This project has received funding from the European Union's Horizon 2021 research and innovation program under the Marie Skłodowska-Curie grant agreement no. 101072427. Views and opinions expressed are however those of the author(s) only and do not necessarily reflect those of the European Union or the European Research Executive Agency. Neither the European Union nor the granting authority can be held responsible for them.

Data Availability Statement: No new data were created.

Acknowledgments: We thank Mia Melhuish for the critical review and proofreading of this manuscript.

Conflicts of Interest: The authors declare the following competing interests: M.M. and A.S. are employees and shareholders of AstraZeneca. A.S. is supported by AstraZeneca/Promega Postodoc Program. A.M. is an industrial PhD student at AstraZeneca.

References

1. Christian, M.; Cermak, T.; Doyle, E.L.; Schmidt, C.; Zhang, F.; Hummel, A.; Bogdanove, A.J.; Voytas, D.F. Targeting DNA Double-Strand Breaks with TAL Effector Nucleases. *Genetics* **2010**, *186*, 756–761. [CrossRef] [PubMed]
2. Kim, Y.-G.; Cha, J.; Chandrasegaran, S. Hybrid Restriction Enzymes: Zinc Finger Fusions to Fok I Cleavage Domain (Flavobacterium Okeanokoites/Chimeric Restriction Endonuclease/Protein Engineering/Recognition and Cleavage Domains). *Proc. Natl. Acad. Sci. USA* **1996**, *93*, 1156–1160. [CrossRef] [PubMed]
3. Jinek, M.; Chylinski, K.; Fonfara, I.; Hauer, M.; Doudna, J.A.; Charpentier, E. A Programmable Dual RNA-Guided DNA Endonuclease in Adaptive Bacterial Immunity One-Sentence Summary. *Science* **2012**, *337*, 816–821. [CrossRef] [PubMed]
4. Gasiunas, G.; Barrangou, R.; Horvath, P.; Siksnys, V. Cas9-CrRNA Ribonucleoprotein Complex Mediates Specific DNA Cleavage for Adaptive Immunity in Bacteria. *Proc. Natl. Acad. Sci. USA* **2012**, *109*, E2579–E2586. [CrossRef]
5. Makarova, K.S.; Wolf, Y.I.; Iranzo, J.; Shmakov, S.A.; Alkhnbashi, O.S.; Brouns, S.J.J.; Charpentier, E.; Cheng, D.; Haft, D.H.; Horvath, P.; et al. Evolutionary Classification of CRISPR–Cas Systems: A Burst of Class 2 and Derived Variants. *Nat. Rev. Microbiol.* **2020**, *18*, 67–83. [CrossRef]
6. A Phase 3b Study to Evaluate Efficacy and Safety of a Single Dose of Autologous CRISPR Cas9 Modified CD34+ Human Hematopoietic Stem and Progenitor Cells (CTX001) in Subjects with Transfusion-Dependent β-Thalassemia or Severe Sickle Cell Disease (2021-006390-37). Available online: https://www.clinicaltrialsregister.eu/ctr-search/trial/2021-006390-37/DE. (accessed on 24 January 2025).
7. A Phase 1/2/3 Study to Evaluate the Safety and Efficacy of a Single Dose of Autologous CRISPR-Cas9 Modified CD34+ Human Hematopoietic Stem and Progenitor Cells (CTX001) in Subjects With Severe Sickle Cell Disease (NCT03745287). Available online: https://clinicaltrials.gov/study/NCT03745287 (accessed on 24 January 2025).
8. Frangoul, H.; Altshuler, D.; Cappellini, M.D.; Chen, Y.-S.; Domm, J.; Eustace, B.K.; Foell, J.; de la Fuente, J.; Grupp, S.; Handgretinger, R.; et al. CRISPR-Cas9 Gene Editing for Sickle Cell Disease and β-Thalassemia. *N. Engl. J. Med.* **2021**, *384*, 252–260. [CrossRef]
9. Exagamglogene Autotemcel (STN:125787). Available online: https://www.fda.gov/vaccines-blood-biologics/casgevy (accessed on 24 January 2025).
10. Scully, R.; Panday, A.; Elango, R.; Willis, N.A. DNA Double-Strand Break Repair-Pathway Choice in Somatic Mammalian Cells. *Nat. Rev. Mol. Cell Biol.* **2019**, *20*, 698–714. [CrossRef]
11. Cejka, P.; Symington, L.S. DNA End Resection: Mechanisms and Control. *Annu. Rev. Genet.* **2021**, 285–307. [CrossRef]
12. Chen, C.C.; Feng, W.; Lim, P.X.; Kass, E.M.; Jasin, M. Homology-Directed Repair and the Role of BRCA1, BRCA2, and Related Proteins in Genome Integrity and Cancer. *Annu. Rev. Cancer Biol.* **2018**, *2*, 313–336. [CrossRef]
13. Zhao, B.; Rothenberg, E.; Ramsden, D.A.; Lieber, M.R. The Molecular Basis and Disease Relevance of Non-Homologous DNA End Joining. *Nat. Rev. Mol. Cell Biol.* **2020**, *21*, 765–781. [CrossRef]
14. Xue, C.; Greene, E.C. DNA Repair Pathway Choices in CRISPR-Cas9-Mediated Genome Editing. *Trends Genet.* **2021**, *37*, 639–656. [CrossRef] [PubMed]
15. Shibata, A.; Moiani, D.; Arvai, A.S.; Perry, J.; Harding, S.M.; Genois, M.M.; Maity, R.; van Rossum-Fikkert, S.; Kertokalio, A.; Romoli, F.; et al. DNA Double-Strand Break Repair Pathway Choice Is Directed by Distinct MRE11 Nuclease Activities. *Mol. Cell* **2014**, *53*, 7–18. [CrossRef] [PubMed]
16. Buis, J.; Wu, Y.; Deng, Y.; Leddon, J.; Westfield, G.; Eckersdorff, M.; Sekiguchi, J.A.M.; Chang, S.; Ferguson, D.O. Mre11 Nuclease Activity Has Essential Roles in DNA Repair and Genomic Stability Distinct from ATM Activation. *Cell* **2008**, *135*, 85–96. [CrossRef] [PubMed]
17. Ceppi, I.; Dello Stritto, M.R.; Mütze, M.; Braunshier, S.; Mengoli, V.; Reginato, G.; Võ, H.M.P.; Jimeno, S.; Acharya, A.; Roy, M.; et al. Mechanism of BRCA1–BARD1 Function in DNA End Resection and DNA Protection. *Nature* **2024**, *634*, 492–500. [CrossRef]
18. Salunkhe, S.; Daley, J.M.; Kaur, H.; Tomimatsu, N.; Xue, C.; Raina, V.B.; Jasper, A.M.; Rogers, C.M.; Li, W.; Zhou, S.; et al. Promotion of DNA End Resection by BRCA1–BARD1 in Homologous Recombination. *Nature* **2024**, *634*, 482–491. [CrossRef]
19. Greenberg, R.A.; Sobhian, B.; Pathania, S.; Cantor, S.B.; Nakatani, Y.; Livingston, D.M. Multifactorial Contributions to an Acute DNA Damage Response by BRCA1/BARD1-Containing Complexes. *Genes. Dev.* **2006**, *20*, 34–46. [CrossRef]
20. Ceppi, I.; Howard, S.M.; Kasaciunaite, K.; Pinto, C.; Anand, R.; Seidel, R.; Cejka, P. CtIP Promotes the Motor Activity of DNA2 to Accelerate Long-Range DNA End Resection. *Proc. Natl. Acad. Sci. USA* **2001**, *117*, 8859–8869. [CrossRef]
21. Myler, L.R.; Gallardo, I.F.; Soniat, M.M.; Deshpande, R.A.; Gonzalez, X.B.; Kim, Y.; Paull, T.T.; Finkelstein, I.J. Single-Molecule Imaging Reveals How Mre11-Rad50-Nbs1 Initiates DNA Break Repair. *Mol. Cell* **2017**, *67*, 891–898.e4. [CrossRef]
22. Sartori, A.A.; Lukas, C.; Coates, J.; Mistrik, M.; Fu, S.; Bartek, J.; Baer, R.; Lukas, J.; Jackson, S.P. Human CtIP Promotes DNA End Resection. *Nature* **2007**, *450*, 509–514. [CrossRef]
23. Wang, H.; Shi, L.Z.; Wong, C.C.L.; Han, X.; Hwang, P.Y.H.; Truong, L.N.; Zhu, Q.; Shao, Z.; Chen, D.J.; Berns, M.W.; et al. The Interaction of CtIP and Nbs1 Connects CDK and ATM to Regulate HR-Mediated Double-Strand Break Repair. *PLoS Genet.* **2013**, *9*, 1003277. [CrossRef]

24. Chen, L.; Nievera, C.J.; Lee, A.Y.L.; Wu, X. Cell Cycle-Dependent Complex Formation of BRCA1·CtIP·MRN Is Important for DNA Double-Strand Break Repair. *J. Biol. Chem.* **2008**, *283*, 7713–7720. [CrossRef] [PubMed]
25. Yun, M.H.; Hiom, K. CtIP-BRCA1 Modulates the Choice of DNA Double-Strand-Break Repair Pathway throughout the Cell Cycle. *Nature* **2009**, *459*, 460–463. [CrossRef] [PubMed]
26. Anand, R.; Ranjha, L.; Cannavo, E.; Cejka, P. Phosphorylated CtIP Functions as a Co-Factor of the MRE11-RAD50-NBS1 Endonuclease in DNA End Resection. *Mol. Cell* **2016**, *64*, 940–950. [CrossRef]
27. Paull, T.T.; Gellert, M. The 3′ to 5′ Exonuclease Activity of Mre11 Facilitates Repair of DNA Double-Strand Breaks. *Mol. Cell* **1998**, *1*, 969–979. [CrossRef] [PubMed]
28. Sturzenegger, A.; Burdova, K.; Kanagaraj, R.; Levikova, M.; Pinto, C.; Cejka, P.; Janscak, P. DNA2 Cooperates with the WRN and BLM RecQ Helicases to Mediate Long-Range DNA End Resection in Human Cells. *J. Biol. Chem.* **2014**, *289*, 27314–27326. [CrossRef]
29. Nimonkar, A.V.; Genschel, J.; Kinoshita, E.; Polaczek, P.; Campbell, J.L.; Wyman, C.; Modrich, P.; Kowalczykowski, S.C. BLM-DNA2-RPA-MRN and EXO1-BLM-RPA-MRN Constitute Two DNA End Resection Machineries for Human DNA Break Repair. *Genes. Dev.* **2011**, *25*, 350–362. [CrossRef]
30. Gravel, S.; Chapman, J.R.; Magill, C.; Jackson, S.P. DNA Helicases Sgs1 and BLM Promote DNA Double-Strand Break Resection. *Genes Dev.* **2008**, *22*, 2767–2772. [CrossRef]
31. Escribano-Díaz, C.; Orthwein, A.; Fradet-Turcotte, A.; Xing, M.; Young, J.T.F.; Tkáč, J.; Cook, M.A.; Rosebrock, A.P.; Munro, M.; Canny, M.D.; et al. A Cell Cycle-Dependent Regulatory Circuit Composed of 53BP1-RIF1 and BRCA1-CtIP Controls DNA Repair Pathway Choice. *Mol. Cell* **2013**, *49*, 872–883. [CrossRef]
32. Isono, M.; Niimi, A.; Oike, T.; Hagiwara, Y.; Sato, H.; Sekine, R.; Yoshida, Y.; Isobe, S.Y.; Obuse, C.; Nishi, R.; et al. BRCA1 Directs the Repair Pathway to Homologous Recombination by Promoting 53BP1 Dephosphorylation. *Cell Rep.* **2017**, *18*, 520–532. [CrossRef]
33. Benson, F.E.; Stasiak, A.; West, S.C. Purification and Characterization of the Human Rad5l Protein, an Analogue of E. Coli RecA. *EMBO J.* **1994**, *13*, 5764–5771.
34. Zelensky, A.; Kanaar, R.; Wyman, C. Mediators of Homologous DNA Pairing. *Cold Spring Harb. Perspect. Biol.* **2014**, *6*, a016451. [CrossRef] [PubMed]
35. Jensen, R.B.; Carreira, A.; Kowalczykowski, S.C. Purified Human BRCA2 Stimulates RAD51-Mediated Recombination. *Nature* **2010**, *467*, 678–683. [CrossRef] [PubMed]
36. Krais, J.J.; Wang, Y.; Patel, P.; Basu, J.; Bernhardy, A.J.; Johnson, N. RNF168-Mediated Localization of BARD1 Recruits the BRCA1-PALB2 Complex to DNA Damage. *Nat. Commun.* **2021**, *12*, 5016. [CrossRef] [PubMed]
37. Sy, S.M.H.; Huen, M.S.Y.; Chen, J.; Livingston, D.M. PALB2 Is an Integral Component of the BRCA Complex Required for Homologous Recombination Repair. *Proc. Natl. Acad. Sci. USA* **2009**, *106*, 7155–7160. [CrossRef] [PubMed]
38. Zhang, F.; Ma, J.; Wu, J.; Ye, L.; Cai, H.; Xia, B.; Yu, X. PALB2 Links BRCA1 and BRCA2 in the DNA-Damage Response. *Curr. Biol.* **2009**, *19*, 524–529. [CrossRef]
39. Baumann, P.; Benson, F.E.; West, S.C. Human Rad51 Protein Promotes ATP-Dependent Homologous Pairing and Strand Transfer Reactions In Vitro. *Cell* **1996**, *87*, 757–766. [CrossRef]
40. Mazina, O.M.; Mazin, A.V. Human Rad54 Protein Stimulates DNA Strand Exchange Activity of HRad51 Protein in the Presence of Ca2+. *J. Biol. Chem.* **2004**, *279*, 52042–52051. [CrossRef]
41. Mazin, A.V.; Alexeev, A.A.; Kowalczykowski, S.C. A Novel Function of Rad54 Protein: Stabilization of the Rad51 Nucleoprotein Filament. *J. Biol. Chem.* **2003**, *278*, 14029–14036. [CrossRef]
42. Sigurdsson, S.; Van Komen, S.; Petukhova, G.; Sung, P. Homologous DNA Pairing by Human Recombination Factors Rad51 and Rad54. *J. Biol. Chem.* **2002**, *277*, 42790–42794. [CrossRef]
43. Buisson, R.; Dion-Côté, A.M.; Coulombe, Y.; Launay, H.; Cai, H.; Stasiak, A.Z.; Stasiak, A.; Xia, B.; Masson, J.Y. Cooperation of Breast Cancer Proteins PALB2 and Piccolo BRCA2 in Stimulating Homologous Recombination. *Nat. Struct. Mol. Biol.* **2010**, *17*, 1247–1254. [CrossRef]
44. Dray, E.; Etchin, J.; Wiese, C.; Saro, D.; Williams, G.J.; Hammel, M.; Yu, X.; Galkin, V.E.; Liu, D.; Tsai, M.S.; et al. Enhancement of RAD51 Recombinase Activity by the Tumor Suppressor PALB2. *Nat. Struct. Mol. Biol.* **2010**, *17*, 1255–1259. [CrossRef] [PubMed]
45. Zhao, W.; Steinfeld, J.B.; Liang, F.; Chen, X.; Maranon, D.G.; Jian Ma, C.; Kwon, Y.; Rao, T.; Wang, W.; Sheng, C.; et al. BRCA1-BARD1 Promotes RAD51-Mediated Homologous DNA Pairing. *Nature* **2017**, *550*, 360–365. [CrossRef] [PubMed]
46. Liang, F.; Longerich, S.; Miller, A.S.; Tang, C.; Buzovetsky, O.; Xiong, Y.; Maranon, D.G.; Wiese, C.; Kupfer, G.M.; Sung, P. Promotion of RAD51-Mediated Homologous DNA Pairing by the RAD51AP1-UAF1 Complex. *Cell Rep.* **2016**, *15*, 2118–2126. [CrossRef]
47. Kim, K.P.; Mirkin, E.V. So Similar yet so Different: The Two Ends of a Double Strand Break. *Mutat. Res.* **2018**, *809*, 70–80. [CrossRef]

48. McVey, M.; Khodaverdian, V.Y.; Meyer, D.; Cerqueira, P.G.; Heyer, W.D. Eukaryotic DNA Polymerases in Homologous Recombination. *Annu. Rev. Genet.* **2016**, *50*, 393–421. [CrossRef]
49. Zapotoczny, G.; Sekelsky, J. Human Cell Assays for Synthesis-Dependent Strand Annealing and Crossing over during Double-Strand Break Repair. *G3-Genes Genomes Genet.* **2017**, *7*, 1191–1199. [CrossRef]
50. Nassif, N.; Penney, J.; Pal, S.; Engels, W.R.; Gloor, G.B. Efficient Copying of Nonhomologous Sequences from Ectopic Sites via P-Element-Induced Gap Repair. *Mol. Cell. Biol.* **1994**, *14*, 1613–1625.
51. Szostak, J.W.; Orr-Weaver, T.L.; Rothstein, R.J.; Stahl, F.W. The Double-Strand-Break Repair Model for Recombination. *Cell* **1983**, *33*, 25–35. [CrossRef]
52. Costantino, L.; Sotiriou, S.K.; Rantala, J.K.; Magin, S.; Mladenov, E.; Helleday, T.; Haber, J.E.; Iliakis, G.; Kallioniemi, O.P.; Halazonetis, T.D. Break-Induced Replication Repair of Damaged Forks Induces Genomic Duplications in Human Cells. *Science* **2014**, *343*, 88–91. [CrossRef]
53. Kockler, Z.W.; Osia, B.; Lee, R.; Musmaker, K.; Malkova, A. Repair of DNA Breaks by Break-Induced Replication. *Annu. Rev. Biochem.* **2021**, 165–191. [CrossRef]
54. Johnson, R.D.; Jasin, M. Sister Chromatid Gene Conversion Is a Prominent Double-Strand Break Repair Pathway in Mammalian Cells. *EMBO J.* **2000**, *19*, 3398–3407. [CrossRef] [PubMed]
55. Bader, A.S.; Bushell, M. IMUT-Seq: High-Resolution DSB-Induced Mutation Profiling Reveals Prevalent Homologous-Recombination Dependent Mutagenesis. *Nat. Commun.* **2023**, *14*, 8419. [CrossRef] [PubMed]
56. Kieffer, S.R.; Lowndes, N.F. Immediate-Early, Early, and Late Responses to DNA Double Stranded Breaks. *Front. Genet.* **2022**, *13*, 793884. [CrossRef]
57. Koike, M.; Yutoku, Y.; Koike, A. Nuclear Localization of Mouse Ku70 in Interphase Cells and Focus Formation of Mouse Ku70 at DNA Damage Sites Immediately after Irradiation. *J. Vet. Med. Sci.* **2015**, *77*, 1137–1142. [CrossRef]
58. Mari, P.-O.; Florea, B.I.; Persengiev, S.P.; Verkaik, N.S.; Brü, H.T.; Modesti, M.; Giglia-Mari, G.; Bezstarosti, K.; Demmers, J.A.A.; Luider, T.M.; et al. Dynamic Assembly of End-Joining Complexes Requires Interaction between Ku70/80 and XRCC4. *Proc. Natl. Acad. Sci. USA* **2006**, *103*, 18597–18602. [CrossRef]
59. Walker, J.R.; Corpina, R.A.; Goldberg, J. Structure of the Ku Heterodimer Bound to DNA and Its Implications for Double-Strand Break Repair. *Nature* **2001**, *412*, 607–614. [CrossRef]
60. Gottlieb, T.M.; Jackson, S.P. The DNA-Dependent Protein Ki for DNA Ends and Association with Ku Antigen. *Cell* **1993**, *72*, 131–142. [CrossRef]
61. Jette, N.; Lees-Miller, S.P. The DNA-Dependent Protein Kinase: A Multifunctional Protein Kinase with Roles in DNA Double Strand Break Repair and Mitosis. *Prog. Biophys. Mol. Biol.* **2015**, *117*, 194–205. [CrossRef]
62. Graham, T.G.W.; Walter, J.C.; Loparo, J.J. Two-Stage Synapsis of DNA Ends during Non-Homologous End Joining. *Mol. Cell* **2016**, *61*, 850–858. [CrossRef]
63. Chen, S.; Lee, L.; Naila, T.; Fishbain, S.; Wang, A.; Tomkinson, A.E.; Lees-Miller, S.P.; He, Y. Structural Basis of Long-Range to Short-Range Synaptic Transition in NHEJ. *Nature* **2021**, *593*, 294–298. [CrossRef]
64. Grawunder, U.; Wilm, M.; Wu, X.; Kulesza, P.; Wilson, T.E.; Mann, M.; Lieber, M.R. Activity of DNA Ligase IV Stimulated by Complex Formation with XRCC4 Protein in Mammalian Cells. *Nature* **1997**, *388*, 492–495. [CrossRef] [PubMed]
65. Stinson, B.M.; Carney, S.M.; Walter, J.C.; Loparo, J.J. Structural Role for DNA Ligase IV in Promoting the Fidelity of Non-Homologous End Joining. *Nat. Commun.* **2024**, *15*, 1250. [CrossRef]
66. Gu, J.; Lu, H.; Tsai, A.G.; Schwarz, K.; Lieber, M.R. Single-Stranded DNA Ligation and XLF-Stimulated Incompatible DNA End Ligation by the XRCC4-DNA Ligase IV Complex: Influence of Terminal DNA Sequence. *Nucleic Acids Res.* **2007**, *35*, 5755–5762. [CrossRef] [PubMed]
67. Tsai, C.J.; Kim, S.A.; Chu, G.; Lehman, I.R. CernunnosXLF Promotes the Ligation of Mismatched and Noncohesive DNA Ends. *Proc. Natl. Acad. Sci. USA* **2007**, *104*, 7851–7856. [CrossRef] [PubMed]
68. Prabhu, K.S.; Kuttikrishnan, S.; Ahamad, N.; Habeeba, U.; Mariyam, Z.; Suleman, M.; Bhat, A.A.; Uddin, S. H2AX: A Key Player in DNA Damage Response and a Promising Target for Cancer Therapy. *Biomed. Pharmacother.* **2024**, *175*, 116663. [CrossRef]
69. Riballo, E.; Kühne, M.; Rief, N.; Doherty, A.; Smith, G.C.M.; Recio, M.J.; Reis, C.; Dahm, K.; Fricke, A.; Krempler, A.; et al. A Pathway of Double-Strand Break Rejoining Dependent upon ATM, Artemis, and Proteins Locating to γ-H2AX Foci. *Mol. Cell* **2004**, *16*, 715–724. [CrossRef]
70. Fradet-Turcotte, A.; Canny, M.D.; Escribano-Díaz, C.; Orthwein, A.; Leung, C.C.Y.; Huang, H.; Landry, M.C.; Kitevski-Leblanc, J.; Noordermeer, S.M.; Sicheri, F.; et al. 53BP1 Is a Reader of the DNA-Damage-Induced H2A Lys 15 Ubiquitin Mark. *Nature* **2013**, *499*, 50–54. [CrossRef]
71. Chapman, J.R.; Barral, P.; Vannier, J.B.; Borel, V.; Steger, M.; Tomas-Loba, A.; Sartori, A.A.; Adams, I.R.; Batista, F.D.; Boulton, S.J. RIF1 Is Essential for 53BP1-Dependent Nonhomologous End Joining and Suppression of DNA Double-Strand Break Resection. *Mol. Cell* **2013**, *49*, 858–871. [CrossRef]

72. Zimmermann, M.; Lottersberger, F.; Buonomo, S.B.; Sfeir, A.; De Lange, T. 53BP1 Regulates DSB Repair Using Rif1 to Control 5′ End Resection. *Science* **2013**, *339*, 700–704. [CrossRef]
73. Dev, H.; Chiang, T.W.W.; Lescale, C.; de Krijger, I.; Martin, A.G.; Pilger, D.; Coates, J.; Sczaniecka-Clift, M.; Wei, W.; Ostermaier, M.; et al. Shieldin Complex Promotes DNA End-Joining and Counters Homologous Recombination in BRCA1-Null Cells. *Nat. Cell Biol.* **2018**, *20*, 954–965. [CrossRef]
74. Noordermeer, S.M.; Adam, S.; Setiaputra, D.; Barazas, M.; Pettitt, S.J.; Ling, A.K.; Olivieri, M.; Álvarez-Quilón, A.; Moatti, N.; Zimmermann, M.; et al. The Shieldin Complex Mediates 53BP1-Dependent DNA Repair. *Nature* **2018**, *560*, 117–121. [CrossRef] [PubMed]
75. Barazas, M.; Annunziato, S.; Pettitt, S.J.; de Krijger, I.; Ghezraoui, H.; Roobol, S.J.; Lutz, C.; Frankum, J.; Song, F.F.; Brough, R.; et al. The CST Complex Mediates End Protection at Double-Strand Breaks and Promotes PARP Inhibitor Sensitivity in BRCA1-Deficient Cells. *Cell Rep.* **2018**, *23*, 2107–2118. [CrossRef] [PubMed]
76. Mirman, Z.; Lottersberger, F.; Takai, H.; Kibe, T.; Gong, Y.; Takai, K.; Bianchi, A.; Zimmermann, M.; Durocher, D.; de Lange, T. 53BP1–RIF1–Shieldin Counteracts DSB Resection through CST- and Polα-Dependent Fill-In. *Nature* **2018**, *560*, 112–116. [CrossRef] [PubMed]
77. Gupta, R.; Somyajit, K.; Narita, T.; Maskey, E.; Stanlie, A.; Kremer, M.; Typas, D.; Lammers, M.; Mailand, N.; Nussenzweig, A.; et al. DNA Repair Network Analysis Reveals Shieldin as a Key Regulator of NHEJ and PARP Inhibitor Sensitivity. *Cell* **2018**, *173*, 972–988.e23. [CrossRef]
78. Callen, E.; Di Virgilio, M.; Kruhlak, M.J.; Nieto-Soler, M.; Wong, N.; Chen, H.T.; Faryabi, R.B.; Polato, F.; Santos, M.; Starnes, L.M.; et al. 53BP1 Mediates Productive and Mutagenic DNA Repair through Distinct Phosphoprotein Interactions. *Cell* **2013**, *153*, 1266–1280. [CrossRef]
79. Paiano, J.; Zolnerowich, N.; Wu, W.; Pavani, R.; Wang, C.; Li, H.; Zheng, L.; Shen, B.; Sleckman, B.P.; Chen, B.R.; et al. Role of 53BP1 in End Protection and DNA Synthesis at DNA Breaks. *Genes. Dev.* **2021**, *35*, 1356–1368. [CrossRef]
80. Mirman, Z.; Sasi, N.K.; King, A.; Chapman, J.R.; de Lange, T. 53BP1–Shieldin-Dependent DSB Processing in BRCA1-Deficient Cells Requires CST–Polα–Primase Fill-in Synthesis. *Nat. Cell Biol.* **2022**, *24*, 51–61. [CrossRef]
81. Bunting, S.F.; Callén, E.; Wong, N.; Chen, H.T.; Polato, F.; Gunn, A.; Bothmer, A.; Feldhahn, N.; Fernandez-Capetillo, O.; Cao, L.; et al. 53BP1 Inhibits Homologous Recombination in Brca1-Deficient Cells by Blocking Resection of DNA Breaks. *Cell* **2010**, *141*, 243–254. [CrossRef]
82. Callen, E.; Zong, D.; Wu, W.; Wong, N.; Stanlie, A.; Ishikawa, M.; Pavani, R.; Dumitrache, L.C.; Byrum, A.K.; Mendez-Dorantes, C.; et al. 53BP1 Enforces Distinct Pre- and Post-Resection Blocks on Homologous Recombination. *Mol. Cell* **2020**, *77*, 26–38.e7. [CrossRef]
83. Mirman, Z.; De Lange, T. 53BP1: A DSB Escort. *Genes Dev.* **2020**, *34*, 7–23. [CrossRef]
84. Wang, J.; Aroumougame, A.; Lobrich, M.; Li, Y.; Chen, D.; Chenj, J.; Gong, Z. PTIP Associates with Artemis to Dictate DNA Repair Pathway Choice. *Genes Dev.* **2014**, *28*, 2693–2698. [CrossRef] [PubMed]
85. Ma, Y.; Pannicke, U.; Schwarz, K.; Lieber, M.R. Hairpin Opening and Overhang Processing by an Artemis/DNA-Dependent Protein Kinase Complex in Nonhomologous End Joining and V(D)J Recombination. *Cell* **2002**, *108*, 781–794. [CrossRef] [PubMed]
86. Goodarzi, A.A.; Yu, Y.; Riballo, E.; Douglas, P.; Walker, S.A.; Ye, R.; Härer, C.; Marchetti, C.; Morrice, N.; Jeggo, P.A.; et al. DNA-PK Autophosphorylation Facilitates Artemis Endonuclease Activity. *EMBO J.* **2006**, *25*, 3880–3889. [CrossRef] [PubMed]
87. Chang, H.H.Y.; Lieber, M.R. Structure-Specific Nuclease Activities of Artemis and the Artemis: DNA-PKcs Complex. *Nucleic Acids Res.* **2016**, *44*, 4991–4997. [CrossRef]
88. Lieber, M.R. Pol X DNA Polymerases Contribute to NHEJ Flexibility. *Nat. Struct. Mol. Biol.* **2023**, *30*, 5–8. [CrossRef]
89. Sfeir, A.; Symington, L.S. Microhomology-Mediated End Joining: A Back-up Survival Mechanism or Dedicated Pathway? *Trends Biochem. Sci.* **2015**, *40*, 701–714. [CrossRef]
90. Patterson-Fortin, J.; D'Andrea, A.D. Exploiting the Microhomology-Mediated End-Joining Pathway in Cancer Therapy. *Cancer Res.* **2020**, *80*, 4593–4600. [CrossRef]
91. Hussain, S.S.; Majumdar, R.; Moore, G.M.; Narang, H.; Buechelmaier, E.S.; Bazil, M.J.; Ravindran, P.T.; Leeman, J.E.; Li, Y.; Jalan, M.; et al. Measuring Nonhomologous End-Joining, Homologous Recombination and Alternative End-Joining Simultaneously at an Endogenous Locus in Any Transfectable Human Cell. *Nucleic Acids Res.* **2021**, *49*, e74. [CrossRef]
92. Simsek, D.; Brunet, E.; Wong, S.Y.W.; Katyal, S.; Gao, Y.; McKinnon, P.J.; Lou, J.; Zhang, L.; Li, J.; Rebar, E.J.; et al. DNA Ligase III Promotes Alternative Nonhomologous End-Joining during Chromosomal Translocation Formation. *PLoS Genet.* **2011**, *7*, e1002080. [CrossRef]
93. Howard, S.M.; Yanez, D.A.; Stark, J.M. DNA Damage Response Factors from Diverse Pathways, Including DNA Crosslink Repair, Mediate Alternative End Joining. *PLoS Genet.* **2015**, *11*, e1004943. [CrossRef]
94. Mateos-Gomez, P.A.; Gong, F.; Nair, N.; Miller, K.M.; Lazzerini-Denchi, E.; Sfeir, A. Mammalian Polymerase θ Promotes Alternative NHEJ and Suppresses Recombination. *Nature* **2015**, *518*, 254–257. [CrossRef]

95. Sharma, S.; Javadekar, S.M.; Pandey, M.; Srivastava, M.; Kumari, R.; Raghavan, S.C. Homology and Enzymatic Requirements of Microhomology-Dependent Alternative End Joining. *Cell Death Dis.* **2015**, *6*, e1697. [CrossRef] [PubMed]
96. Wang, M.; Wu, W.; Wu, W.; Rosidi, B.; Zhang, L.; Wang, H.; Iliakis, G. PARP-1 and Ku Compete for Repair of DNA Double Strand Breaks by Distinct NHEJ Pathways. *Nucleic Acids Res.* **2006**, *34*, 6170–6182. [CrossRef]
97. Haince, J.F.; McDonald, D.; Rodrigue, A.; Déry, U.; Masson, J.Y.; Hendzel, M.J.; Poirier, G.G. PARP1-Dependent Kinetics of Recruitment of MRE11 and NBS1 Proteins to Multiple DNA Damage Sites. *J. Biol. Chem.* **2008**, *283*, 1197–1208. [CrossRef]
98. Audebert, M.; Salles, B.; Calsou, P. Involvement of Poly(ADP-Ribose) Polymerase-1 and XRCC1/DNA Ligase III in an Alternative Route for DNA Double-Strand Breaks Rejoining. *J. Biol. Chem.* **2004**, *279*, 55117–55126. [CrossRef]
99. Vekariya, U.; Minakhin, L.; Chandramouly, G.; Tyagi, M.; Kent, T.; Sullivan-Reed, K.; Atkins, J.; Ralph, D.; Nieborowska-Skorska, M.; Kukuyan, A.M.; et al. PARG Is Essential for Polθ-Mediated DNA End-Joining by Removing Repressive Poly-ADP-Ribose Marks. *Nat. Commun.* **2024**, *15*, 5822. [CrossRef]
100. Seki, M.; Marini, F.; Wood, R.D. POLQ (Pol θ), a DNA Polymerase and DNA-Dependent ATPase in Human Cells. *Nucleic Acids Res.* **2003**, *31*, 6117–6126. [CrossRef]
101. Mateos-Gomez, P.A.; Kent, T.; Deng, S.K.; Mcdevitt, S.; Kashkina, E.; Hoang, T.M.; Pomerantz, R.T.; Sfeir, A. The Helicase Domain of Polθ Counteracts RPA to Promote Alt-NHEJ. *Nat. Struct. Mol. Biol.* **2017**, *24*, 1116–1123. [CrossRef]
102. Ozdemir, A.Y.; Rusanov, T.; Kent, T.; Siddique, L.A.; Pomerantz, R.T. Polymerase θ-Helicase Efficiently Unwinds DNA and RNA-DNA Hybrids. *J. Biol. Chem.* **2018**, *293*, 5259–5269. [CrossRef]
103. Hogg, M.; Sauer-Eriksson, A.E.; Johansson, E. Promiscuous DNA Synthesis by Human DNA Polymerase θ. *Nucleic Acids Res.* **2012**, *40*, 2611–2622. [CrossRef]
104. Zahn, K.E.; Averill, A.M.; Aller, P.; Wood, R.D.; Doublié, S. Human DNA Polymerase θ Grasps the Primer Terminus to Mediate DNA Repair. *Nat. Struct. Mol. Biol.* **2015**, *22*, 304–311. [CrossRef] [PubMed]
105. Mengwasser, K.E.; Adeyemi, R.O.; Leng, Y.; Choi, M.Y.; Clairmont, C.; D'Andrea, A.D.; Elledge, S.J. Genetic Screens Reveal FEN1 and APEX2 as BRCA2 Synthetic Lethal Targets. *Mol. Cell* **2019**, *73*, 885–899.e6. [CrossRef] [PubMed]
106. Fleury, H.; MacEachern, M.K.; Stiefel, C.M.; Anand, R.; Sempeck, C.; Nebenfuehr, B.; Maurer-Alcalá, K.; Ball, K.; Proctor, B.; Belan, O.; et al. The APE2 Nuclease Is Essential for DNA Double-Strand Break Repair by Microhomology-Mediated End Joining. *Mol. Cell* **2023**, *83*, 1429–1445.e8. [CrossRef] [PubMed]
107. Tomkinson, A.E.; Sallmyr, A. Structure and Function of the DNA Ligases Encoded by the Mammalian LIG3 Gene. *Gene* **2013**, *531*, 150–157. [CrossRef]
108. Okano, S.; Lan, L.; Caldecott, K.W.; Mori, T.; Yasui, A. Spatial and Temporal Cellular Responses to Single-Strand Breaks in Human Cells. *Mol. Cell. Biol.* **2003**, *23*, 3974–3981. [CrossRef]
109. Lin, F.-L.; Sperle, K.; Sternberg, N. Model for Homologous Recombination During Transfer of DNA into Mouse L Cells: Role for DNA Ends in the Recombination Process. *Mol. Cell. Biol.* **1984**, *4*, 1020–1034.
110. Bennardo, N.; Cheng, A.; Huang, N.; Stark, J.M. Alternative-NHEJ Is a Mechanistically Distinct Pathway of Mammalian Chromosome Break Repair. *PLoS Genet.* **2008**, *4*, e1000110. [CrossRef]
111. Bhargava, R.; Onyango, D.O.; Stark, J.M. Regulation of Single-Strand Annealing and Its Role in Genome Maintenance. *Trends Genet.* **2016**, *32*, 566–575. [CrossRef]
112. van de Kooij, B.; Kruswick, A.; van Attikum, H.; Yaffe, M.B. Multi-Pathway DNA-Repair Reporters Reveal Competition between End-Joining, Single-Strand Annealing and Homologous Recombination at Cas9-Induced DNA Double-Strand Breaks. *Nat. Commun.* **2022**, *13*, 5295. [CrossRef]
113. Grimme, J.M.; Honda, M.; Wright, R.; Okuno, Y.; Rothenberg, E.; Mazin, A.V.; Ha, T.; Spies, M. Human Rad52 Binds and Wraps Single-Stranded DNA and Mediates Annealing via Two HRad52-SsDNA Complexes. *Nucleic Acids Res.* **2010**, *38*, 2917–2930. [CrossRef]
114. Kagawa, W.; Kurumizaka, H.; Ikawa, S.; Yokoyama, S.; Shibata, T. Homologous Pairing Promoted by the Human Rad52 Protein. *J. Biol. Chem.* **2001**, *276*, 35201–35208. [CrossRef] [PubMed]
115. Liang, C.C.; Greenhough, L.A.; Masino, L.; Maslen, S.; Bajrami, I.; Tuppi, M.; Skehel, M.; Taylor, I.A.; West, S.C. Mechanism of Single-Stranded DNA Annealing by RAD52–RPA Complex. *Nature* **2024**, *629*, 697–703. [CrossRef] [PubMed]
116. Stasiak, A.Z.; Larquet, E.; Stasiak, A.; Müller, S.; Engel, A.; Van Dyck, E.; West, S.C.; Egelman, E.H. The Human Rad52 Protein Exists as a Heptameric Ring. *Curr. Biol.* **2000**, *10*, 337–340. [CrossRef]
117. Balboni, B.; Marotta, R.; Rinaldi, F.; Milordini, G.; Varignani, G.; Girotto, S.; Cavalli, A. An Integrative Structural Study of the Human Full-Length RAD52 at 2.2 Å Resolution. *Commun. Biol.* **2024**, *7*, 956. [CrossRef]
118. Kagawa, W.; Kurumizaka, H.; Ishitani, R.; Fukai, S.; Nureki, O.; Shibata, T.; Yokoyama, S. Crystal Structure of the Homologous-Pairing Domain from the Human Rad52 Recombinase in the Undecameric Form. *Mol. Cell* **2002**, *10*, 359–371. [CrossRef] [PubMed]
119. Ranatunga, W.; Jackson, D.; Lloyd, J.A.; Forget, A.L.; Knight, K.L.; Borgstahl, G.E.O. Human RAD52 Exhibits Two Modes of Self-Association. *J. Biol. Chem.* **2001**, *276*, 15876–15880. [CrossRef]

120. Kinoshita, C.; Takizawa, Y.; Saotome, M.; Ogino, S.; Kurumizaka, H.; Kagawa, W. The Cryo-EM Structure of Full-Length RAD52 Protein Contains an Undecameric Ring. *FEBS Open Bio* **2023**, *13*, 408–418. [CrossRef]
121. Sargent, R.G.; Meservy, J.L.; Perkins, B.D.; Kilburn, A.E.; Intody, Z.; Adair, G.M.; Nairn, R.S.; Wilson, J.H. Role of the Nucleotide Excision Repair Gene ERCC1 in Formation of Recombination-Dependent Rearrangements in Mammalian Cells. *Nucleic Acids Res.* **2000**, *28*, 3771–3778. [CrossRef]
122. Al-minawi, A.Z.; Saleh-gohari, N.; Helleday, T. The ERCC1/XPF Endonuclease Is Required for Efficient Single-Strand Annealing and Gene Conversion in Mammalian Cells. *Nucleic Acids Res.* **2008**, *36*, 1–9. [CrossRef]
123. Motycka, T.A.; Bessho, T.; Post, S.M.; Sung, P.; Tomkinson, A.E. Physical and Functional Interaction between the XPF/ERCC1 Endonuclease and HRad52. *J. Biol. Chem.* **2004**, *279*, 13634–13639. [CrossRef]
124. Zhao, X.; Wei, C.; Li, J.; Xing, P.; Li, J.; Zheng, S.; Chen, X. Cell Cycle-Dependent Control of Homologous Recombination. *Acta Biochim. Biophys. Sin. (Shanghai)* **2017**, *49*, 655–668. [CrossRef] [PubMed]
125. Beucher, A.; Birraux, J.; Tchouandong, L.; Barton, O.; Shibata, A.; Conrad, S.; Goodarzi, A.A.; Krempler, A.; Jeggo, P.A.; Löbrich, M. ATM and Artemis Promote Homologous Recombination of Radiation-Induced DNA Double-Strand Breaks in G2. *EMBO J.* **2009**, *28*, 3413–3427. [CrossRef]
126. Rothkamm, K.; Krüger, I.; Thompson, L.H.; Lübrich, M. Pathways of DNA Double-Strand Break Repair during the Mammalian Cell Cycle. *Mol. Cell. Biol.* **2003**, *23*, 5706–5715. [CrossRef] [PubMed]
127. Hinz, J.M.; Yamada, N.A.; Salazar, E.P.; Tebbs, R.S.; Thompson, L.H. Influence of Double-Strand-Break Repair Pathways on Radiosensitivity throughout the Cell Cycle in CHO Cells. *DNA Repair* **2005**, *4*, 782–792. [CrossRef]
128. van Sluis, M.; McStay, B. A Localized Nucleolar DNA Damage Response Facilitates Recruitment of the Homology-Directed Repair Machinery Independent of Cell Cycle Stage. *Genes. Dev.* **2015**, *29*, 1151–1163. [CrossRef]
129. Yilmaz, D.; Furst, A.; Meaburn, K.; Lezaja, A.; Wen, Y.; Altmeyer, M.; Reina-San-Martin, B.; Soutoglou, E. Activation of Homologous Recombination in G1 Preserves Centromeric Integrity. *Nature* **2021**, *600*, 748–753. [CrossRef]
130. Ait Saada, A.; Lambert, S.A.E.; Carr, A.M. Preserving Replication Fork Integrity and Competence via the Homologous Recombination Pathway. *DNA Repair* **2018**, *71*, 135–147. [CrossRef]
131. Löbrich, M.; Jeggo, P. A Process of Resection-Dependent Nonhomologous End Joining Involving the Goddess Artemis. *Trends Biochem. Sci.* **2017**, *42*, 690–701. [CrossRef]
132. Shibata, A.; Jeggo, P.A. Roles for the DNA-PK Complex and 53BP1 in Protecting Ends from Resection during DNA Double-Strand Break Repair. *J. Radiat. Res.* **2020**, *61*, 718–726. [CrossRef]
133. Biehs, R.; Steinlage, M.; Barton, O.; Juhász, S.; Künzel, J.; Spies, J.; Shibata, A.; Jeggo, P.A.; Löbrich, M. DNA Double-Strand Break Resection Occurs during Non-Homologous End Joining in G1 but Is Distinct from Resection during Homologous Recombination. *Mol. Cell* **2017**, *65*, 671–684.e5. [CrossRef]
134. Shibata, A.; Conrad, S.; Birraux, J.; Geuting, V.; Barton, O.; Ismail, A.; Kakarougkas, A.; Meek, K.; Taucher-Scholz, G.; Löbrich, M.; et al. Factors Determining DNA Double-Strand Break Repair Pathway Choice in G2 Phase. *EMBO J.* **2011**, *30*, 1079–1092. [CrossRef]
135. Brinkman, E.K.; Chen, T.; de Haas, M.; Holland, H.A.; Akhtar, W.; van Steensel, B. Kinetics and Fidelity of the Repair of Cas9-Induced Double-Strand DNA Breaks. *Mol. Cell* **2018**, *70*, 801–813.e6. [CrossRef] [PubMed]
136. Sternberg, S.H.; Redding, S.; Jinek, M.; Greene, E.C.; Doudna, J.A. DNA Interrogation by the CRISPR RNA-Guided Endonuclease Cas9. *Nature* **2014**, *507*, 62–67. [CrossRef] [PubMed]
137. Richardson, C.D.; Ray, G.J.; DeWitt, M.A.; Curie, G.L.; Corn, J.E. Enhancing Homology-Directed Genome Editing by Catalytically Active and Inactive CRISPR-Cas9 Using Asymmetric Donor DNA. *Nat. Biotechnol.* **2016**, *34*, 339–344. [CrossRef] [PubMed]
138. Stark, J.M.; Pierce, A.J.; Oh, J.; Pastink, A.; Jasin, M. Genetic Steps of Mammalian Homologous Repair with Distinct Mutagenic Consequences. *Mol. Cell. Biol.* **2004**, *24*, 9305–9316. [CrossRef]
139. Muñoz, M.C.; Laulier, C.; Gunn, A.; Cheng, A.; Robbiani, D.F.; Nussenzweig, A.; Stark, J.M. Ring Finger Nuclear Factor RNF168 Is Important for Defects in Homologous Recombination Caused by Loss of the Breast Cancer Susceptibility Factor BRCA1. *J. Biol. Chem.* **2012**, *287*, 40618–40628. [CrossRef]
140. Anantha, R.W.; Simhardi, S.; Foo, T.K.; Miao, S.; Liu, J.; Shen, Z.; Ganesan, S.; Xia, B. Functional and Mutational Landscapes of BRCA1 for Homology-Directed Repair and Therapy Resistance. *eLife* **2017**, *6*, e21350. [CrossRef]
141. Truong, L.N.; Li, Y.; Shi, L.Z.; Hwang, P.Y.H.; He, J.; Wang, H.; Razavian, N.; Berns, M.W.; Wu, X. Microhomology-Mediated End Joining and Homologous Recombination Share the Initial End Resection Step to Repair DNA Double-Strand Breaks in Mammalian Cells. *Proc. Natl. Acad. Sci. USA* **2013**, *110*, 7720–7725. [CrossRef]
142. Llorens-Agost, M.; Ensminger, M.; Le, H.P.; Gawai, A.; Liu, J.; Cruz-García, A.; Bhetawal, S.; Wood, R.D.; Heyer, W.D.; Löbrich, M. POLθ-Mediated End Joining Is Restricted by RAD52 and BRCA2 until the Onset of Mitosis. *Nat. Cell Biol.* **2021**, *23*, 1095–1104. [CrossRef]
143. Brambati, A.; Sacco, O.; Porcella, S.; Heyza, J.; Kareh, M.; Schmidt, J.C.; Sfeir, A. RHINO Directs MMEJ to Repair DNA Breaks in Mitosis. *Science* **2023**, *381*, 653–660. [CrossRef]

144. DeWitt, M.A.; Magis, W.; Bray, N.L.; Wang, T.; Berman, J.R.; Urbinati, F.; Heo, S.J.; Mitros, T.; Muñoz, D.P.; Boffelli, D.; et al. Selection-Free Genome Editing of the Sickle Mutation in Human Adult Hematopoietic Stem/Progenitor Cells. *Sci. Transl. Med.* **2016**, *8*, 360ra134. [CrossRef]
145. Yang, L.; Guell, M.; Byrne, S.; Yang, J.L.; De Los Angeles, A.; Mali, P.; Aach, J.; Kim-Kiselak, C.; Briggs, A.W.; Rios, X.; et al. Optimization of Scarless Human Stem Cell Genome Editing. *Nucleic Acids Res.* **2013**, *41*, 9049–9061. [CrossRef]
146. Fu, Y.W.; Dai, X.Y.; Wang, W.T.; Yang, Z.X.; Zhao, J.J.; Zhang, J.P.; Wen, W.; Zhang, F.; Oberg, K.C.; Zhang, L.; et al. Dynamics and Competition of CRISPR-Cas9 Ribonucleoproteins and AAV Donor-Mediated NHEJ, MMEJ and HDR Editing. *Nucleic Acids Res.* **2021**, *49*, 969–985. [CrossRef] [PubMed]
147. Weterings, E.; Gallegos, A.C.; Dominick, L.N.; Cooke, L.S.; Bartels, T.N.; Vagner, J.; Matsunaga, T.O.; Mahadevan, D. A Novel Small Molecule Inhibitor of the DNA Repair Protein Ku70/80. *DNA Repair* **2016**, *43*, 98–106. [CrossRef] [PubMed]
148. Sun, P.; Zhao, W.; Li, H.; Feng, Y.; Chen, L.; Cao, H. STL127705 Synergize with Olaparib in Castration-Resistant Prostate Cancer by Inhibiting Homologous Recombination and Non-Homologous End-Joining Repair. *Am. J. Cancer Res.* **2023**, *13*, 2030–2040. [PubMed]
149. Srivastava, M.; Nambiar, M.; Sharma, S.; Karki, S.S.; Goldsmith, G.; Hegde, M.; Kumar, S.; Pandey, M.; Singh, R.K.; Ray, P.; et al. An Inhibitor of Nonhomologous End-Joining Abrogates Double-Strand Break Repair and Impedes Cancer Progression. *Cell* **2012**, *151*, 1474–1487. [CrossRef]
150. Maruyama, T.; Dougan, S.K.; Truttmann, M.C.; Bilate, A.M.; Ingram, J.R.; Ploegh, H.L. Increasing the Efficiency of Precise Genome Editing with CRISPR-Cas9 by Inhibition of Nonhomologous End Joining. *Nat. Biotechnol.* **2015**, *33*, 538–542. [CrossRef]
151. Chu, V.T.; Weber, T.; Wefers, B.; Wurst, W.; Sander, S.; Rajewsky, K.; Kühn, R. Increasing the Efficiency of Homology-Directed Repair for CRISPR-Cas9-Induced Precise Gene Editing in Mammalian Cells. *Nat. Biotechnol.* **2015**, *33*, 543–548. [CrossRef]
152. Pinder, J.; Salsman, J.; Dellaire, G. Nuclear Domain "knock-in" Screen for the Evaluation and Identification of Small Molecule Enhancers of CRISPR-Based Genome Editing. *Nucleic Acids Res.* **2015**, *43*, 9379–9392. [CrossRef]
153. Riesenberg, S.; Maricic, T. Targeting Repair Pathways with Small Molecules Increases Precise Genome Editing in Pluripotent Stem Cells. *Nat Commun* **2018**, *9*, 2164. [CrossRef]
154. Yang, D.; Scavuzzo, M.A.; Chmielowiec, J.; Sharp, R.; Bajic, A.; Borowiak, M. Enrichment of G2/M Cell Cycle Phase in Human Pluripotent Stem Cells Enhances HDR-Mediated Gene Repair with Customizable Endonucleases. *Sci. Rep.* **2016**, *6*, 21264. [CrossRef] [PubMed]
155. Greco, G.E.; Matsumoto, Y.; Brooks, R.C.; Lu, Z.; Lieber, M.R.; Tomkinson, A.E. SCR7 Is Neither a Selective nor a Potent Inhibitor of Human DNA Ligase IV. *DNA Repair* **2016**, *43*, 18–23. [CrossRef] [PubMed]
156. Song, J.; Yang, D.; Xu, J.; Zhu, T.; Chen, Y.E.; Zhang, J. RS-1 Enhances CRISPR/Cas9- and TALEN-Mediated Knock-in Efficiency. *Nat. Commun.* **2016**, *7*, 10548. [CrossRef] [PubMed]
157. Wimberger, S.; Akrap, N.; Firth, M.; Brengdahl, J.; Engberg, S.; Schwinn, M.K.; Slater, M.R.; Lundin, A.; Hsieh, P.P.; Li, S.; et al. Simultaneous Inhibition of DNA-PK and Polθ Improves Integration Efficiency and Precision of Genome Editing. *Nat. Commun.* **2023**, *14*, 4761. [CrossRef]
158. Munck, J.M.; Batey, M.A.; Zhao, Y.; Jenkins, H.; Richardson, C.J.; Cano, C.; Tavecchio, M.; Barbeau, J.; Bardos, J.; Cornell, L.; et al. Chemosensitization of Cancer Cells by KU-0060648, a Dual Inhibitor of DNA-PK and PI-3K. *Mol. Cancer Ther.* **2012**, *11*, 1789–1798. [CrossRef]
159. Leahy, J.J.J.; Golding, B.T.; Griffin, R.J.; Hardcastle, I.R.; Richardson, C.; Rigoreau, L.; Smith, G.C.M. Identification of a Highly Potent and Selective DNA-Dependent Protein Kinase (DNA-PK) Inhibitor (NU7441) by Screening of Chromenone Libraries. *Bioorg Med. Chem. Lett.* **2004**, *14*, 6083–6087. [CrossRef]
160. Veuger, S.J.; Curtin, N.J.; Richardson, C.J.; Smith, G.C.M.; Durkacz, B.W. Radiosensitization and DNA Repair Inhibition by the Combined Use of Novel Inhibitors of DNA-Dependent Protein Kinase and Poly(ADP-Ribose) Polymerase-1. *Cancer Res.* **2003**, *63*, 6008–6015.
161. Fok, J.H.L.; Ramos-Montoya, A.; Vazquez-Chantada, M.; Wijnhoven, P.W.G.; Follia, V.; James, N.; Farrington, P.M.; Karmokar, A.; Willis, S.E.; Cairns, J.; et al. AZD7648 Is a Potent and Selective DNA-PK Inhibitor That Enhances Radiation, Chemotherapy and Olaparib Activity. *Nat. Commun.* **2019**, *10*, 5065. [CrossRef]
162. Riesenberg, S.; Chintalapati, M.; Macak, D.; Kanis, P.; Maricic, T.; Pääbo, S. Simultaneous Precise Editing of Multiple Genes in Human Cells. *Nucleic Acids Res.* **2019**, *47*, E116. [CrossRef]
163. Zenke, F.T.; Zimmermann, A.; Sirrenberg, C.; Dahmen, H.; Kirkin, V.; Pehl, U.; Grombacher, T.; Wilm, C.; Fuchss, T.; Amendt, C.; et al. Pharmacologic Inhibitor of DNA-PK, M3814, Potentiates Radiotherapy and Regresses Human Tumors in Mouse Models. *Mol. Cancer Ther.* **2020**, *19*, 1091–1101. [CrossRef]
164. Maresca, M.; Lin, V.G.; Guo, N.; Yang, Y. Obligate Ligation-Gated Recombination (ObLiGaRe): Custom-Designed Nuclease-Mediated Targeted Integration through Nonhomologous End Joining. *Genome Res.* **2013**, *23*, 539–546. [CrossRef] [PubMed]

165. Suzuki, K.; Tsunekawa, Y.; Hernandez-Benitez, R.; Wu, J.; Zhu, J.; Kim, E.J.; Hatanaka, F.; Yamamoto, M.; Araoka, T.; Li, Z.; et al. In Vivo Genome Editing via CRISPR/Cas9 Mediated Homology-Independent Targeted Integration. *Nature* **2016**, *540*, 144–149. [CrossRef] [PubMed]
166. Robert, F.; Barbeau, M.; Éthier, S.; Dostie, J.; Pelletier, J. Pharmacological Inhibition of DNA-PK Stimulates Cas9-Mediated Genome Editing. *Genome Med.* **2015**, *7*, 93. [CrossRef]
167. Zhang, J.P.; Li, X.L.; Li, G.H.; Chen, W.; Arakaki, C.; Botimer, G.D.; Baylink, D.; Zhang, L.; Wen, W.; Fu, Y.W.; et al. Efficient Precise Knockin with a Double Cut HDR Donor after CRISPR/Cas9-Mediated Double-Stranded DNA Cleavage. *Genome Biol.* **2017**, *18*, 35. [CrossRef]
168. Harnor, S.J.; Brennan, A.; Cano, C. Targeting DNA-Dependent Protein Kinase for Cancer Therapy. *ChemMedChem* **2017**, *12*, 895–900. [CrossRef]
169. Riesenberg, S.; Kanis, P.; Macak, D.; Wollny, D.; Düsterhöft, D.; Kowalewski, J.; Helmbrecht, N.; Maricic, T.; Pääbo, S. Efficient High-Precision Homology-Directed Repair-Dependent Genome Editing by HDRobust. *Nat Methods* **2023**, *20*, 1388–1399. [CrossRef]
170. Selvaraj, S.; Feist, W.N.; Viel, S.; Vaidyanathan, S.; Dudek, A.M.; Gastou, M.; Rockwood, S.J.; Ekman, F.K.; Oseghale, A.R.; Xu, L.; et al. High-Efficiency Transgene Integration by Homology-Directed Repair in Human Primary Cells Using DNA-PKcs Inhibition. *Nat. Biotechnol.* **2024**, *42*, 731–744. [CrossRef]
171. Feng, L.; Fong, K.W.; Wang, J.; Wang, W.; Chen, J. RIF1 Counteracts BRCA1-Mediated End Resection during DNA Repair. *J. Biol. Chem.* **2013**, *288*, 11135–11143. [CrossRef]
172. Canny, M.D.; Moatti, N.; Wan, L.C.K.; Fradet-Turcotte, A.; Krasner, D.; Mateos-Gomez, P.A.; Zimmermann, M.; Orthwein, A.; Juang, Y.C.; Zhang, W.; et al. Inhibition of 53BP1 Favors Homology-Dependent DNA Repair and Increases CRISPR-Cas9 Genome-Editing Efficiency. *Nat. Biotechnol.* **2018**, *36*, 95–102. [CrossRef]
173. Nambiar, T.S.; Billon, P.; Diedenhofen, G.; Hayward, S.B.; Taglialatela, A.; Cai, K.; Huang, J.W.; Leuzzi, G.; Cuella-Martin, R.; Palacios, A.; et al. Stimulation of CRISPR-Mediated Homology-Directed Repair by an Engineered RAD18 Variant. *Nat. Commun.* **2019**, *10*, 3395. [CrossRef]
174. Jayavaradhan, R.; Pillis, D.M.; Goodman, M.; Zhang, F.; Zhang, Y.; Andreassen, P.R.; Malik, P. CRISPR-Cas9 Fusion to Dominant-Negative 53BP1 Enhances HDR and Inhibits NHEJ Specifically at Cas9 Target Sites. *Nat. Commun.* **2019**, *10*, 2866. [CrossRef] [PubMed]
175. Zatreanu, D.; Robinson, H.M.R.; Alkhatib, O.; Boursier, M.; Finch, H.; Geo, L.; Grande, D.; Grinkevich, V.; Heald, R.A.; Langdon, S.; et al. Polθ Inhibitors Elicit BRCA-Gene Synthetic Lethality and Target PARP Inhibitor Resistance. *Nat. Commun.* **2021**, *12*, 3636. [CrossRef] [PubMed]
176. Stockley, M.L.; Ferdinand, A.; Benedetti, G.; Blencowe, P.; Boyd, S.M.; Calder, M.; Charles, M.D.; Edwardes, L.V.; Ekwuru, T.; Finch, H.; et al. Discovery, Characterization, and Structure-Based Optimization of Small-Molecule in Vitro and in Vivo Probes for Human DNA Polymerase Theta. *J. Med. Chem.* **2022**, *65*, 13879–13891. [CrossRef]
177. Schimmel, J.; Muñoz-Subirana, N.; Kool, H.; van Schendel, R.; van der Vlies, S.; Kamp, J.A.; de Vrij, F.; Kushner, S.A.; Smith, G.C.M.; Boulton, S.J.; et al. Modulating Mutational Outcomes and Improving Precise Gene Editing at CRISPR-Cas9-Induced Breaks by Chemical Inhibition of End-Joining Pathways. *Cell Rep.* **2023**, *42*, 112019. [CrossRef]
178. Lee, J.-S. Activation of ATM-Dependent DNA Damage Signal Pathway by a Histone Deacetylase Inhibitor, Trichostatin A. *Cancer Res. Treat.* **2007**, *3*, 125–130. [CrossRef]
179. Jimeno, S.; Fernández-Ávila, M.J.; Cruz-García, A.; Cepeda-García, C.; Gómez-Cabello, D.; Huertas, P. Neddylation Inhibits CtIP-Mediated Resection and Regulates DNA Double Strand Break Repair Pathway Choice. *Nucleic Acids Res.* **2015**, *43*, 987–999. [CrossRef]
180. Glanzer, J.G.; Carnes, K.A.; Soto, P.; Liu, S.; Parkhurst, L.J.; Oakley, G.G. A Small Molecule Directly Inhibits the P53 Transactivation Domain from Binding to Replication Protein A. *Nucleic Acids Res.* **2013**, *41*, 2047–2059. [CrossRef]
181. Glanzer, J.G.; Liu, S.; Oakley, G.G. Small Molecule Inhibitor of the RPA70 N-Terminal Protein Interaction Domain Discovered Using in Silico and in Vitro Methods. *Bioorg Med. Chem.* **2011**, *19*, 2589–2595. [CrossRef]
182. Kosicki, M.; Tomberg, K.; Bradley, A. Repair of Double-Strand Breaks Induced by CRISPR–Cas9 Leads to Large Deletions and Complex Rearrangements. *Nat. Biotechnol.* **2018**, *36*, 765–771. [CrossRef]
183. Turchiano, G.; Andrieux, G.; Klermund, J.; Blattner, G.; Pennucci, V.; el Gaz, M.; Monaco, G.; Poddar, S.; Mussolino, C.; Cornu, T.I.; et al. Quantitative Evaluation of Chromosomal Rearrangements in Gene-Edited Human Stem Cells by CAST-Seq. *Cell Stem Cell* **2021**, *28*, 1136–1147.e5. [CrossRef]
184. Leibowitz, M.L.; Papathanasiou, S.; Doerfler, P.A.; Blaine, L.J.; Sun, L.; Yao, Y.; Zhang, C.-Z.; Weiss, M.J.; Pellman, D. Chromothripsis as an On-Target Consequence of CRISPR-Cas9 Genome Editing. *Nature Genetics* **2021**, *53*, 895–905. [CrossRef] [PubMed]
185. Komor, A.C.; Kim, Y.B.; Packer, M.S.; Zuris, J.A.; Liu, D.R. Programmable Editing of a Target Base in Genomic DNA without Double-Stranded DNA Cleavage. *Nature* **2016**, *533*, 420–424. [CrossRef] [PubMed]

186. Gaudelli, N.M.; Komor, A.C.; Rees, H.A.; Packer, M.S.; Badran, A.H.; Bryson, D.I.; Liu, D.R. Programmable Base Editing of T to G C in Genomic DNA without DNA Cleavage. *Nature* **2017**, *551*, 464–471. [CrossRef]
187. Anzalone, A.V.; Randolph, P.B.; Davis, J.R.; Sousa, A.A.; Koblan, L.W.; Levy, J.M.; Chen, P.J.; Wilson, C.; Newby, G.A.; Raguram, A.; et al. Search-and-Replace Genome Editing without Double-Strand Breaks or Donor DNA. *Nature* **2019**, *576*, 149–157. [CrossRef]
188. Arezi, B.; Hogrefe, H. Novel Mutations in Moloney Murine Leukemia Virus Reverse Transcriptase Increase Thermostability through Tighter Binding to Template-Primer. *Nucleic Acids Res.* **2009**, *37*, 473–481. [CrossRef]
189. Baranauskas, A.; Paliksa, S.; Alzbutas, G.; Vaitkevicius, M.; Lubiene, J.; Letukiene, V.; Burinskas, S.; Sasnauskas, G.; Skirgaila, R. Generation and Characterization of New Highly Thermostable and Processive M-MuLV Reverse Transcriptase Variants. *Protein Eng. Des. Sel.* **2012**, *25*, 657–668. [CrossRef]
190. Chen, P.J.; Hussmann, J.A.; Yan, J.; Knipping, F.; Ravisankar, P.; Chen, P.F.; Chen, C.; Nelson, J.W.; Newby, G.A.; Sahin, M.; et al. Enhanced Prime Editing Systems by Manipulating Cellular Determinants of Editing Outcomes. *Cell* **2021**, *184*, 5635–5652.e29. [CrossRef]
191. Yan, J.; Oyler-Castrillo, P.; Ravisankar, P.; Ward, C.C.; Levesque, S.; Jing, Y.; Simpson, D.; Zhao, A.; Li, H.; Yan, W.; et al. Improving Prime Editing with an Endogenous Small RNA-Binding Protein. *Nature* **2024**, *628*, 639–647. [CrossRef]
192. Doman, J.L.; Pandey, S.; Neugebauer, M.E.; An, M.; Davis, J.R.; Randolph, P.B.; McElroy, A.; Gao, X.D.; Raguram, A.; Richter, M.F.; et al. Phage-Assisted Evolution and Protein Engineering Yield Compact, Efficient Prime Editors. *Cell* **2023**, *186*, 3983–4002.e26. [CrossRef]
193. Li, X.; Zhou, L.; Gao, B.Q.; Li, G.; Wang, X.; Wang, Y.; Wei, J.; Han, W.; Wang, Z.; Li, J.; et al. Highly Efficient Prime Editing by Introducing Same-Sense Mutations in PegRNA or Stabilizing Its Structure. *Nat. Commun.* **2022**, *13*, 1669. [CrossRef]
194. Zhang, G.; Liu, Y.; Huang, S.; Qu, S.; Cheng, D.; Yao, Y.; Ji, Q.; Wang, X.; Huang, X.; Liu, J. Enhancement of Prime Editing via XrRNA Motif-Joined PegRNA. *Nat. Commun.* **2022**, *13*, 1856. [CrossRef] [PubMed]
195. Nelson, J.W.; Randolph, P.B.; Shen, S.P.; Everette, K.A.; Chen, P.J.; Anzalone, A.V.; An, M.; Newby, G.A.; Chen, J.C.; Hsu, A.; et al. Engineered PegRNAs Improve Prime Editing Efficiency. *Nat. Biotechnol.* **2022**, *40*, 402–410. [CrossRef]
196. Liu, B.; Dong, X.; Cheng, H.; Zheng, C.; Chen, Z.; Rodríguez, T.C.; Liang, S.Q.; Xue, W.; Sontheimer, E.J. A Split Prime Editor with Untethered Reverse Transcriptase and Circular RNA Template. *Nat. Biotechnol.* **2022**, *40*, 1388–1393. [CrossRef] [PubMed]
197. Feng, Y.; Liu, S.; Mo, Q.; Liu, P.; Xiao, X.; Ma, H. Enhancing Prime Editing Efficiency and Flexibility with Tethered and Split PegRNAs. *Protein Cell* **2023**, *14*, 304–308. [CrossRef]
198. Hendel, A.; Bak, R.O.; Clark, J.T.; Kennedy, A.B.; Ryan, D.E.; Roy, S.; Steinfeld, I.; Lunstad, B.D.; Kaiser, R.J.; Wilkens, A.B.; et al. Chemically Modified Guide RNAs Enhance CRISPR-Cas Genome Editing in Human Primary Cells. *Nat. Biotechnol.* **2015**, *33*, 985–989. [CrossRef]
199. Chen, R.; Cao, Y.; Liu, Y.; Zhao, D.; Li, J.; Cheng, Z.; Bi, C.; Zhang, X. Enhancement of a Prime Editing System via Optimal Recruitment of the Pioneer Transcription Factor P65. *Nat. Commun.* **2023**, *14*, 257. [CrossRef]
200. Park, S.J.; Jeong, T.Y.; Shin, S.K.; Yoon, D.E.; Lim, S.Y.; Kim, S.P.; Choi, J.; Lee, H.; Hong, J.I.; Ahn, J.; et al. Targeted Mutagenesis in Mouse Cells and Embryos Using an Enhanced Prime Editor. *Genome Biol.* **2021**, *22*, 170. [CrossRef]
201. Song, M.; Lim, J.M.; Min, S.; Oh, J.S.; Kim, D.Y.; Woo, J.S.; Nishimasu, H.; Cho, S.R.; Yoon, S.; Kim, H.H. Generation of a More Efficient Prime Editor 2 by Addition of the Rad51 DNA-Binding Domain. *Nat. Commun.* **2021**, *12*, 5617. [CrossRef]
202. Truong, D.J.J.; Geilenkeuser, J.; Wendel, S.V.; Wilming, J.C.H.; Armbrust, N.; Binder, E.M.H.; Santl, T.H.; Siebenhaar, A.; Gruber, C.; Phlairaharn, T.; et al. Exonuclease-Enhanced Prime Editors. *Nat. Methods* **2024**, *21*, 455–464. [CrossRef]
203. Velimirovic, M.; Zanetti, L.C.; Shen, M.W.; Fife, J.D.; Lin, L.; Cha, M.; Akinci, E.; Barnum, D.; Yu, T.; Sherwood, R.I. Peptide Fusion Improves Prime Editing Efficiency. *Nat. Commun.* **2022**, *13*, 3512. [CrossRef]
204. Liu, P.; Liang, S.Q.; Zheng, C.; Mintzer, E.; Zhao, Y.G.; Ponnienselvan, K.; Mir, A.; Sontheimer, E.J.; Gao, G.; Flotte, T.R.; et al. Improved Prime Editors Enable Pathogenic Allele Correction and Cancer Modelling in Adult Mice. *Nat. Commun.* **2021**, *12*, 2121. [CrossRef] [PubMed]
205. Nishimasu, H.; Shi, X.; Ishiguro, S.; Gao, L.; Hirano, S.; Okazaki, S.; Noda, T.; Abudayyeh, O.O.; Gootenberg, J.S.; Mori, H.; et al. Engineered CRISPR-Cas9 Nuclease with Expanded Targeting Space. *Science* **2018**, *361*, 1259–1262. [CrossRef] [PubMed]
206. Walton, R.T.; Christie, K.A.; Whittaker, M.N.; Kleinstiver, B.P. Unconstrained Genome Targeting with Near-PAMless Engineered CRISPR-Cas9 Variants. *Science* **2020**, *368*, 290–296. [CrossRef] [PubMed]
207. Kweon, J.; Yoon, J.K.; Jang, A.H.; Shin, H.R.; See, J.E.; Jang, G.; Kim, J.I.; Kim, Y. Engineered Prime Editors with PAM Flexibility. *Mol. Ther.* **2021**, *29*, 2001–2007. [CrossRef]
208. Kleinstiver, B.P.; Prew, M.S.; Tsai, S.Q.; Topkar, V.V.; Nguyen, N.T.; Zheng, Z.; Gonzales, A.P.W.; Li, Z.; Peterson, R.T.; Yeh, J.R.J.; et al. Engineered CRISPR-Cas9 Nucleases with Altered PAM Specificities. *Nature* **2015**, *523*, 481–485. [CrossRef]
209. Kleinstiver, B.P.; Pattanayak, V.; Prew, M.S.; Tsai, S.Q.; Nguyen, N.T.; Zheng, Z.; Joung, J.K. High-Fidelity CRISPR-Cas9 Nucleases with No Detectable Genome-Wide off-Target Effects. *Nature* **2016**, *529*, 490–495. [CrossRef]
210. Böck, D.; Rothgangl, T.; Villiger, L.; Schmidheini, L.; Matsushita, M.; Mathis, N.; Ioannidi, E.; Rimann, N.; Man Grisch-Chan, H.; Kreutzer, S.; et al. In Vivo Prime Editing of a Metabolic Liver Disease in Mice. *Sci. Transl. Med.* **2022**, *14*, eabl9238. [CrossRef]

211. Oh, Y.; Lee, W.J.; Hur, J.K.; Song, W.J.; Lee, Y.; Kim, H.; Gwon, L.W.; Kim, Y.H.; Park, Y.H.; Kim, C.H.; et al. Expansion of the Prime Editing Modality with Cas9 from Francisella Novicida. *Genome Biol.* **2022**, *23*, 92. [CrossRef]
212. Liang, R.; He, Z.; Zhao, K.T.; Zhu, H.; Hu, J.; Liu, G.; Gao, Q.; Liu, M.; Zhang, R.; Qiu, J.L.; et al. Prime Editing Using CRISPR-Cas12a and Circular RNAs in Human Cells. *Nat. Biotechnol.* **2024**, *42*, 1867–1875. [CrossRef]
213. Liu, B.; Dong, X.; Zheng, C.; Keener, D.; Chen, Z.; Cheng, H.; Watts, J.K.; Xue, W.; Sontheimer, E.J. Targeted Genome Editing with a DNA-Dependent DNA Polymerase and Exogenous DNA-Containing Templates. *Nat. Biotechnol.* **2024**, *42*, 1039–1045. [CrossRef]
214. Ferreira da Silva, J.; Tou, C.J.; King, E.M.; Eller, M.L.; Rufino-Ramos, D.; Ma, L.; Cromwell, C.R.; Metovic, J.; Benning, F.M.C.; Chao, L.H.; et al. Click Editing Enables Programmable Genome Writing Using DNA Polymerases and HUH Endonucleases. *Nat. Biotechnol.* **2024**. [CrossRef]
215. Nguyen, L.T.; Rakestraw, N.R.; Pizzano, B.L.M.; Young, C.B.; Huang, Y.; Beerensson, K.T.; Fang, A.; Antal, S.G.; Anamisis, K.V.; Peggs, C.M.D.; et al. Efficient Genome Editing with Chimeric Oligonucleotide-Directed Editing. *bioRxiv* **2024**. [CrossRef]
216. Jiang, K.; Yan, Z.; Di Bernardo, M.; Sgrizzi, S.R.; Villiger, L.; Kayabolen, A.; Kim, B.; Carscadden, J.K.; Hiraizumi, M.; Nishimasu, H.; et al. Rapid Protein Evolution by Few-Shot Learning with a Protein Language Model. *bioRxiv* **2024**. [CrossRef]
217. Anzalone, A.V.; Gao, X.D.; Podracky, C.J.; Nelson, A.T.; Koblan, L.W.; Raguram, A.; Levy, J.M.; Mercer, J.A.M.; Liu, D.R. Programmable Deletion, Replacement, Integration and Inversion of Large DNA Sequences with Twin Prime Editing. *Nat. Biotechnol.* **2022**, *40*, 731–740. [CrossRef] [PubMed]
218. Wang, J.; He, Z.; Wang, G.; Zhang, R.; Duan, J.; Gao, P.; Lei, X.; Qiu, H.; Zhang, C.; Zhang, Y.; et al. Efficient Targeted Insertion of Large DNA Fragments without DNA Donors. *Nat. Methods* **2022**, *19*, 331–340. [CrossRef]
219. Choi, J.; Chen, W.; Suiter, C.C.; Lee, C.; Chardon, F.M.; Yang, W.; Leith, A.; Daza, R.M.; Martin, B.; Shendure, J. Precise Genomic Deletions Using Paired Prime Editing. *Nat. Biotechnol.* **2022**, *40*, 218–226. [CrossRef]
220. Zhuang, Y.; Liu, J.; Wu, H.; Zhu, Q.; Yan, Y.; Meng, H.; Chen, P.R.; Yi, C. Increasing the Efficiency and Precision of Prime Editing with Guide RNA Pairs. *Nat. Chem. Biol.* **2022**, *18*, 29–37. [CrossRef]
221. Tao, R.; Wang, Y.; Jiao, Y.; Hu, Y.; Li, L.; Jiang, L.; Zhou, L.; Qu, J.; Chen, Q.; Yao, S. Bi-PE: Bi-Directional Priming Improves CRISPR/Cas9 Prime Editing in Mammalian Cells. *Nucleic Acids Res.* **2022**, *50*, 6423–6434. [CrossRef]
222. Zheng, C.; Liu, B.; Dong, X.; Gaston, N.; Sontheimer, E.J.; Xue, W. Template-Jumping Prime Editing Enables Large Insertion and Exon Rewriting in Vivo. *Nat. Commun.* **2023**, *14*, 3369. [CrossRef]
223. Merrick, C.A.; Zhao, J.; Rosser, S.J. Serine Integrases: Advancing Synthetic Biology. *ACS Synth. Biol.* **2018**, *7*, 299–310. [CrossRef]
224. Thomason, L.C.; Calendar, R.; Ow, D.W. Gene Insertion and Replacement in Schizosaccharomyces Pombe Mediated by the Streptomyces Bacteriophage ΦC31 Site-Specific Recombination System. *Mol. Genet. Genom.* **2001**, *265*, 1031–1038. [CrossRef] [PubMed]
225. Pandey, S.; Gao, X.D.; Krasnow, N.A.; McElroy, A.; Tao, Y.A.; Duby, J.E.; Steinbeck, B.J.; McCreary, J.; Pierce, S.E.; Tolar, J.; et al. Efficient Site-Specific Integration of Large Genes in Mammalian Cells via Continuously Evolved Recombinases and Prime Editing. *Nat. Biomed. Eng.* **2025**, *9*, 22–39. [CrossRef] [PubMed]
226. Yarnall, M.T.N.; Ioannidi, E.I.; Schmitt-Ulms, C.; Krajeski, R.N.; Lim, J.; Villiger, L.; Zhou, W.; Jiang, K.; Garushyants, S.K.; Roberts, N.; et al. Drag-and-Drop Genome Insertion of Large Sequences without Double-Strand DNA Cleavage Using CRISPR-Directed Integrases. *Nat. Biotechnol.* **2023**, *41*, 500–512. [CrossRef]
227. Lin, Q.; Jin, S.; Zong, Y.; Yu, H.; Zhu, Z.; Liu, G.; Kou, L.; Wang, Y.; Qiu, J.L.; Li, J.; et al. High-Efficiency Prime Editing with Optimized, Paired PegRNAs in Plants. *Nat. Biotechnol.* **2021**, *39*, 923–927. [CrossRef]
228. Durrant, M.G.; Fanton, A.; Tycko, J.; Hinks, M.; Chandrasekaran, S.S.; Perry, N.T.; Schaepe, J.; Du, P.P.; Lotfy, P.; Bassik, M.C.; et al. Systematic Discovery of Recombinases for Efficient Integration of Large DNA Sequences into the Human Genome. *Nat. Biotechnol.* **2023**, *41*, 488–499. [CrossRef]
229. Acharya, S.; Wilson, T.; Gradia, S.; Kane, M.F.; Guerrette, S.; Marsischky, G.T.; Kolodner, R.; Fishel, R. HMSH2 Forms Specific Mispair-Binding Complexes with HMSH3 and HMSH6. *Proc. Natl. Acad. Sci. USA* **1996**, *93*, 13629–13634. [CrossRef]
230. Drummond, J.T.; Li, G.-M.; Longley, M.J.; Modrich, P. Isolation of an HMSH2-P160 Heterodimer That Restores DNA Mismatch Repair to Tumor Cells. *Science* **1995**, *268*, 1909–1912. [CrossRef]
231. Genschel, J.; Littman, S.J.; Drummond, J.T.; Modrich, P. Isolation of MutSβ from Human Cells and Comparison of the Mismatch Repair Specificities of MutSβ and MutSα. *J. Biol. Chem.* **1998**, *273*, 19895–19901. [CrossRef]
232. Palombo, F.; Gallinari, P.; Iaccarino, I.; Lettieri, T.; Huges, M.; D'Arrigo, A.; Truong, O.; Hsuan, J.J.; Jiricny, J. GTPB, a 160-Kilodalton Protein Essential for Mitsmatch-Binding Activity in Human Cells. *Science* **1995**, *268*, 1912–1914. [CrossRef]
233. Palombo, F.; Iaccarino, I.; Nakajima, E.; Ikejima, M.; Shimada, T.; Jiricny, J. HMutSbeta, a Heterodimer of HMSH2 and HMSH3, Binds to Insertion/Deletion Loops in DNA. *Curr. Biol.* **1996**, *6*, 1181–1184. [CrossRef]
234. Papadopoulos, N.; Nicolaides, N.C.; Liu, B.; Parsons, R.; Lengauer, C.; Palombo, F.; D'Arrigo, A.; Markowitz, S.; Willson, J.K.; Kinzler, K.W. Mutations of GTBP in Genetically Unstable Cells. *Science* **1995**, *268*, 1915–1917. [CrossRef] [PubMed]

235. Li, G.-M.; Modrich, P. Restoration of Mismatch Repair to Nuclear Extracts of H6 Colorectal Tumor Cells by a Heterodimer of Human MutL Homologs (Cancer/Genetic Instability). *Proc. Natl. Acad. Sci. USA* **1995**, *92*, 1950–1954. [CrossRef] [PubMed]
236. Blackwell, L.J.; Wang, S.; Modrich, P. DNA Chain Length Dependence of Formation and Dynamics of HMutSα·hMutLα·Heteroduplex Complexes. *J. Biol. Chem.* **2001**, *276*, 33233–33240. [CrossRef] [PubMed]
237. Kadyrov, F.A.; Dzantiev, L.; Constantin, N.; Modrich, P. Endonucleolytic Function of MutLα in Human Mismatch Repair. *Cell* **2006**, *126*, 297–308. [CrossRef]
238. Kadyrov, F.A.; Genschel, J.; Fang, Y.; Penland, E.; Edelmann, W.; Modrich, P. A Possible Mechanisms for Exonuclease 1-Dependent Eukaryotic Mismatch Repair. *Proc. Natl. Acad. Sci. USA* **2009**, *106*, 8495–8500. [CrossRef]
239. Genschel, J.; Modrich, P. Mechanism of 5′-Directed Excision in Human Mismatch Repair. *Mol. Cell* **2003**, *12*, 1077–1086. [CrossRef]
240. Blanko, E.R.; Kadyrova, L.Y.; Kadyrov, F.A. DNA Mismatch Repair Interacts with CAF-1- and ASF1A-H3-H4-Dependent Histone (H3-H4)2 Tetramer Deposition. *J. Biol. Chem.* **2016**, *291*, 9203–9217. [CrossRef]
241. Longley, M.J.; Pierce, A.J.; Modrich, P. DNA Polymerase δ Is Required for Human Mismatch Repair in Vitro. *Journal of Biological Chemistry* **1997**, *272*, 10917–10921. [CrossRef]
242. Constantin, N.; Dzantiev, L.; Kadyrov, F.A.; Modrich, P. Human Mismatch Repair: RECONSTITUTION OF A NICK-DIRECTED BIDIRECTIONAL REACTION. *J. Biol. Chem.* **2005**, *280*, 39752–39761. [CrossRef]
243. Kratz, K.; Artola-Borán, M.; Kobayashi-Era, S.; Koh, G.; Oliveira, G.; Kobayashi, S.; Oliveira, A.; Zou, X.; Richter, J.; Tsuda, M.; et al. FANCD2-Associated Nuclease 1 Partially Compensates for the Lack of Exonuclease 1 in Mismatch Repair. *Mol Cell Biol* **2021**, *41*, e0030321. [CrossRef]
244. Kadyrova, L.Y.; Dahal, B.K.; Gujar, V.; Daley, J.M.; Sung, P.; Kadyrov, F.A. The Nuclease Activity of DNA2 Promotes Exonuclease 1-Independent Mismatch Repair. *J. Biol. Chem.* **2022**, *298*, 101831. [CrossRef] [PubMed]
245. Zhang, Y.; Yuan, F.; Presnell, S.R.; Tian, K.; Gao, Y.; Tomkinson, A.E.; Gu, L.; Li, G.M. Reconstitution of 5′-Directed Human Mismatch Repair in a Purified System. *Cell* **2005**, *122*, 693–705. [CrossRef] [PubMed]
246. Sallmyr, A.; Rashid, I.; Bhandari, S.K.; Naila, T.; Tomkinson, A.E. Human DNA Ligases in Replication and Repair. *DNA Repair* **2020**, *93*, 102908. [CrossRef]
247. Habib, O.; Habib, G.; Hwang, G.-H.; Bae, S. Comprehensive Analysis of Prime Editing Outcomes in Human Embryonic Stem Cells. *Nucleic Acids Res.* **2022**, *50*, 1187–1197. [CrossRef]
248. Ferreira da Silva, J.; Oliveira, G.P.; Arasa-Verge, E.A.; Kagiou, C.; Moretton, A.; Timelthaler, G.; Jiricny, J.; Loizou, J.I. Prime Editing Efficiency and Fidelity Are Enhanced in the Absence of Mismatch Repair. *Nat. Commun.* **2022**, *13*, 760. [CrossRef]
249. Park, J.C.; Kim, Y.J.; Han, J.H.; Kim, D.; Park, M.J.; Kim, J.; Jang, H.K.; Bae, S.; Cha, H.J. MutSα and MutSβ as Size-Dependent Cellular Determinants for Prime Editing in Human Embryonic Stem Cells. *Mol. Ther. Nucleic Acids* **2023**, *32*, 914–922. [CrossRef]
250. Li, X.; Chen, W.; Martin, B.K.; Calderon, D.; Lee, C.; Choi, J.; Chardon, F.M.; McDiarmid, T.A.; Daza, R.M.; Kim, H.; et al. Chromatin Context-Dependent Regulation and Epigenetic Manipulation of Prime Editing. *Cell* **2024**, *187*, 2411–2427.e25. [CrossRef]
251. McCulloch, S.D.; Gu, L.; Li, G.M. Bi-Directional Processing of DNA Loops by Mismatch Repair-Dependent and -Independent Pathways in Human Cells. *J. Biol. Chem.* **2003**, *278*, 3891–3896. [CrossRef]
252. Koeppel, J.; Weller, J.; Peets, E.M.; Pallaseni, A.; Kuzmin, I.; Raudvere, U.; Peterson, H.; Liberante, F.G.; Parts, L. Prediction of Prime Editing Insertion Efficiencies Using Sequence Features and DNA Repair Determinants. *Nat. Biotech.* **2023**, *41*, 1446–1456. [CrossRef]
253. Thomas, D.C.; Roberts, J.D.; Kunkel, T.A. Heteroduplex Repair in Extracts of Human HeLa Cells. *J. Biol. Chem.* **1991**, *266*, 3744–3751. [CrossRef]
254. Balakrishnan, L.; Bambara, R.A. Okazaki Fragment Metabolism. *Cold Spring Harb. Perspect. Biol.* **2013**, *5*, a010173. [CrossRef] [PubMed]
255. Khatib, J.B.; Nicolae, C.M.; Moldovan, G.L. Role of Translesion DNA Synthesis in the Metabolism of Replication-Associated Nascent Strand Gaps. *J. Mol. Biol.* **2024**, *436*, 168275. [CrossRef] [PubMed]
256. Kondratick, C.M.; Washington, M.T.; Spies, M. Making Choices: DNA Replication Fork Recovery Mechanisms. *Semin. Cell Dev. Biol.* **2021**, *113*, 27–37. [CrossRef]
257. Kannouche, P.L.; Wing, J.; Lehmann, A.R. Interaction of Human DNA Polymerase η with Monoubiquitinated PCNA: A Possible Mechanism for the Polymerase Switch in Response to DNA Damage. *Mol. Cell* **2004**, *14*, 491–500. [CrossRef]
258. Motegi, A.; Liaw, H.-J.; Lee, K.-Y.; Roest, H.P.; Maas, A.; Wu, X.; Moinova, H.; Markowitz, S.D.; Ding, H.; J Hoeijmakers, J.H.; et al. Polyubiquitination of Proliferating Cell Nuclear Antigen by HLTF and SHPRH Prevents Genomic Instability from Stalled Replication Forks. *Proc. Natl. Acad. Sci. USA* **2008**, *105*, 12411–12416. [CrossRef]
259. Chiu, R.K.; Brun, J.; Ramaekers, C.; Theys, J.; Weng, L.; Lambin, P.; Gray, D.A.; Wouters, B.G. Lysine 63-Polyubiquitination Guards against Translesion Synthesis-Induced Mutations. *PLoS Genet.* **2006**, *2*, e116. [CrossRef]
260. Motegi, A.; Sood, R.; Moinova, H.; Markowitz, S.D.; Liu, P.P.; Myung, K. Human SHPRH Suppresses Genomic Instability through Proliferating Cell Nuclear Antigen Polyubiquitination. *J. Cell Biol.* **2006**, *175*, 703–708. [CrossRef]

261. Joseph, S.A.; Taglialatela, A.; Leuzzi, G.; Huang, J.W.; Cuella-Martin, R.; Ciccia, A. Time for Remodeling: SNF2-Family DNA Translocases in Replication Fork Metabolism and Human Disease. *DNA Repair* **2020**, *95*, 102943. [CrossRef]
262. Blastyák, A.; Hajdú, I.; Unk, I.; Haracska, L. Role of Double-Stranded DNA Translocase Activity of Human HLTF in Replication of Damaged DNA. *Mol. Cell. Biol.* **2009**, *30*, 684–693. [CrossRef]
263. Hishiki, A.; Hara, K.; Ikegaya, Y.; Yokoyama, H.; Shimizu, T.; Sato, M.; Hashimoto, H. Structure of a Novel DNA-Binding Domain of Helicase-like Transcription Factor (HLTF) and Its Functional Implication in DNA Damage Tolerance. *J. Biol. Chem.* **2015**, *290*, 13215–13223. [CrossRef]
264. Miller, A.K.; Mao, G.; Knicely, B.G.; Daniels, H.G.; Rahal, C.; Putnam, C.D.; Kolodner, R.D.; Goellner, E.M. Rad5 and Its Human Homologs, HLTF and SHPRH, Are Novel Interactors of Mismatch Repair. *Front. Cell Dev. Biol.* **2022**, *10*, 843121. [CrossRef] [PubMed]
265. Mazur, D.J.; Perrino, F.W. Excision of 3′ Termini by the Trex1 and TREX2 3′–5′ Exonucleases. Characterization of the Recombinant Proteins. *J. Biol. Chem.* **2001**, *276*, 17022–17029. [CrossRef]
266. Mazur, D.J.; Perrino, F.W. Identification and Expression of the TREX1 and TREX2 CDNA Sequences Encoding Mammalian 3′–5′ Exonucleases. *J. Biol. Chem.* **1999**, *274*, 19655–19660. [CrossRef] [PubMed]
267. Höss, M.; Robins, P.; Naven, T.J.; Pappin, D.J.; Sgouros, J.; Lindahl, T. A Human DNA Editing Enzyme Homologous to the Escherichia Coli DnaQ/MutD Protein. *EMBO J.* **1999**, *18*, 3868–3875. [CrossRef] [PubMed]
268. Gao, D.; Li, T.; Li, X.D.; Chen, X.; Li, Q.Z.; Wight-Carter, M.; Chen, Z.J. Activation of Cyclic GMP-AMP Synthase by Self-DNA Causes Autoimmune Diseases. *Proc. Natl. Acad. Sci. USA* **2015**, *112*, E5699–E5705. [CrossRef] [PubMed]
269. Ahn, J.; Ruiz, P.; Barber, G.N. Intrinsic Self-DNA Triggers Inflammatory Disease Dependent on STING. *J. Immunol.* **2014**, *193*, 4634–4642. [CrossRef]
270. Richards, A.; van den Maagdenberg, A.M.J.M.; Jen, J.C.; Kavanagh, D.; Bertram, P.; Spitzer, D.; Liszewski, M.K.; Barilla-Labarca, M.-L.; Terwindt, G.M.; Kasai, Y.; et al. C-Terminal Truncations in Human 3′-5′ DNA Exonuclease TREX1 Cause Autosomal Dominant Retinal Vasculopathy with Cerebral Leukodystrophy. *Nat. Genet.* **2007**, *39*, 1068–1070. [CrossRef]
271. Lee-Kirsch, M.A.; Gong, M.; Chowdhury, D.; Senenko, L.; Engel, K.; Lee, Y.-A.; de Silva, U.; Bailey, S.L.; Witte, T.; Vyse, T.J.; et al. Mutations in the Gene Encoding the 3′-5′ DNA Exonuclease TREX1 Are Associated with Systemic Lupus Erythematosus. *Nat. Genet.* **2007**, *39*, 1065–1067. [CrossRef]
272. Chowdhury, D.; Beresford, P.J.; Zhu, P.; Zhang, D.; Sung, J.-S.; Demple, B.; Perrino, F.W.; Lieberman, J. The Exonuclease TREX1 Is in the SET Complex and Acts in Concert with NM23-H1 to Degrade DNA during Granzyme A-Mediated Cell Death. *Mol. Cell.* **2006**, *23*, 133–142. [CrossRef]
273. Yang, Y.-G.; Lindahl, T.; Barnes, D.E. Trex1 Exonuclease Degrades SsDNA to Prevent Chronic Checkpoint Activation and Autoimmune Disease. *Cell* **2007**, *131*, 873–886. [CrossRef]
274. Christmann, M.; Tomicic, M.T.; Aasland, D.; Berdelle, N.; Kaina, B. Three Prime Exonuclease I (TREX1) Is Fos/AP-1 Regulated by Genotoxic Stress and Protects against Ultraviolet Light and Benzo(a)Pyrene-Induced DNA Damage. *Nucleic Acids Res.* **2010**, *38*, 6418–6432. [CrossRef] [PubMed]
275. Marple, T.; Son, M.Y.; Cheng, X.; Ko, J.H.; Sung, P.; Hasty, P. TREX2 Deficiency Suppresses Spontaneous and Genotoxin-Associated Mutagenesis. *Cell Rep.* **2024**, *43*, 113637. [CrossRef] [PubMed]
276. Ko, J.H.; Son, M.Y.; Zhou, Q.; Molnarova, L.; Song, L.; Mlcouskova, J.; Jekabsons, A.; Montagna, C.; Krejci, L.; Hasty, P. TREX2 Exonuclease Causes Spontaneous Mutations and Stress-Induced Replication Fork Defects in Cells Expressing RAD51K133A. *Cell Rep.* **2020**, *33*, 108543. [CrossRef]
277. Adikusuma, F.; Lushington, C.; Arudkumar, J.; Godahewa, G.I.; Chey, Y.C.J.; Gierus, L.; Piltz, S.; Geiger, A.; Jain, Y.; Reti, D.; et al. Optimized Nickase- and Nuclease-Based Prime Editing in Human and Mouse Cells. *Nucleic Acids Res.* **2021**, *49*, 10785–10795. [CrossRef]
278. Peterka, M.; Akrap, N.; Li, S.; Wimberger, S.; Hsieh, P.P.; Degtev, D.; Bestas, B.; Barr, J.; van de Plassche, S.; Mendoza-Garcia, P.; et al. Harnessing DSB Repair to Promote Efficient Homology-Dependent and -Independent Prime Editing. *Nat. Commun.* **2022**, *13*, 1240. [CrossRef]
279. Jiang, T.; Zhang, X.-O.; Weng, Z.; Xue, W. Deletion and Replacement of Long Genomic Sequences Using Prime Editing. *Nat. Biotechnol.* **2021**, *40*, 227–234. [CrossRef]
280. Tao, R.; Wang, Y.; Hu, Y.; Jiao, Y.; Zhou, L.; Jiang, L.; Li, L.; He, X.; Li, M.; Yu, Y.; et al. WT-PE: Prime Editing with Nuclease Wild-Type Cas9 Enables Versatile Large-Scale Genome Editing. *Signal Transduct. Target. Ther.* **2022**, *7*, 108. [CrossRef]
281. Li, X.; Zhang, G.; Huang, S.; Liu, Y.; Tang, J.; Zhong, M.; Wang, X.; Sun, W.; Yao, Y.; Ji, Q.; et al. Development of a Versatile Nuclease Prime Editor with Upgraded Precision. *Nat Commun* **2023**, *14*, 305. [CrossRef]
282. Antoniou, P.; Dacquay, L.; Selfjord, N.; Madeyski-Bengtson, K.; Loyd, A.-L.; Gordon, E.; Thom, G.; Hsieh, P.-P.; Wimberger, S.; Šviković, S.; et al. Improved Nuclease-Based Prime Editing by DNA Repair Modulation and PegRNA Engineering. *bioRxiv* **2024**. [CrossRef]

283. Fu, Y.; He, X.; Gao, X.D.; Li, F.; Ge, S.; Yang, Z.; Fan, X. Prime Editing: Current Advances and Therapeutic Opportunities in Human Diseases. *Sci. Bull.* **2023**, *68*, 3278–3291. [CrossRef]
284. Venditti, C.P. Safety Questions for AAV Gene Therapy. *Nat. Biotechnol.* **2021**, *39*, 24–26. [CrossRef] [PubMed]
285. Wang, D.; Tai, P.W.L.; Gao, G. Adeno-Associated Virus Vector as a Platform for Gene Therapy Delivery. *Nat. Rev. Drug Discov.* **2019**, *18*, 358–378. [CrossRef] [PubMed]
286. Raguram, A.; Banskota, S.; Liu, D.R. Therapeutic in Vivo Delivery of Gene Editing Agents. *Cell* **2022**, *185*, 2806–2827. [CrossRef]
287. Davis, J.R.; Banskota, S.; Levy, J.M.; Newby, G.A.; Wang, X.; Anzalone, A.V.; Nelson, A.T.; Chen, P.J.; Hennes, A.D.; An, M.; et al. Efficient Prime Editing in Mouse Brain, Liver and Heart with Dual AAVs. *Nat. Biotechnol.* **2024**, *42*, 253–264. [CrossRef]
288. Zhi, S.; Chen, Y.; Wu, G.; Wen, J.; Wu, J.; Liu, Q.; Li, Y.; Kang, R.; Hu, S.; Wang, J.; et al. Dual-AAV Delivering Split Prime Editor System for in Vivo Genome Editing. *Mol. Ther.* **2022**, *30*, 283–294. [CrossRef]
289. Zheng, C.; Liang, S.Q.; Liu, B.; Liu, P.; Kwan, S.Y.; Wolfe, S.A.; Xue, W. A Flexible Split Prime Editor Using Truncated Reverse Transcriptase Improves Dual-AAV Delivery in Mouse Liver. *Mol. Ther.* **2022**, *30*, 1343–1351. [CrossRef]
290. She, K.; Liu, Y.; Zhao, Q.; Jin, X.; Yang, Y.; Su, J.; Li, R.; Song, L.; Xiao, J.; Yao, S.; et al. Dual-AAV Split Prime Editor Corrects the Mutation and Phenotype in Mice with Inherited Retinal Degeneration. *Signal Transduct. Target. Ther.* **2023**, *8*, 57. [CrossRef]
291. Grünewald, J.; Miller, B.R.; Szalay, R.N.; Cabeceiras, P.K.; Woodilla, C.J.; Holtz, E.J.B.; Petri, K.; Joung, J.K. Engineered CRISPR Prime Editors with Compact, Untethered Reverse Transcriptases. *Nat. Biotechnol.* **2023**, *41*, 337–343. [CrossRef]
292. Qin, H.; Zhang, W.; Zhang, S.; Feng, Y.; Xu, W.; Qi, J.; Zhang, Q.; Xu, C.; Liu, S.; Zhang, J.; et al. Vision Rescue via Unconstrained in Vivo Prime Editing in Degenerating Neural Retinas. *J. Exp. Med.* **2023**, *220*, e20220776. [CrossRef]
293. Lan, T.; Chen, H.; Tang, C.; Wei, Y.; Liu, Y.; Zhou, J.; Zhuang, Z.; Zhang, Q.; Chen, M.; Zhou, X.; et al. Mini-PE, a Prime Editor with Compact Cas9 and Truncated Reverse Transcriptase. *Mol. Ther. Nucleic Acids* **2023**, *33*, 890–897. [CrossRef]
294. Yamada, M.; Watanabe, Y.; Gootenberg, J.S.; Hirano, H.; Ran, F.A.; Nakane, T.; Ishitani, R.; Zhang, F.; Nishimasu, H.; Nureki, O. Crystal Structure of the Minimal Cas9 from Campylobacter Jejuni Reveals the Molecular Diversity in the CRISPR-Cas9 Systems. *Mol. Cell* **2017**, *65*, 1109–1121.e3. [CrossRef] [PubMed]
295. Kim, E.; Koo, T.; Park, S.W.; Kim, D.; Kim, K.; Cho, H.Y.; Song, D.W.; Lee, K.J.; Jung, M.H.; Kim, S.; et al. In Vivo Genome Editing with a Small Cas9 Orthologue Derived from Campylobacter Jejuni. *Nat. Commun.* **2017**, *8*, 14500. [CrossRef] [PubMed]
296. Schmidheini, L.; Mathis, N.; Marquart, K.F.; Rothgangl, T.; Kissling, L.; Böck, D.; Chanez, C.; Wang, J.P.; Jinek, M.; Schwank, G. Continuous Directed Evolution of a Compact CjCas9 Variant with Broad PAM Compatibility. *Nat. Chem. Biol.* **2024**, *20*, 333–343. [CrossRef]
297. Jang, H.; Jo, D.H.; Cho, C.S.; Shin, J.H.; Seo, J.H.; Yu, G.; Gopalappa, R.; Kim, D.; Cho, S.R.; Kim, J.H.; et al. Application of Prime Editing to the Correction of Mutations and Phenotypes in Adult Mice with Liver and Eye Diseases. *Nat. Biomed. Eng.* **2022**, *6*, 181–194. [CrossRef]
298. Wei, R.; Yu, Z.; Ding, L.; Lu, Z.; Yao, K.; Zhang, H.; Huang, B.; He, M.; Ma, L. Improved Split Prime Editors Enable Efficient in Vivo Genome Editing. *Cell Rep.* **2025**, *44*, 115144. [CrossRef]
299. Masarwy, R.; Stotsky-Oterin, L.; Elisha, A.; Hazan-Halevy, I.; Peer, D. Delivery of Nucleic Acid Based Genome Editing Platforms via Lipid Nanoparticles: Clinical Applications. *Adv. Drug Deliv. Rev.* **2024**, *211*, 115359. [CrossRef]
300. Chen, Z.; Kelly, K.; Cheng, H.; Dong, X.; Hedger, A.K.; Li, L.; Sontheimer, E.J.; Watts, J.K. In Vivo Prime Editing by Lipid Nanoparticle Co-Delivery of Chemically Modified PegRNA and Prime Editor MRNA. *GEN Biotechnol.* **2023**, *2*, 490–502. [CrossRef]
301. Herrera-Barrera, M.; Gautam, M.; Lokras, A.; Vlasova, K.; Foged, C.; Sahay, G. Lipid Nanoparticle-Enabled Intracellular Delivery of Prime Editors. *AAPS J.* **2023**, *25*, 65. [CrossRef]
302. Hosseini, S.Y.; Mallick, R.; Mäkinen, P.; Ylä-Herttuala, S. Insights into Prime Editing Technology: A Deep Dive into Fundamentals, Potentials, and Challenges. *Hum. Gene Ther.* **2024**, *35*, 649–668. [CrossRef]
303. An, M.; Raguram, A.; Du, S.W.; Banskota, S.; Davis, J.R.; Newby, G.A.; Chen, P.Z.; Palczewski, K.; Liu, D.R. Engineered Virus-like Particles for Transient Delivery of Prime Editor Ribonucleoprotein Complexes in Vivo. *Nat. Biotechnol.* **2024**, *42*, 1526–1537. [CrossRef]
304. Halegua, T.; Risson, V.; Carras, J.; Rouyer, M.; Coudert, L.; Jacquier, A.; Schaeffer, L.; Ohlmann, T.; Mangeot, P.E. Delivery of Prime Editing in Human Stem Cells Using Pseudoviral NanoScribes Particles. *Nat. Commun.* **2025**, *16*, 397. [CrossRef] [PubMed]
305. Liang, S.Q.; Liu, P.; Ponnienselvan, K.; Suresh, S.; Chen, Z.; Kramme, C.; Chatterjee, P.; Zhu, L.J.; Sontheimer, E.J.; Xue, W.; et al. Genome-Wide Profiling of Prime Editor off-Target Sites in Vitro and in Vivo Using PE-Tag. *Nat. Methods* **2023**, *20*, 898–907. [CrossRef] [PubMed]
306. Kwon, J.; Kim, M.; Bae, S.; Jo, A.; Kim, Y.; Lee, J.K. TAPE-Seq Is a Cell-Based Method for Predicting Genome-Wide off-Target Effects of Prime Editor. *Nat. Commun.* **2022**, *13*, 7975. [CrossRef]
307. Kosicki, M.; Allen, F.; Steward, F.; Tomberg, K.; Pan, Y.; Bradley, A. Cas9-Induced Large Deletions and Small Indels Are Controlled in a Convergent Fashion. *Nat. Commun.* **2022**, *13*, 3422. [CrossRef]

308. Wen, W.; Quan, Z.J.; Li, S.A.; Yang, Z.X.; Fu, Y.W.; Zhang, F.; Li, G.H.; Zhao, M.; Yin, M.D.; Xu, J.; et al. Effective Control of Large Deletions after Double-Strand Breaks by Homology-Directed Repair and DsODN Insertion. *Genome Biol.* **2021**, *22*, 236. [CrossRef]
309. Cullot, G.; Aird, E.J.; Schlapansky, M.F.; Yeh, C.D.; van de Venn, L.; Vykhlyantseva, I.; Kreutzer, S.; Mailänder, D.; Lewków, B.; Klermund, J.; et al. Genome Editing with the HDR-Enhancing DNA-PKcs Inhibitor AZD7648 Causes Large-Scale Genomic Alterations. *Nat. Biotechnol.* **2024**. [CrossRef]

cells MDPI

Review

Regulation of Precise DNA Repair by Nuclear Actin Polymerization: A Chance for Improving Gene Therapy?

Xiubin He and Cord Brakebusch *

Biotech Research and Innovation Centre (BRIC), University of Copenhagen, Ole Maaløes Vej 5, 2200 Copenhagen, Denmark; xiubin.he@bric.ku.dk
* Correspondence: cord.brakebusch@bric.ku.dk

Abstract: Although more difficult to detect than in the cytoplasm, it is now clear that actin polymerization occurs in the nucleus and that it plays a role in the specific processes of the nucleus such as transcription, replication, and DNA repair. A number of studies suggest that nuclear actin polymerization is promoting precise DNA repair by homologous recombination, which could potentially be of help for precise genome editing and gene therapy. This review summarizes the findings and describes the challenges and chances in the field.

Keywords: nuclear actin; actin dynamics; actin polymerization; double-strand DNA breaks; DNA repair; DNA damage response; homologous recombination; chromatin remodeling; genome integrity

1. Introduction

Rare diseases are diseases with a frequency of less than 0.05%. More than 7000 rare diseases are known to be affecting about 350 million people worldwide [1]. Most rare diseases are caused by mutations, and gene therapy is therefore the only curative treatment option for these patients. CRISPR gene editing is an exciting possibility for therapeutic repair of defective genes in rare diseases [2,3]. Here, a complex of a Cas9 protein with a short guide RNA function as a genomic "scissor", which precisely cuts the genome at a single place close to the mutation, introducing a double-strand DNA break (DSB). In the presence of a repair template, the defective mutation is repaired precisely without leaving any traces in the genome. In the absence of a repair template and in competition with precise gene editing, an error-prone repair takes place, which introduces further mutations at the DSB. To increase the efficiency of gene therapy and reduce costs, a large amount of research is currently trying to increase the efficiency of precise repair during CRISPR gene editing. Nuclear actin polymerization is a candidate pathway to promote precise gene editing.

Actin is expressed in all cells and contributes by polymerization to the cytoskeleton and cell morphology. Regulation of the actin cytoskeleton is crucial for cell migration and adhesion by controlling membrane protrusions and cell contraction [4,5]. It also plays a role in endocytosis and intracellular transport of vesicles and organelles [6]. In recent years, it has been revealed that actin polymerization can also occur in the nucleus, mostly in response to stress such as serum stimulation or irradiation. This was reported to influence transcription, translation, replication, and DNA repair [7,8]. With respect to DNA repair, nuclear actin polymerization was often linked to the promotion of precise repair by homologous recombination (HR) [9]. In this review, we describe the basics of actin polymerization and the repair of DSBs, and we present and discuss the current knowledge of how nuclear actin polymerization and DNA repair might be linked.

2. Regulation of Actin Polymerization

Actin exists in two forms: the monomeric actin (G-actin) and the filamentous actin (F-actin). These forms convert to each other in a dynamic equilibrium that is controlled by a large number of actin-binding proteins (ABPs) (Figure 1) [8]. Above a critical concentration, G-actin can polymerize spontaneously to F-actin, but G-actin-binding molecules such as profilin reduce the concentration of free G-actin in the cell below the critical level, preventing uncontrolled polymerization. Profilin also promotes the exchange of ADP bound to actin by ATP, which is a key step for actin polymerization (Figure 1). At low concentrations of free actin, actin polymerization is triggered by actin nucleation-promoting factors (NPFs). Formins such as mDia and Formin 2 (FMN2) promote the linear polymerization of actin filaments, whereas the Arp2/3 complex, together with Wasp family proteins, initiates the growth of branched actin networks (Figure 1) [10,11]. A third group of NPFs are tandem-monomer-binding proteins such as Spire. Force generated by actin polymerization can mediate membrane protrusions or intracellular movements of bacteria in mammalian cells [4]. Crosslinking proteins like filamin, fimbrin, and α-actinin organize F-actin into networks and bundles, providing structural integrity to the cytoskeleton [12–14]. Within the actin filaments, actin-bound ATP is hydrolyzed to ADP, which allows binding and severing of the filaments by cofilin and stimulates depolymerization at the pointed end of the filaments (Figure 1) [5]. Barbed-end capping proteins like CapZ and gelsolin bind to the fast-growing, barbed end of actin filaments, inhibiting elongation, while the pointed-end capping protein tropomodulin can inhibit depolymerization. Myosins can "walk" on F-actin in an ATP-dependent manner, which is used for cell contraction or for the intracellular transport of vesicles and organelles [15]. Finally, many signaling pathways regulate the activity of ABPs by protein–protein interaction, post-translational modification, or by regulating their expression [15].

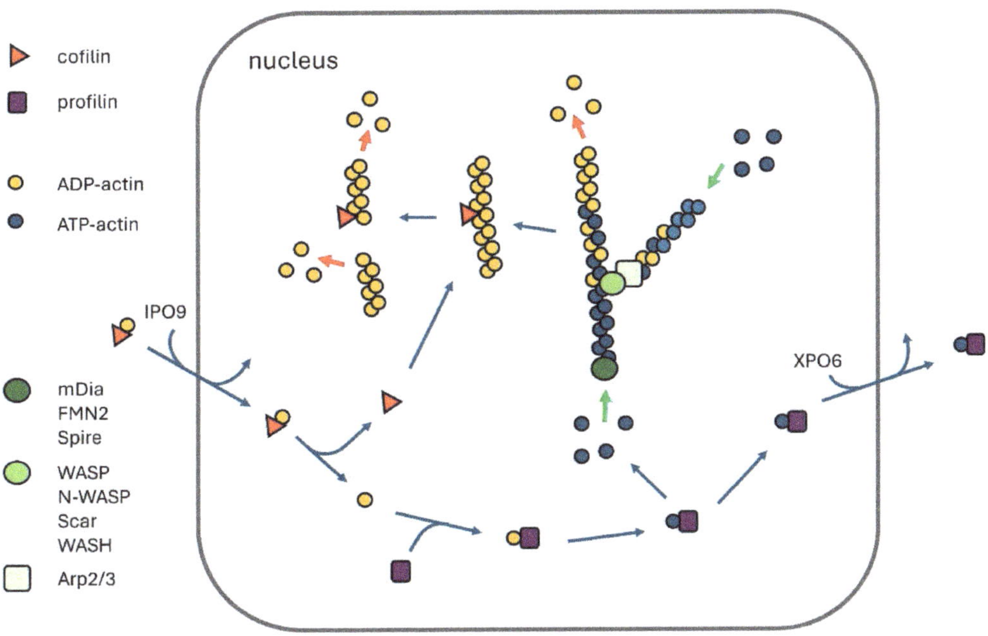

Figure 1. Regulation of nuclear actin polymerization. ADP-actin-cofilin is imported into the nucleus by IPO9, while profilin exports ATP-actin to the cytoplasm with the help of XPO6. Profilin binding to actin promotes the exchange of ADP for ATP, which facilitates actin polymerization. Formin proteins (mDia, FMN2) contribute to linear F-actin formation, which is accelerated by Spire. Arp2/3, together with Wasp family proteins, mediates branched F-actin formation. Cofilin facilitates actin depolymerization by severing ADP-bound F-actin.

3. Actin Polymerization in the Nucleus

Monomeric and filamentous actin are not only found in the cytoplasm but also in the nucleus. In addition, many ABPs are present in the nucleus, suggesting that regulated actin polymerization in the nucleus might be of physiological importance for nuclear processes [8]. Although monomeric, unbound actin, based on its small size of 42 kD, should be able to freely diffuse to the nucleus, it uses energy-dependent transport processes to quickly move in and out of the nucleus. Import of G-actin to the nucleus is mediated by unphosphorylated cofilin binding to preferentially ADP-bound G-actin and importin 9 (IPO9) (Figure 1) [16,17]. Phosphorylation of cofilin will prevent its binding to actin and, consequently, nuclear import. Regulation of cofilin phosphorylation by LIM kinase and slingshot phosphatase appears therefore as a major physiological pathway to modulate the nuclear import of G-actin [18,19]. Indeed, phosphorylation of cofilin during mitotic exit corresponded to the formation of nuclear F-actin [20], and cofilin regulators were identified in a genome-wide screen for inhibitors of nuclear actin localization and polymerization [21].

Export from the nucleus requires binding of G-actin to profilin and exportin 6 (XPO6) [22]. Inhibition of the nuclear export of actin by knockdown (KD) of XPO6 increases nuclear G-actin levels and promotes nuclear F-actin formation [23,24].

Interestingly, unphosphorylated cofilin not only promotes the nuclear import of G-actin but also severs F-actin [25]. Thus, active cofilin in the cytoplasm will promote both depolymerization of F-actin in the cytoplasm and transport of G-actin to the nucleus (Figure 1), suggesting that nuclear G-actin levels are increasing when cytoplasmic actin depolymerization is high. In addition, active cofilin mediating nuclear transport of G-actin might promote the severing and depolymerization of ADP-bound F-actin in the nucleus, resulting in a short lifetime of nuclear actin filaments induced by cofilin dephosphorylation. On the other hand, profilin not only promotes the nuclear export of actin but also facilitates the polymerization of ATP-actin by binding to NPFs [5]. These contrasting relationships might suggest that actin polymerization is less likely in the nucleus than in the cytoplasm, and that normally, nuclear F-actin in unstressed mammalian cells is much more difficult to detect than cytoplasmic F-actin. Whether the relatively low amounts of nuclear actin polymerization are due to evolutionary pressure, trying to avoid toxically high levels of nuclear actin polymerization, is unclear.

4. Repair Mechanisms of Double-Strand DNA Breaks

DSBs can be repaired by error-prone mechanisms, including classical nonhomologous end joining (cNHEJ) and microhomology-mediated end joining (MMEJ), or by precise homologous recombination (HR) (Figure 2).

cNHEJ is started by the binding of Ku70/Ku80 dimers to the DSBs, which triggers the association with other repair-associated proteins [26]. The free DNA ends are either ligated directly or after some processing by nucleases or polymerases, which often results in insertions or deletions (indels).

On the other hand, binding of CtIP and the MRE11-Rad50-NBS1 (MRN) complex, which contains the nuclease MRE11, to DSBs is initiating end resection, which is the generation of single-stranded overhangs on both sides of the DSBs [27]. This process occurs only during the S and G2 phases [28,29], while cNHEJ is possible during all phases of the cell cycle [26].

If the overhangs are about 20 bps, a short stretch of complementary base-pairing can associate the opposite sides of the break, which is then sealed by the combined action of nucleases and polymerases [30]. This repair is called MMEJ and results in a short deletion of DNA.

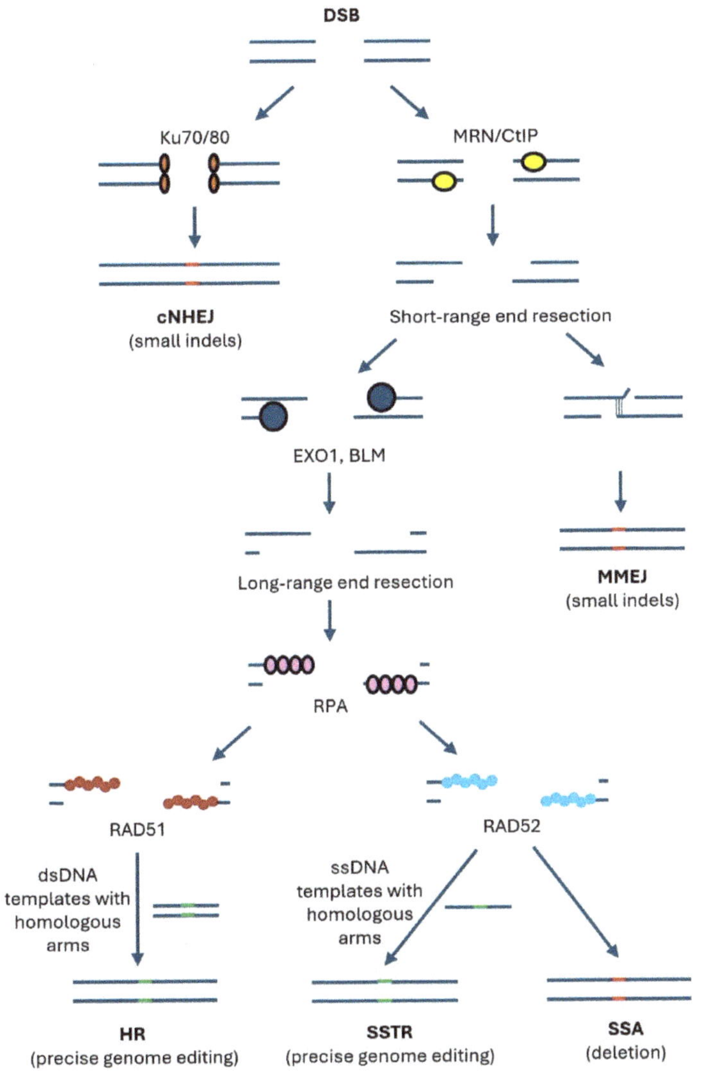

Figure 2. Repair pathways of DSB. In the cNHEJ pathway, Ku70/80 proteins bind to DSBs, which results in small deletions and insertions. When the MRN-CtIP complex binds to DSBs, short-range end resection occurs and ssDNA overhangs are generated, which can lead to the MMEJ pathway. DSBs can further enter long-range end resection mediated by proteins such as EXO1 or BLM, where the long ssDNA overhangs are bound by RPA. RPA is then replaced by RAD51 or RAD52, which direct repair to HR or SSA and SSTR, respectively. Indels are indicated in red, and precise genome editing is indicated in green.

Longer end resections, mediated by helicases and nucleases such as EXO1 and BLM, will enable the binding of replication protein A (RPA) to the single strand ends, which is later replaced by RAD52 or RAD51 [31–33]. In the absence of a template, complementary base pairing of RAD52-bound stretches longer than 20 bp can trigger deletion-causing DNA repair. This repair pathway is called single-strand annealing (SSA). In the presence of a single-stranded DNA (ssDNA) template with homologous sequences on both sides of the DSBs, precise genome editing can be performed (single-stranded template repair,

SSTR). RAD51-bound single strands can undergo HR with the help of double-stranded DNA (dsDNA) templates with homologous sequences on both sides of the DSB. Templates could be sister chromatids or exogenously provided DNA.

Precise genome editing via HR is a hope for patients suffering from rare diseases that are caused by mutations. Genome editing by CRISPR/Cas9 introduces a DSB close to the mutation, which is then repaired with the help of a template. Unfortunately, error-prone repairs by cNHEJ and MMEJ are mostly outcompeting the precise edits by SSTR and HR, resulting in a low percentage of correctly repaired genes. For the clinical application of CRISPR genome editing, it is therefore essential to maximize the percentage of cells that are repaired by precise DNA repair. Mechanisms that can contribute to this goal should be investigated for their potential use in gene therapy, and nuclear actin polymerization is a candidate pathway for promoting HR.

5. Methodological Challenges

Investigating the role of nuclear actin polymerization in DSB repair faces several methodological problems.

Firstly, the effect of nuclear actin polymerization might be dependent on the method applied to induce DSBs. DSBs can be induced by different methods. Often used are irradiation or chemical reagents [34,35] that induce multiple DSBs but also have side effects such as the production of reactive oxygen species [36]. In addition, different DSB-inducing agents have different preferences with respect to the target site [37]. A more subtle method for DSB induction is the expression of sequence-specific nucleases that cut the genome at single or multiple DSBs [38,39]. Yet, only in the latter case will it be possible to detect effects based on the clustering of multiple DSBs.

Secondly, the method used for detection of nuclear actin polymerization might be either not nucleus-specific or include the risk of directly affecting actin polymerization. A common way to detect F-actin in fixed cells is by staining them with fluorescently labeled phalloidin [20]. However, the high amount of cytoplasmic F-actin makes it, in most cases, difficult to clearly identify potential nuclear F-actin. In corresponding wide-field images, stress fibers can appear to extend from the cytoplasm into the nuclear area, although they are in fact located above the nucleus. In confocal or super-resolution microscopy, these false positive nuclear stress fibers appear significantly weaker in the apparent nuclear area and can be identified as extranuclear in 3D stack analyses.

Detection of nuclear F-actin by transfection of cells with fluorescent protein probes binding to F-actin and targeted to the nucleus by a nuclear localization sequence (NLS) has two advantages: They largely avoid the detection of cytoplasmic F-actin, and they allow the monitoring of nuclear F-actin in living cells. Some of these probes were shown to promote the non-physiological formation of F-actin, while others appeared to be neutral [40]. However, depending on the expression level and the cellular system used, artifact formation can never be completely excluded [41]. Actin chromobodies are intracellularly expressed fluorescent proteins that bind to actin via an antibody-derived protein domain. Cobb et al. reported that following expression of actin chromobodies in U2OS cells, about 20% of the cells showed clearly visible nuclear nodules or fibrils, which in some cases were thick, long, and bended [42]. Palumbieri et al. stably expressed a FLAG-tagged NLS-actin fusion protein at low levels in U2OS cells, which was detected after fixation by immunofluoresent staining for the FLAG tag [43]. With this system, they detected nuclear foci and fibrils in about 30% of the untreated cells. Such structures have not been described before in phalloidin-stained, untransfected U2OS cells. One possible explanation for this discrepancy would be that actin chromobodies and FLAG-tagged actin might be much more sensitive in detecting nuclear F-actin than phalloidin. Another explanation would be that these probes stabilize F-actin structures and, by this mechanism, increase nuclear F-actin.

In MCF7 cells, transient expression of actin chromobodies detected long and bended nuclear F-actin only following inhibition of p53 [44]. Importantly, these F-actin structures could easily be detected by phalloidin staining, but only in cells transfected with actin chromobodies. Similarly, nLifeact-GFP expression resulted in phalloidin-stainable, thick nuclear actin filaments, while cells not expressing the probe showed no nuclear F-actin detectable by phalloidin. These data suggest that actin-binding probes expressed in living cells indeed have the potential to induce nuclear F-actin.

As a final methodological challenge, it is difficult to specifically interfere with nuclear actin polymerization. Nuclear and cytoplasmic actin pools are connected in a dynamic equilibrium, and nuclear–cytoplasmic transport is regulated by the amounts of actin transport molecules and by their post-translational modifications [16]. Molecules regulating actin polymerization are often present both in the cytoplasm and the nucleus. Targeting molecules in the nucleus with an NLS sequence might lead to unphysiologically high levels of this protein in the nucleus. Targeting a non-polymerizable actin mutant in the nucleus reduces nuclear actin polymerization but might have effects on nuclear proteins that bind G-actin. Inhibitors, such as KD or knockout (KO), will efficiently work on the nuclear protein but also on cytoplasmic proteins, which indirectly might affect DNA repair.

Torii et al. explored the effect of the overexpression of nuclear-targeted and GFP-coupled wild-type and S14C actin, an actin polymerization-promoting mutant [44]. With wild-type actin, inhibition of p53 and treatment with doxorubicin promoted the formation of thin nuclear actin filaments not detectable by phalloidin. In contrast, S14C actin resulted in actin nodules with very thin, short fibrillar extensions, all of which could be stained by phalloidin. While the molecular reason for these differences is not clear, they indicate that results obtained with NLS-actin-GFP fusions as well as with actin mutants need to be interpreted very carefully.

6. Nuclear Actin Polymerization Promotes DNA Repair by HR

Several studies have demonstrated that actin and ABPs bind to DSBs and promote actin polymerization and clustering of DNA repair foci, which are membrane-less condensates that form around sites of DNA damage [45]. Importantly, nuclear actin polymerization is often correlated with increased precise genome editing.

The first evidence of nuclear actin potentially being involved in DSB repair came from an F-actin disruption experiment in human MCF-7 breast cancer cells [46]. Shin et al. uncovered that protein levels of phosphorylated histone H2AX (γH2AX), a marker of DSB, were dramatically increased after latrunculin B (LatB)-induced actin disruption [46]. Subsequently, Andrin et al. confirmed a direct association between polymerized actin and DSB repair proteins, including Ku70, Ku80, Mre11, and Rad51, through in vitro F-actin co-precipitation [47]. Of note, these proteins include repair proteins required for cNHEJ as well as for resection and HR. In addition, the authors showed by different assays that inhibition of actin polymerization inhibits the repair of irradiation-induced DSBs, particularly cNHEJ [47]. However, neither of the above evidence could prove a direct interaction between nuclear F-actin and DNA repair proteins, as the binding might be mediated by ABPs. Since expression of an NLS-tagged non-polymerizable actin mutant (G13R) reduced retention of Ku80-GFP at laser micro-irradiation-induced DNA defects similar to treatment with the actin polymerization inhibitor cytochalasin D [47], cytoplasmic actin polymerization appeared not to be involved in the observed effects.

Later, Belin et al. detected an increased amount of both long and short actin filaments in the nucleus induced by DNA-damaging agents such as methyl-methanesulfonate (MMS) using phalloidin staining and live-cell actin probes in Hela cells [23]. The experiments showed that DNA damage-induced nuclear actin assembly requires the nucleation factors FMN2 and Spire-1/Spire-2 but not mDia1/2, which earlier were shown to trigger nuclear actin polymerization (Figure 3) [23,48]. Depletion of either FMN2 or IPO9 increased the number of DSBs detected by 53BP1 or γH2AX as markers [23]. The authors considered this observation to be inefficient DSB clearance in the absence of nuclear actin filaments, but it cannot be excluded that the clustering of DSBs contributes to the decrease in DSB foci. The mechanism by which nuclear F-actin DNA promotes DNA repair was not studied.

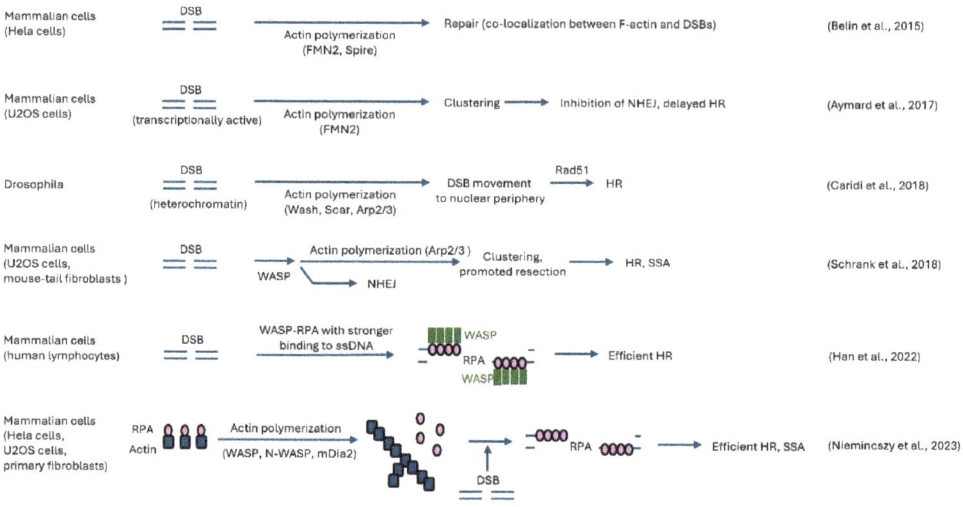

Figure 3. Nuclear actin polymerization-dependent regulation of DNA repair. Presented are different studies indicating the involvement of nuclear actin polymerization in the repair of DSBs with the corresponding suggested molecular mechanisms. The clustering and movement of DSBs were reported to depend in some models on FMN2 [42,44], but in other systems on Arp2/3-dependent actin polymerization [45,46]. WASP was found to release RPA bound to G-actin [48], but also to bind to DSBs and facilitate end resection by nuclear F-actin formation [46]. Independent of actin polymerization, WASP binding to RPA was demonstrated to increase RPA binding to ssDNA [47].

Utilizing a high-throughput chromosome conformation capture assay (capture Hi-C), Aymard et al. revealed that DSBs at transcriptionally active genes cluster in the G1 cell cycle phase, which coincides with delayed repair [49]. Clustering depended on the MRN complex, the actin nucleation promoting formin FMN2, and the linker of the nuclear and cytoplasmic skeleton (LINC) complex, which connects structural proteins close to the internal nuclear membrane with the actin, microtubule, and intermediate filament networks of the cytoplasm. The researchers proposed that DSB clustering may help to sequester DSBs and prevent error-prone cNHEJ repair in G1, enabling delayed but precise DSB repair by HR in G2 [49]. This study suggested that nuclear actin polymerization is required in DSB clustering-driven repair by HR (Figure 3). However, the specific mechanism of how actin polymerization triggers DSB clustering and whether indeed nuclear actin polymerization is involved were not tested.

In *Drosophila*, repair of DSBs in heterochromatic regions by HR involves the DSB movement to the nuclear periphery before Rad51 recruitment, probably to avoid aberrant recombination with other DNA sequences than the sister chromosome [50]. Caridi et al. demonstrated that this DSB movement is dependent on Arp2/3-mediated nuclear actin polymerization and independent of the actin nucleators Spire and mDia (Figure 3) [50]. Furthermore, siRNA experiments demonstrated that Arp2/3 is activated by Wash or Scar but not by WASP or Whamy. Following irradiation, Arp2/3 co-precipitated with the heterochromatin repair complex Smc5/6, suggesting a close coupling of DSB repair with actin polymerization. Heterochromatic DSB movement also required myosin and the myosin activator Unc45. In contrast to heterochromatic DSBs, euchromatic DSBs rarely moved to the periphery, and undamaged heterochromatic DNA hardly moved at all. Interestingly, Arp2/3 was not important for clustering of early repair foci in heterochromatin, while it was essential for euchromatic early repair foci clustering.

Immunofluorescence staining of endogenously expressed WASP, an Arp2/3 stimulating NPF, revealed its localization at all DSBs in U2OS sarcoma cells and mouse-tail fibroblasts (MTFs) [51]. However, only at DSBs undergoing resection in the G2 cell cycle phase did Arp2/3 dependent actin polymerization take place, resulting in DSB clustering. These clustered DSBs were positive for RPA, RAD51, or RAD52, suggesting that clustering might affect DNA repair by HR and SSA (Figure 3). Indeed, inhibitors of Arp2/3-dependent actin polymerization or of WASP reduced HR and SSA, as shown by I-SceI assays, where restriction enzyme-induced DSBs are repaired in a template-dependent or independent manner [51]. Neither FMN2 nor other formins such as mDia1 and mDia2 were involved in this effect, as shown by KD and inhibitors. Interestingly, MMEJ or cNHEJ, neither of which require substantial resection, were not impaired by inhibiting WASP or Arp2/3, implying a resection-promoting role of actin polymerization in HR repair. The authors presented a model where resection enhances DSB movement, which, in a positive feedback-loop, increases resection [51]. How WASP is activated after recruitment to DSBs only at DSBs undergoing resection was not clarified.

Recently, Lamm et al. showed that aphidicolin-induced replication stress stimulated nuclear actin polymerization involving IQGAP1, WASP, Arp2/3, and inhibition of cofilin by phosphorylation [52]. Nuclear actin polymerization was found to contribute to replication fork repair. Here, myosin II was required for the movement of the replication foci. This suggests that the machinery required for the movement of DSBs and replication foci is partially overlapping.

A different role for WASP was suggested by Han et al. [53]. They observed that WASP interacts with RPA, which increases its interaction with ssDNA (Figure 3). WASP KO leads to less RPA binding to ssDNA and increased γH2AX staining. This could indicate the accumulation of unrepaired DSBs due to insufficient HR and SSA, which are both dependent on RPA. In yeast, SSA-induced deletion was not significantly influenced by loss of the WASP orthologue Las17, but DSB repair by SSA was even improved by Las17 deficiency. WASP-deficient B and T cells showed differential cell cycle changes to genotoxic stress dependent on the genotoxin, which is of interest as resection-dependent repair only occurs in S/G2 [53].

A completely different way of regulating RPA binding to ssDNA was proposed by Nieminuszczy and colleagues [54]. They showed that RPA can bind to actin and that this interaction prevents RPA binding to ssDNA. Actin polymerization releases RPA from actin, allowing RPA to bind to ssDNA and promote resection dependent DNA repair (HR and SSA) (Figure 3). KD of the actin nucleators WASP, N-WASP, or mDia1 all showed a similar increase in DSBs in the S/G2 phase identified by RBP1 foci following HU treatment. In this case, no direct interaction of the actin polymerization machinery with the DSB appeared necessary.

A principally similar indirect type of regulation was earlier described for the MAL/MRTF transcription factor. In the cytoplasm, MAL/MRTF is bound to G-actin [55]. Actin polymerization releases MAL/MRTF, which can then travel to the nucleus, where it stimulates transcription in collaboration with the SRF transcription factor [56]. Potentially, nuclear actin polymerization might influence the nuclear activity of MAL/MRTF [57]. Also, for the regulation of MAL/MRTF signaling, the exact type of NPF triggering actin polymerization is not relevant.

Related to this principle of regulation, nuclear RNA polymerases and several chromatin-modifying complexes, such as INO80 or TIP60, were shown to bind to monomeric actin, which appears to be important for the stability and function of these complexes [58–60]. If nuclear actin polymerization reduces the availability of G-actin, the stability of these chromatin-modifying complexes might be decreased. Importantly, these actin monomer-binding complexes might affect DNA repair, as was shown, for example, for TIP60 and INO80 [60–65].

Taken together, most of these studies indicate that nuclear actin polymerization promotes DSB repair by HR and SSA by facilitating resection. The exact molecular mechanisms underlying it, however, are not entirely clear. Several, maybe parallel, mechanisms were suggested involving different types of actin NPFs. To further investigate this process, it will be helpful to induce DSBs by methods with fewer side effects on other cellular processes than irradiation or chemical reagents. Secondly, the use of gene KO in the cell lines of interest could complement the published siRNA and small molecule inhibitor studies, which have a higher chance for partial as well as off-target effects. Rescue of the KO models by re-expression of the wild-type molecule would safely confirm that the observed effect is indeed related to the mutated gene. Thirdly, performing the rescue with molecules fused to nuclear export sequences (NES) or NLS and directing the proteins to the cytoplasm or nucleus, respectively, might help to distinguish between cytoplasmic and nuclear functions.

7. DSB Movement and Repair

The molecular details of how DSB movement or clustering support DNA repair by HR or SSA are not clear. One option would be that DSBs are moved from a region with a low likelihood of HR to a region with a high likelihood. Similarly, DSB clustering might increase the local concentration of HR-related proteins, facilitating recombination. On the other hand, increased DSB concentration could also increase the competition for HR-related proteins.

Several lines of evidence have suggested chromatin mobility changes in response to DSBs [49,66–70]. The pattern of chromatin mobility after DNA damage could be different depending on the DSB location, repair sites, and time [66,71,72]. Chromatin motions proximal to DSB sites retain higher mobility, whereas the motions of chromatin distal to DSBs are globally reduced [66]. In yeast and human cells, the motion of damaged DNA requires HR machinery [49,73], but it is not clear how DSB clusters direct to repair-conducive sites, how they interact with HR regulators, and in which role nuclear actin is involved. Another open question is how the directionality of the movement is controlled.

However, the clustering and spatial compartmentalization of DSBs triggered by WASP-Arp2/3-dependent nuclear actin polymerization also increase the risk of chromosomal translocations [74–76]. A study found that the location of DSBs proximal or distal to the centromere exhibits different mechanisms in DNA repair [71]. DSB mobility near the centromere relied on γH2AX, independently of the Rad9-dependent checkpoint and the Rad51 nucleofilament. The movement of sub-telomeric distal DSBs, on the other hand, contingent on the presence of a homologous sequence close to the centromere, required both Rad9 and the Rad51 nucleofilament essential for HR [71,77].

8. Is Nuclear Actin Polymerization Required for Genome Maintenance?

Defects in DNA repair caused by the alterations of nuclear actin polymerization could promote the frequency of mutations and contribute to the development of cancer.

Wiskott–Aldrich syndrome patients lack a functional WASP-encoding gene and suffer from severe immunosuppression and a predisposition to non-Hodgkin lymphoma and leukemia [78]. Lymphocytes from WAS patients displayed defective resection, implying impaired DNA repair by HR [51]. Whether this predisposition to leukemia and lymphoma is caused by defective HR repair, however, has not been shown yet. Since all nuclear actin polymerization regulators identified also have cytoplasmic functions, mutations in these genes will most likely influence both nuclear and cytoplasmic functions, making it difficult to prove a correlation between defective nuclear function and cancer development.

9. Can Facilitated Actin Polymerization Promote Genome Editing by HR?

While the inhibition of actin polymerization was shown by many groups to interfere with DSB repair by HR, it is still not clear whether it will be possible to facilitate nuclear actin polymerization in a way that promotes DSB repair by HR or SSTR, which would be of interest for ex vivo gene therapy of stem cells. If the release of RPA from monomeric actin is a rate-limiting step, all treatments that elevate nuclear actin polymerization would be supportive. However, transient knockdown of the nuclear actin exporter XPO6 in

HeLa cells promoted nuclear actin polymerization but did not alter the frequency of MMS-induced DSBs as detected by 53BP1 foci [47].

The movement of DSBs was shown by several groups to occur exclusively at DSBs undergoing HR [50,51]. It would be interesting to test whether it is possible to increase the fraction of DSB undergoing movement by altering the regulation of nuclear actin polymerization and whether this would increase the fraction of DSBs that are repaired by HR.

Irradiation was reported to cause nuclear activation of RhoA [79], a Rho GTPase that in the cytoplasm induces actin polymerization via the mDia formins and via activation of myosin-dependent contraction. It was reported that RhoA inhibition increased the number of DSBs by irradiation and impaired the NHEJ pathway in glioma cells in a p53-dependent way, while HR was only affected in p53-mutant cells [80]. However, the effect of nuclear RhoA activation or inhibition on HR has not yet been investigated.

When in a closed conformation, WASP is neither able to bind to Arp2/3 nor to induce actin polymerization. The binding of GTP-bound Cdc42 to WASP induces a conformational shift and enables actin polymerization. Whether this WASP activation scheme could be integrated with the proposed binding of WASP to all DSBs, while Arp2/3 is binding only to those DSBs undergoing resection, needs to be studied. In any case, the regulation of the actin polymerization activity of nuclear WASP will be of interest.

Cofilin controls the nuclear import of actin as well as the severing of nuclear F-actin. Cofilin dephosphorylation, promoting its binding to ADP-actin, was reported to transiently increase nuclear F-actin [20]. Whether this increase affects DSB repair was not tested. On the other hand, siRNA against cofilin promoted nuclear F-actin during mitotic exit [20], while it prevented nuclear actin accumulation induced by XPO6 siRNA [16]. This suggests that the regulation of cofilin needs to be carefully regulated to obtain the desired result of more nuclear F-actin due to reduced severing.

10. Conclusions

The current data do not identify strong candidates for actin polymerization regulating genes that could be transiently tuned to promote precise genome editing. However, an improved molecular understanding of how nuclear actin polymerization is controlled might provide such targets. For future experiments, phalloidin staining appears to be the gold standard for monitoring nuclear F-actin. If live cell actin probes are used, it should always be controlled whether their expression affects nuclear F-actin as detectable by phalloidin staining. If nuclear actin polymerization is manipulated by inhibition, overexpression, or knockout of actin regulators controlling, for example, nuclear transport, polymerization, or severing of F-actin, it needs to be demonstrated that these actin regulators are altered in activity or expression under physiological circumstances. Moreover, under physiological conditions, changes in actin regulators should correspond to changes in nuclear F-actin. Distinguishing the effects of nuclear actin polymerization from effects on actin monomer-binding chromatin modifier complexes will remain a challenge. Perhaps testing the consequences of nuclear actin polymerization on DSB repair in cell lines with KO of one or several genes of chromatin modifier complexes could be an option. DSB induction by CRISPR will be a subtle method to trigger DNA repair, avoiding the unwanted side effects associated with irradiation or chemical carcinogens, which indirectly might affect DSB repair. Finally, testing the consequences of altered actin polymerization directly on the HDR and NHEJ efficiency of CRISPR gene editing will increase the chances for novel findings that improve the efficiency of gene therapy for rare diseases.

Author Contributions: X.H. and C.B. wrote the manuscript and prepared the figures. All authors have read and agreed to the published version of the manuscript.

Funding: This study was supported by the European Union (Project: 101,072,427—GetRadi—HORIZON-MSCA-DN-2021).

Acknowledgments: The views and opinions expressed are, however, those of the authors only and do not necessarily reflect those of the European Union or the European Research Executive Agency. Neither the European Union nor the European Research Executive Agency can be held responsible for them.

Conflicts of Interest: The authors declare no conflicts of interest.

References

1. Tambuyzer, E.; Vandendriessche, B.; Austin, C.P.; Brooks, P.J.; Larsson, K.; Miller Needleman, K.I.; Valentine, J.; Davies, K.; Groft, S.C.; Preti, R.; et al. Therapies for rare diseases: Therapeutic modalities, progress and challenges ahead. *Nat. Rev. Drug Discov.* **2020**, *19*, 93–111. [CrossRef] [PubMed]
2. Wang, J.Y.; Doudna, J.A. CRISPR technology: A decade of genome editing is only the beginning. *Science* **2023**, *379*, eadd8643. [CrossRef] [PubMed]
3. Pacesa, M.; Pelea, O.; Jinek, M. Past, present, and future of CRISPR genome editing technologies. *Cell* **2024**, *187*, 1076–1100. [CrossRef] [PubMed]
4. Svitkina, T. The Actin Cytoskeleton and Actin-Based Motility. *Cold Spring Harb. Perspect. Biol.* **2018**, *10*, a018267. [CrossRef] [PubMed]
5. Pollard, T.D. Actin and Actin-Binding Proteins. *Cold Spring Harb. Perspect. Biol.* **2016**, *8*, a018226. [CrossRef] [PubMed]
6. Picco, A.; Kukulski, W.; Manenschijn, H.E.; Specht, T.; Briggs, J.A.G.; Kaksonen, M. The contributions of the actin machinery to endocytic membrane bending and vesicle formation. *Mol. Biol. Cell* **2018**, *29*, 1346–1358. [CrossRef] [PubMed]
7. Ulferts, S.; Prajapati, B.; Grosse, R.; Vartiainen, M.K. Emerging Properties and Functions of Actin and Actin Filaments Inside the Nucleus. *Cold Spring Harb. Perspect. Biol.* **2021**, *13*, a040121. [CrossRef] [PubMed]
8. Hurst, V.; Shimada, K.; Gasser, S.M. Nuclear Actin and Actin-Binding Proteins in DNA Repair. *Trends Cell Biol.* **2019**, *29*, 462–476. [CrossRef]
9. Caridi, C.P.; Plessner, M.; Grosse, R.; Chiolo, I. Nuclear actin filaments in DNA repair dynamics. *Nat. Cell Biol.* **2019**, *21*, 1068–1077. [CrossRef]
10. Le, S.; Yu, M.; Bershadsky, A.; Yan, J. Mechanical regulation of formin-dependent actin polymerization. *Semin. Cell Dev. Biol.* **2020**, *102*, 73–80. [CrossRef]
11. Ding, B.; Narvaez-Ortiz, H.Y.; Singh, Y.; Hocky, G.M.; Chowdhury, S.; Nolen, B.J. Structure of Arp2/3 complex at a branched actin filament junction resolved by single-particle cryo-electron microscopy. *Proc. Natl. Acad. Sci. USA* **2022**, *119*, e2202723119. [CrossRef]
12. Wang, J.; Nakamura, F. Identification of Filamin A Mechanobinding Partner II: Fimbacin Is a Novel Actin Cross-Linking and Filamin A Binding Protein. *Biochemistry* **2019**, *58*, 4737–4743. [CrossRef] [PubMed]
13. Senger, F.; Pitaval, A.; Ennomani, H.; Kurzawa, L.; Blanchoin, L.; Théry, M. Spatial integration of mechanical forces by α-actinin establishes actin network symmetry. *J. Cell Sci.* **2019**, *132*, jcs236604. [CrossRef] [PubMed]
14. Rajan, S.; Kudryashov, D.S.; Reisler, E. Actin Bundles Dynamics and Architecture. *Biomolecules* **2023**, *13*, 450. [CrossRef] [PubMed]
15. Lee, G.; Leech, G.; Rust, M.J.; Das, M.; McGorty, R.J.; Ross, J.L.; Robertson-Anderson, R.M. Myosin-driven actin-microtubule networks exhibit self-organized contractile dynamics. *Sci. Adv.* **2021**, *7*, eabe4334. [CrossRef]
16. Dopie, J.; Skarp, K.P.; Rajakylä, E.K.; Tanhuanpää, K.; Vartiainen, M.K. Active maintenance of nuclear actin by importin 9 supports transcription. *Proc. Natl. Acad. Sci. USA* **2012**, *109*, E544–E552. [CrossRef]
17. Pollard, T.D.; Blanchoin, L.; Mullins, R.D. Molecular mechanisms controlling actin filament dynamics in nonmuscle cells. *Annu. Rev. Biophys. Biomol. Struct.* **2000**, *29*, 545–576. [CrossRef]
18. Bernard, O. Lim kinases, regulators of actin dynamics. *Int. J. Biochem. Cell Biol.* **2007**, *39*, 1071–1076. [CrossRef]
19. Soosairajah, J.; Maiti, S.; Wiggan, O.; Sarmiere, P.; Moussi, N.; Sarcevic, B.; Sampath, R.; Bamburg, J.R.; Bernard, O. Interplay between components of a novel LIM kinase-slingshot phosphatase complex regulates cofilin. *EMBO J.* **2005**, *24*, 473–486. [CrossRef]
20. Baarlink, C.; Plessner, M.; Sherrard, A.; Morita, K.; Misu, S.; Virant, D.; Kleinschnitz, E.M.; Harniman, R.; Alibhai, D.; Baumeister, S.; et al. A transient pool of nuclear F-actin at mitotic exit controls chromatin organization. *Nat. Cell Biol.* **2017**, *19*, 1389–1399. [CrossRef]
21. Dopie, J.; Rajakylä, E.K.; Joensuu, M.S.; Huet, G.; Ferrantelli, E.; Xie, T.; Jäälinoja, H.; Jokitalo, E.; Vartiainen, M.K. Genome-wide RNAi screen for nuclear actin reveals a network of cofilin regulators. *J. Cell Sci.* **2015**, *128*, 2388–2400. [CrossRef]
22. Stüven, T.; Hartmann, E.; Görlich, D. Exportin 6: A novel nuclear export receptor that is specific for profilin.actin complexes. *EMBO J.* **2003**, *22*, 5928–5940. [CrossRef]
23. Belin, B.J.; Lee, T.; Mullins, R.D. Correction: DNA damage induces nuclear actin filament assembly by Formin-2 and Spire-1/2 that promotes efficient DNA repair. *eLife* **2015**, *4*, e11935. [CrossRef] [PubMed]
24. Bohnsack, M.T.; Stüven, T.; Kuhn, C.; Cordes, V.C.; Görlich, D. A selective block of nuclear actin export stabilizes the giant nuclei of *Xenopus* oocytes. *Nat. Cell Biol.* **2006**, *8*, 257–263. [CrossRef] [PubMed]
25. Andrianantoandro, E.; Pollard, T.D. Mechanism of actin filament turnover by severing and nucleation at different concentrations of ADF/cofilin. *Mol. Cell* **2006**, *24*, 13–23. [CrossRef]

26. Scully, R.; Panday, A.; Elango, R.; Willis, N.A. DNA double-strand break repair-pathway choice in somatic mammalian cells. *Nat. Rev. Mol. Cell Biol.* **2019**, *20*, 698–714. [CrossRef]
27. Deshpande, R.A.; Myler, L.R.; Soniat, M.M.; Makharashvili, N.; Lee, L.; Lees-Miller, S.P.; Finkelstein, I.J.; Paull, T.T. DNA-dependent protein kinase promotes DNA end processing by MRN and CtIP. *Sci. Adv.* **2020**, *6*, eaay0922. [CrossRef] [PubMed]
28. Anand, R.; Ranjha, L.; Cannavo, E.; Cejka, P. Phosphorylated CtIP Functions as a Co-factor of the MRE11-RAD50-NBS1 Endonuclease in DNA End Resection. *Mol. Cell* **2016**, *64*, 940–950. [CrossRef]
29. Ceppi, I.; Howard, S.M.; Kasaciunaite, K.; Pinto, C.; Anand, R.; Seidel, R.; Cejka, P. CtIP promotes the motor activity of DNA2 to accelerate long-range DNA end resection. *Proc. Natl. Acad. Sci. USA* **2020**, *117*, 8859–8869. [CrossRef]
30. Kamp, J.A.; Lemmens, B.; Romeijn, R.J.; Changoer, S.C.; van Schendel, R.; Tijsterman, M. Helicase Q promotes homology-driven DNA double-strand break repair and prevents tandem duplications. *Nat. Commun.* **2021**, *12*, 7126. [CrossRef]
31. Mimitou, E.P.; Symington, L.S. Sae2, Exo1 and Sgs1 collaborate in DNA double-strand break processing. *Nature* **2008**, *455*, 770–774. [CrossRef] [PubMed]
32. Cejka, P.; Symington, L.S. DNA End Resection: Mechanism and Control. *Annu. Rev. Genet.* **2021**, *55*, 285–307. [CrossRef] [PubMed]
33. Ranjha, L.; Howard, S.M.; Cejka, P. Main steps in DNA double-strand break repair: An introduction to homologous recombination and related processes. *Chromosoma* **2018**, *127*, 187–214. [CrossRef] [PubMed]
34. Sage, E.; Shikazono, N. Radiation-induced clustered DNA lesions: Repair and mutagenesis. *Free Radic. Biol. Med.* **2017**, *107*, 125–135. [CrossRef]
35. Vítor, A.C.; Huertas, P.; Legube, G.; de Almeida, S.F. Studying DNA Double-Strand Break Repair: An Ever-Growing Toolbox. *Front. Mol. Biosci.* **2020**, *7*, 24. [CrossRef]
36. Kang, M.A.; So, E.Y.; Simons, A.L.; Spitz, D.R.; Ouchi, T. DNA damage induces reactive oxygen species generation through the H2AX-Nox1/Rac1 pathway. *Cell Death Dis.* **2012**, *3*, e249. [CrossRef]
37. Volkova, N.V.; Meier, B.; González-Huici, V.; Bertolini, S.; Gonzalez, S.; Vöhringer, H.; Abascal, F.; Martincorena, I.; Campbell, P.J.; Gartner, A.; et al. Mutational signatures are jointly shaped by DNA damage and repair. *Nat. Commun.* **2020**, *11*, 2169. [CrossRef] [PubMed]
38. Bischoff, N.; Wimberger, S.; Kühn, R.; Laugesen, A.; Turan, V.; Larsen, B.D.; Sørensen, C.S.; Helin, K.; Bennett, E.P.; Maresca, M.; et al. The methylation inhibitor 3DZNep promotes HDR pathway choice during CRISPR-Cas9 genome editing. *Gene Genome Ed.* **2023**, *5*, 100023. [CrossRef]
39. Iacovoni, J.S.; Caron, P.; Lassadi, I.; Nicolas, E.; Massip, L.; Trouche, D.; Legube, G. High-resolution profiling of gammaH2AX around DNA double strand breaks in the mammalian genome. *EMBO J.* **2010**, *29*, 1446–1457. [CrossRef]
40. Melak, M.; Plessner, M.; Grosse, R. Actin visualization at a glance. *J. Cell Sci.* **2017**, *130*, 525–530. [CrossRef]
41. Du, J.; Fan, Y.L.; Chen, T.L.; Feng, X.Q. Lifeact and Utr230 induce distinct actin assemblies in cell nuclei. *Cytoskeleton* **2015**, *72*, 570–575. [CrossRef] [PubMed]
42. Cobb, A.M.; De Silva, S.A.; Hayward, R.; Sek, K.; Ulferts, S.; Grosse, R.; Shanahan, C.M. Filamentous nuclear actin regulation of PML NBs during the DNA damage response is deregulated by prelamin A. *Cell Death Dis.* **2022**, *13*, 1042. [CrossRef]
43. Palumbieri, M.D.; Merigliano, C.; González-Acosta, D.; Kuster, D.; Krietsch, J.; Stoy, H.; von Känel, T.; Ulferts, S.; Welter, B.; Frey, J.; et al. Nuclear actin polymerization rapidly mediates replication fork remodeling upon stress by limiting PrimPol activity. *Nat. Commun.* **2023**, *14*, 7819. [CrossRef] [PubMed]
44. Torii, T.; Sugimoto, W.; Itoh, K.; Kinoshita, N.; Gessho, M.; Goto, T.; Uehara, I.; Nakajima, W.; Budirahardja, Y.; Miyoshi, D.; et al. Loss of p53 function promotes DNA damage-induced formation of nuclear actin filaments. *Cell Death Dis.* **2023**, *14*, 766. [CrossRef]
45. Miné-Hattab, J.; Liu, S.; Taddei, A. Repair Foci as Liquid Phase Separation: Evidence and Limitations. *Genes* **2022**, *13*, 1846. [CrossRef] [PubMed]
46. Shin, I.J.; Ahn, Y.T.; Kim, Y.; Kim, J.M.; An, W.G. Actin disruption agents induce phosphorylation of histone H2AX in human breast adenocarcinoma MCF-7 cells. *Oncol. Rep.* **2011**, *25*, 1313–1319. [CrossRef]
47. Andrin, C.; McDonald, D.; Attwood, K.M.; Rodrigue, A.; Ghosh, S.; Mirzayans, R.; Masson, J.Y.; Dellaire, G.; Hendzel, M.J. A requirement for polymerized actin in DNA double-strand break repair. *Nucleus* **2012**, *3*, 384–395. [CrossRef]
48. Baarlink, C.; Wang, H.; Grosse, R. Nuclear actin network assembly by formins regulates the SRF coactivator MAL. *Science* **2013**, *340*, 864–867. [CrossRef]
49. Aymard, F.; Aguirrebengoa, M.; Guillou, E.; Javierre, B.M.; Bugler, B.; Arnould, C.; Rocher, V.; Iacovoni, J.S.; Biernacka, A.; Skrzypczak, M.; et al. Genome-wide mapping of long-range contacts unveils clustering of DNA double-strand breaks at damaged active genes. *Nat. Struct. Mol. Biol.* **2017**, *24*, 353–361. [CrossRef]
50. Caridi, C.P.; D'Agostino, C.; Ryu, T.; Zapotoczny, G.; Delabaere, L.; Li, X.; Khodaverdian, V.Y.; Amaral, N.; Lin, E.; Rau, A.R.; et al. Nuclear F-actin and myosins drive relocalization of heterochromatic breaks. *Nature* **2018**, *559*, 54–60. [CrossRef]
51. Schrank, B.R.; Aparicio, T.; Li, Y.; Chang, W.; Chait, B.T.; Gundersen, G.G.; Gottesman, M.E.; Gautier, J. Nuclear ARP2/3 drives DNA break clustering for homology-directed repair. *Nature* **2018**, *559*, 61–66. [CrossRef]
52. Lamm, N.; Read, M.N.; Nobis, M.; Van Ly, D.; Page, S.G.; Masamsetti, V.P.; Timpson, P.; Biro, M.; Cesare, A.J. Nuclear F-actin counteracts nuclear deformation and promotes fork repair during replication stress. *Nat. Cell Biol.* **2020**, *22*, 1460–1470. [CrossRef] [PubMed]

53. Han, S.S.; Wen, K.K.; García-Rubio, M.L.; Wold, M.S.; Aguilera, A.; Niedzwiedz, W.; Vyas, Y.M. WASp modulates RPA function on single-stranded DNA in response to replication stress and DNA damage. *Nat. Commun.* **2022**, *13*, 3743. [CrossRef] [PubMed]
54. Nieminuszczy, J.; Martin, P.R.; Broderick, R.; Krwawicz, J.; Kanellou, A.; Mocanu, C.; Bousgouni, V.; Smith, C.; Wen, K.K.; Woodward, B.L.; et al. Actin nucleators safeguard replication forks by limiting nascent strand degradation. *Nucleic Acids Res.* **2023**, *51*, 6337–6354. [CrossRef]
55. Mouilleron, S.; Guettler, S.; Langer, C.A.; Treisman, R.; McDonald, N.Q. Molecular basis for G-actin binding to RPEL motifs from the serum response factor coactivator MAL. *EMBO J.* **2008**, *27*, 3198–3208. [CrossRef] [PubMed]
56. Vartiainen, M.K.; Guettler, S.; Larijani, B.; Treisman, R. Nuclear actin regulates dynamic subcellular localization and activity of the SRF cofactor MAL. *Science* **2007**, *316*, 1749–1752. [CrossRef] [PubMed]
57. Marco, S.; Neilson, M.; Moore, M.; Perez-Garcia, A.; Hall, H.; Mitchell, L.; Lilla, S.; Blanco, G.R.; Hedley, A.; Zanivan, S.; et al. Nuclear-capture of endosomes depletes nuclear G-actin to promote SRF/MRTF activation and cancer cell invasion. *Nat. Commun.* **2021**, *12*, 6829. [CrossRef] [PubMed]
58. Kapoor, P.; Chen, M.; Winkler, D.D.; Luger, K.; Shen, X. Evidence for monomeric actin function in INO80 chromatin remodeling. *Nat. Struct. Mol. Biol.* **2013**, *20*, 426–432. [CrossRef] [PubMed]
59. Hofmann, W.A.; Stojiljkovic, L.; Fuchsova, B.; Vargas, G.M.; Mavrommatis, E.; Philimonenko, V.; Kysela, K.; Goodrich, J.A.; Lessard, J.L.; Hope, T.J.; et al. Actin is part of pre-initiation complexes and is necessary for transcription by RNA polymerase II. *Nat. Cell Biol.* **2004**, *6*, 1094–1101. [CrossRef]
60. Bassi, C.; Li, Y.T.; Khu, K.; Mateo, F.; Baniasadi, P.S.; Elia, A.; Mason, J.; Stambolic, V.; Pujana, M.A.; Mak, T.W.; et al. The acetyltransferase Tip60 contributes to mammary tumorigenesis by modulating DNA repair. *Cell Death Differ.* **2016**, *23*, 1198–1208. [CrossRef]
61. Fréchard, A.; Faux, C.; Hexnerova, R.; Crucifix, C.; Papai, G.; Smirnova, E.; McKeon, C.; Ping, F.L.Y.; Helmlinger, D.; Schultz, P.; et al. The structure of the NuA4-Tip60 complex reveals the mechanism and importance of long-range chromatin modification. *Nat. Struct. Mol. Biol.* **2023**, *30*, 1337–1345. [CrossRef] [PubMed]
62. Morrison, A.J.; Highland, J.; Krogan, N.J.; Arbel-Eden, A.; Greenblatt, J.F.; Haber, J.E.; Shen, X. INO80 and gamma-H2AX interaction links ATP-dependent chromatin remodeling to DNA damage repair. *Cell* **2004**, *119*, 767–775. [CrossRef] [PubMed]
63. Lademann, C.A.; Renkawitz, J.; Pfander, B.; Jentsch, S. The INO80 Complex Removes H2A.Z to Promote Presynaptic Filament Formation during Homologous Recombination. *Cell Rep.* **2017**, *19*, 1294–1303. [CrossRef] [PubMed]
64. Sokolova, V.; Lee, G.; Mullins, A.; Mody, P.; Watanabe, S.; Tan, D. DNA-translocation-independent role of INO80 remodeler in DNA damage repairs. *J. Biol. Chem.* **2023**, *299*, 105245. [CrossRef] [PubMed]
65. Gospodinov, A.; Vaissiere, T.; Krastev, D.B.; Legube, G.; Anachkova, B.; Herceg, Z. Mammalian Ino80 mediates double-strand break repair through its role in DNA end strand resection. *Mol. Cell. Biol.* **2011**, *31*, 4735–4745. [CrossRef] [PubMed]
66. Locatelli, M.; Lawrimore, J.; Lin, H.; Sanaullah, S.; Seitz, C.; Segall, D.; Kefer, P.; Salvador Moreno, N.; Lietz, B.; Anderson, R.; et al. DNA damage reduces heterogeneity and coherence of chromatin motions. *Proc. Natl. Acad. Sci. USA* **2022**, *119*, e2205166119. [CrossRef] [PubMed]
67. García Fernández, F.; Fabre, E. The Dynamic Behavior of Chromatin in Response to DNA Double-Strand Breaks. *Genes* **2022**, *13*, 215. [CrossRef] [PubMed]
68. Lawrimore, J.; Barry, T.M.; Barry, R.M.; York, A.C.; Friedman, B.; Cook, D.M.; Akialis, K.; Tyler, J.; Vasquez, P.; Yeh, E.; et al. Microtubule dynamics drive enhanced chromatin motion and mobilize telomeres in response to DNA damage. *Mol. Biol. Cell* **2017**, *28*, 1701–1711. [CrossRef]
69. Strecker, J.; Gupta, G.D.; Zhang, W.; Bashkurov, M.; Landry, M.C.; Pelletier, L.; Durocher, D. DNA damage signalling targets the kinetochore to promote chromatin mobility. *Nat. Cell Biol.* **2016**, *18*, 281–290. [CrossRef]
70. Amaral, N.; Ryu, T.; Li, X.; Chiolo, I. Nuclear Dynamics of Heterochromatin Repair. *Trends Genet.* **2017**, *33*, 86–100. [CrossRef]
71. García Fernández, F.; Almayrac, E.; Carré Simon, À.; Batrin, R.; Khalil, Y.; Boissac, M.; Fabre, E. Global chromatin mobility induced by a DSB is dictated by chromosomal conformation and defines the HR outcome. *eLife* **2022**, *11*, e78015. [CrossRef] [PubMed]
72. Liu, J.; Vidi, P.A.; Lelièvre, S.A.; Irudayaraj, J.M. Nanoscale histone localization in live cells reveals reduced chromatin mobility in response to DNA damage. *J. Cell Sci.* **2015**, *128*, 599–604. [CrossRef] [PubMed]
73. Dion, V.; Kalck, V.; Horigome, C.; Towbin, B.D.; Gasser, S.M. Increased mobility of double-strand breaks requires Mec1, Rad9 and the homologous recombination machinery. *Nat. Cell Biol.* **2012**, *14*, 502–509. [CrossRef]
74. Piazza, A.; Wright, W.D.; Heyer, W.D. Multi-invasions Are Recombination Byproducts that Induce Chromosomal Rearrangements. *Cell* **2017**, *170*, 760–773.e15. [CrossRef]
75. Zagelbaum, J.; Schooley, A.; Zhao, J.; Schrank, B.R.; Callen, E.; Zha, S.; Gottesman, M.E.; Nussenzweig, A.; Rabadan, R.; Dekker, J.; et al. Author Correction: Multiscale reorganization of the genome following DNA damage facilitates chromosome translocations via nuclear actin polymerization. *Nat. Struct. Mol. Biol.* **2023**, *30*, 1048. [CrossRef]
76. Li, H.; McCord, R.P. Actin up: Shifting chromosomes toward repair, but also translocations. *Nat. Struct. Mol. Biol.* **2023**, *30*, 2–4. [CrossRef] [PubMed]
77. Cho, N.W.; Dilley, R.L.; Lampson, M.A.; Greenberg, R.A. Interchromosomal homology searches drive directional ALT telomere movement and synapsis. *Cell* **2014**, *159*, 108–121. [CrossRef]

78. Candotti, F. Clinical Manifestations and Pathophysiological Mechanisms of the Wiskott-Aldrich Syndrome. *J. Clin. Immunol.* **2018**, *38*, 13–27. [CrossRef]
79. Dubash, A.D.; Guilluy, C.; Srougi, M.C.; Boulter, E.; Burridge, K.; García-Mata, R. The small GTPase RhoA localizes to the nucleus and is activated by Net1 and DNA damage signals. *PLoS ONE* **2011**, *6*, e17380. [CrossRef]
80. Magalhaes, Y.T.; Boell, V.K.; Cardella, G.D.; Forti, F.L. Downregulation of the Rho GTPase pathway abrogates resistance to ionizing radiation in wild-type p53 glioblastoma by suppressing DNA repair mechanisms. *Cell Death Dis.* **2023**, *14*, 283. [CrossRef]

cells MDPI

Review

CRISPR/Cas-Based Ex Vivo Gene Therapy and Lysosomal Storage Disorders: A Perspective Beyond Cas9

Andrés Felipe Leal [1,2,3,*], Luis Eduardo Prieto [2] and Harry Pachajoa [1,2,4,*]

1 Centro de Investigaciones en Anomalías Congénitas y Enfermedades Raras, Universidad Icesi, Cali 760031, Colombia
2 Centro de Investigaciones Clínicas, Fundación Valle de Lili, Cali 760001, Colombia; luis.prieto.va@fvl.org.co
3 Institute for the Study of Inborn Errors of Metabolism, Faculty of Science, Pontificia Universidad Javeriana, Bogotá 110231, Colombia
4 Departamento de Genética Clínica, Fundación Valle de Lili, Cali 760001, Colombia
* Correspondence: lealb.af@javeriana.edu.co (A.F.L.); hmpachajoa@icesi.edu.co (H.P.)

Abstract: Lysosomal storage disorders (LSDs) are inherited metabolic conditions characterized by lysosomal enzyme deficiencies leading to substrate accumulation. As genetic diseases, LSDs can be treated with gene therapies (GT), including the CRISPR/Cas systems. The CRISPR/Cas systems enable precise and programmable genome editing, leading to targeted modifications at specific genomic loci. While the classical CRISPR/Cas9 system has been extensively used to generate LSD disease models and correct disease-associated genetic alterations through homologous recombination (HR), recently described Cas proteins as well as CRISPR/Cas9-derived strategies such as base editing, prime editing, and homology-independent targeted integration (HITI) offer a novel way to develop innovative treatments for LSDs. The direct administration of the CRISPR/Cas9 system remains the primary strategy evaluated in several LSDs; nevertheless, the ex vivo CRISPR/Cas9-based approach has been recently explored, primarily in central nervous system-affecting LSDs. Ex vivo approaches involve genetically modifying, in theory, any patient cells in the laboratory and reintroducing them into the patient to provide a therapeutic effect. This manuscript reviews the molecular aspects of the CRISPR/Cas technology and its implementation in ex vivo strategies for LSDs while discussing novel approaches beyond the classical CRISPR/Cas9 system.

Keywords: CRISPR/Cas; ex vivo; gene therapy; lysosomal storage disorders

1. Introduction

Lysosomal storage disorders (LSDs) are metabolic conditions characterized by overaccumulation of partially or non-degraded substrates due to impaired lysosomal function [1]. Primary LSD care involves enzyme replacement therapy (ERT) and symptom treatment to improve the patient's quality of life [2]. Strategies such as substrate reduction, degradation therapy, and molecular chaperones could become potential treatments in the future [3,4]. As genetic diseases, LSDs can be treated via gene therapy (GT) to correct the mutation causing the disease. Although classical GT enables the delivery of therapeutic genes via adeno-associated virus (AAV) and lentivirus (LV), the vector dilution effect and random integration are significant challenges associated with these strategies [5–7].

Contrary to the AAV- and LV-based GTs, the clustered regularly interspaced short palindromic repeats (CRISPR) and CRISPR-associated proteins (Cas), abbreviated as CRISPR/Cas, lead to modifying the genome through precise DNA cutting and then harnessing natural homologous recombination (HR) DNA repair in the presence of a donor

 https://doi.org/10.3390/cells14151147

template (Figure 1). The absence of a donor template results in indel (insertion or deletion) formation due to the predominant activity of the non-homologous end-joining (NHEJ) pathway in mammalian cells [8,9]. While in vivo CRISPR/Cas9-based GT has demonstrated promising outcomes in several LSD models [10–12], the pre-existing anti-Cas9 immune response could limit its therapeutic efficacy [13,14]. In consequence, ex vivo CRISPR/Cas9-based GT has emerged as a novel alternative for several LSDs [15].

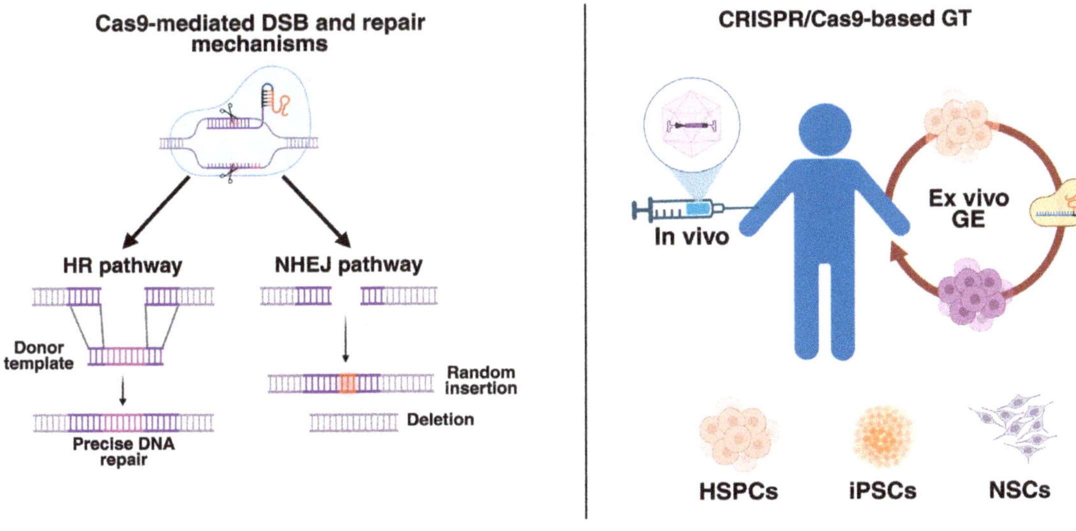

Figure 1. The classical CRISPR/Cas9 system and its use as a gene therapy (GT) approach. (**Left panel**) The classical CRISPR/Cas9 system requires a double-strand break (DSB) within the DNA mediated by an RNA-guided Cas9 protein. Upon DSB, the cell activates two central repair mechanisms: the homologous recombination (HR) and non-homologous end-joining (NHEJ) pathways. While HR is predominant in dividing cells, the NHEJ is active in both dividing and non-dividing cells. Most importantly, HR needs a donor template with homologous arms to be inserted into the host genome. Without a donor template, the NHEJ mediates the DSB repair, producing random insertions or deletions known as indels. (**Right panel**) The CRISPR/Cas9 system can be used for in vivo and ex vivo genome editing (GE). Likewise, current studies have demonstrated the suitability of hematopoietic stem progenitor cells (HSPCs), induced pluripotent stem cells (iPSCs), and neural stem cells (NSCs) for ex vivo CRISPR/Cas9-based GTs. This figure was created with Biorender.com.

Most of the ex vivo CRISPR/Cas9 approaches tested for LSDs rely on modifying hematopoietic stem progenitor cells (HSPCs) in a controlled laboratory environment and later transplanting them into LSD animal models [15]. Although not all LSDs affect the central nervous system (CNS), about two-thirds of patients present neurological compromise [16]. In this regard, HSPCs transplantation (HSCT) has shown encouraging outcomes in CNS-affecting LSDs, such as mucopolysaccharidosis (MPS) I and metachromatic leukodystrophy, as it can induce medullary repopulation [17]. Relevantly, HSPCs can differentiate into all hematological precursors, including macrophages, which can cross the blood–brain barrier (BBB) and transform into microglia in response to the brain's microenvironment [18]. A significant increase in BBB permeability appears to favor the migration of macrophages from the bloodstream into the CNS [19,20]. Likewise, it has been described that HSCT can be successfully implemented in some non-CNS-affecting LSDs, such as MPS IVA, as it improves the 6-minute walk test, increases stature, and enhances respiratory function and hearing in early-treated MPS IVA patients [21,22]. In this case, mature lysosomal enzyme-producing blood cells can distribute throughout the body and mediate cross-correction, thereby ameliorating systemic symptoms in tissues distant from the bone marrow (BM), such as the spleen, liver, and heart, by decreasing substrate accumulation, which ultimately increases patient survival [23].

HSCT is a promising alternative for treating LSDs; however, graft-versus-host disease (GVHD) is a significant concern that limits allogeneic HSCT even in the presence of matched human leukocyte antigen (HLA) [24]. Interestingly, using the CRISPR/Cas system, it is possible to isolate patients' HSPCs, carefully engineer them in the laboratory, and finally reinfuse them into the patient to repopulate the BM (Figure 1) [15]. This autologous HSCT procedure significantly reduces the risk of GVHD and graft rejection [24]. While HSPCs are the most common cells used in ex vivo CRISPR/Cas-based GTs, the feasibility of induced pluripotent stem cells (iPSCs) and neural stem cells (NSCs) has also been demonstrated for some LSDs (Figure 1). Although the CRISPR/Cas9 system is the most common strategy for ex vivo approaches in LSDs, successful gene editing requires the generation of double-strand breaks (DSBs) within the DNA by Cas9 [25], which raises critical safety concerns, including potential unwanted Cas9 cutting. Thus, alternatives such as CRISPR/Cas12a, base editing (BE), prime editing (PE), and homology-independent targeted integration (HITI) could offer innovative options, improving the safety and efficiency of the CRISPR/Cas-based genome editing (GE).

This review summarizes the classical ex vivo CRISPR/Cas9 approaches for LSDs while exploring novel alternatives beyond Cas9 and HR for inducing genomic DNA rewriting. A comprehensive literature search was performed using PubMed, Web of Science, and Google Scholar databases. The search included peer-reviewed articles published between January 2015 and May 2025, with a focus on the most recent developments (2020–2025), using the Boolean operators ("CRISPR/Cas" OR "CRISPR-based editing") AND ("ex vivo gene therapy" OR "hematopoietic stem cells" OR "induced pluripotent stem cells") AND ("lysosomal storage disorders" OR "MPS" OR "Gaucher" OR "Fabry" OR "Krabbe") AND ("base editing" OR "prime editing" OR "Cas12" OR "Cas13" OR "HITI").

2. The CRISPR/Cas Systems

The CRISPR/Cas system is an adaptive immune system in prokaryotes [8,9]. The locus CRISPR harbors short repetitive elements separated by spacers and a series of CRISPR-associated (Cas) genes that encode for Cas proteins. During an invasion by bacteriophages, transposons, or plasmids, the CRISPR/Cas locus is activated to induce Cas-mediated cutting of foreign DNA [8]. The released DNA fragments are later inserted into the CRISPR locus as new spacers, conferring variability to the system. Upon re-infection by the same invader, a CRISPR-associated RNA (crRNA) recognizes the foreign genetic material and mediates Cas-dependent cleavage, thereby protecting the host [26]. The CRISPR/Cas systems are classified into two classes, six types, thirty-three subtypes, and seventeen variants (Figure 2) [27].

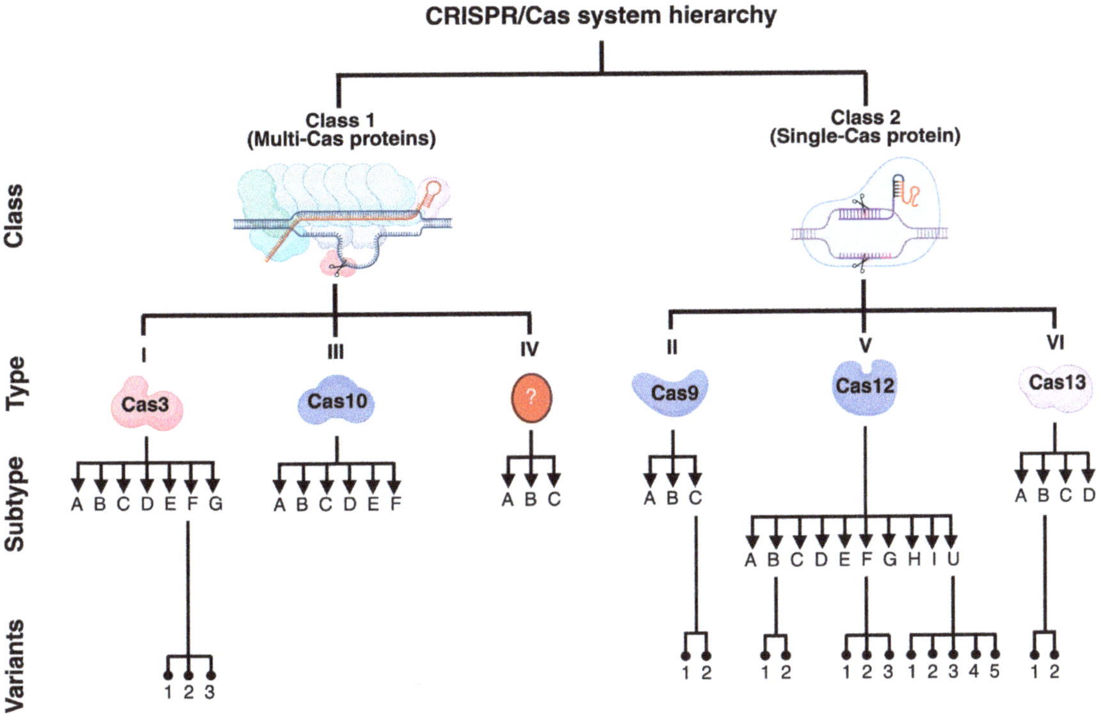

Figure 2. Classification of the CRISPR/Cas systems. The CRISPR/Cas system is primarily classified based on the presence of multi- or single-Cas proteins. Due to its simplicity, Cas proteins belonging to class 2 are the most explored for GE purposes; nevertheless, some class I proteins (i.e., Cas3) are also potential tools for treating human diseases. This figure was created with Biorender.com.

2.1. DNA-Targeting Cas Enzymes

While Cas9 remains the most popular DNA-binding endonuclease used in CRISPR/Cas, several other Cas proteins have been described and could be considered for GE purposes. The main features of Cas proteins are detailed in Table 1.

Table 1. Main features of Cas proteins.

* Cas Variant	MW (kDa)	Substrate Preference	PAM Requirement	Collateral Activity	Editing Mechanism
Cas9	160	dsDNA	5′-NGG-3′	No	Blunt-end DSBs
Cas3	120	ssDNA	Varies (Cascade-dependent)	No	Processive DNA degradation
Cas12a (Cpf1)	130	dsDNA	5′-TTTV-3′	Yes (ssDNA)	Staggered DSBs
Cas 12e (CasX)	112	dsDNA	5′-TTCN-3′	No	ssDNA cleavage
Cas12j (CasΦ)	95	ssDNA	PAM-independent	Yes (ssDNA)	Staggered DSBs
Cas12f1	60	ssDNA	T-rich	Yes (ssDNA)	ssDNA cleavage
Cas14	50	ssDNA	PAM-independent	Yes (ssDNA)	ssDNA cleavage
Cas7-11	120	ssRNA	PFS-like motifs	No	RNA cleavage
Cas13	150	ssRNA	PFS	Yes (ssRNA)	RNA cleavage

* Alternative names given to Cas variants are presented in parentheses. **MW**, molecular weight; **PAM**, protospacer-adjacent motif; **PFS**, protospacer-flanking site.

2.1.1. Cas9

The Cas9 enzyme is the most commonly used endonuclease in CRISPR/Cas. The Cas9 enzyme is composed of a recognition and a nuclease lobe. Two catalytic domains, termed HNH (histidine–asparagine–histidine) and RuvC, mediate the DSB of dsDNA upon protospacer-adjacent motif (PAM) recognition [8,9,28]. Several modifications of the Cas9 protein have been reported, including high-fidelity Cas9 (HF-Cas9), Cas9 nickase (nCas9, lacking endonuclease activity from HNH or RuvC), and dead Cas9 (dCas9) as well as NHEJ-inhibiting proteins such as Brex27, CtIP, DN1S, and RAD52, among others, which are fused to Cas9 for enhancing the HR pathway [8,9]. We have published a comprehensive review of Cas9 variants and encourage readers to consult it [8].

2.1.2. Cas3

The Cas3 enzyme is an effector enzyme with both endonuclease and helicase activity, predominantly binding ssDNA in a 3′-to-5′ direction upon exerting its helicase activity [29,30]. Contrary to Cas9, which induces small deletions [31], Cas3 "shreds" DNA, producing large deletions up to multiple kilobases from the starting point. The Cas3-mediated DNA processing results in fewer off-target effects compared to those produced by Cas9 [32,33]. Cas3 has been previously tested by Morisaka et al. (2019) in an exon-skipping strategy using Duchenne muscular dystrophy (DMD)-induced pluripotent stem cells (iPSCs) [34]. DMD is a genetic disorder caused by pathogenic variant mutations in the dystrophin gene, resulting in a deficiency of the dystrophin protein, which is essential for maintaining muscle function [35]. Results from this study demonstrated the recovery of the dystrophin protein through CRISPR/Cas3-mediated skipping of exon 45 [34], providing proof of concept for the therapeutic applicability of the CRISPR/Cas3 system.

2.1.3. Cas12

Cas12a, initially classified as Cpf1, is a DNA endonuclease that generates staggered breaks resulting in 5 bp overhangs [36]. This is particularly relevant for HR, as it generates ssDNA, leading to access to the HR machinery. Most interestingly, Cas12a is approximately one-third the size of Cas9, making it more suitable for therapeutic applications where package size is limited, such as AAV vectors [37]. Although Cas12a has been used for the diagnosis of infection by SARS-CoV-2 [38,39], it has also been tested for GE applications. For instance, Cas12a has demonstrated a therapeutic impact in restoring the expression of the *CFTR* gene in an in vitro model of cystic fibrosis, resulting in an efficiency of up to 8%. Still, Cas12 resulted in lower efficacy compared to Cas9 [40]. Although Cas12e [41–43] and Cas12j [43,44] have been reported, Cas12f1 has garnered greater attention due to its smaller size compared to its counterparts [45,46]. Cas12f1 was recently tested by Cui et al. (2024) [47] to knock out the Nr2e3 gene in a mouse model of retinitis pigmentosa using AAV vectors. Upon a single subretinal injection in two-week-old mice, the authors found over 70% transduction efficiency in retinal cells and about 41% indel frequency [47]. Notably, mice showed improved cone function at one month post-treatment and an increased outer nuclear layer [47], suggesting that retinal degeneration was rescued upon CRISPR/Cas12f1 treatment.

2.1.4. Cas14

Cas14 is the smallest Cas protein that can only bind and cleave ssDNA in a PAM-independent way [48]. Unlike Cas12a, Cas14 recognizes ssDNA with high specificity, a feature that has been a key factor in detecting single-nucleotide polymorphisms and differentiating between pathogens [49]. Contrary to other Cas proteins, Cas14 is susceptible to mismatches in the middle of the target region (~6 bp) [50], making it unsuitable for GT purposes.

2.2. RNA-Targeting Cas Enzymes

Most approaches observed in CRISPR/Cas utilize Cas proteins that can bind and cut DNA. Nevertheless, RNA-targeting Cas enzymes can also be implemented in clinical practice. Since these Cas enzymes do not target DNA but RNA, they can be used for transient knockdown.

2.2.1. Cas7-11

Cas7-11 is a single effector that cuts ssRNA with minimal cell toxicity and lower off-target effects in mammalian cells compared to its counterpart, Cas13 [51,52]. Cas7-11 comprises a putative Cas11 domain and several Cas7 subunits [52,53]. As a recently discovered Cas enzyme, only a few pieces of evidence are available showing its potential in a transient knockdown. Moreno-Sánchez et al. (2025) [54] reported using CRISPR/Cas7-11 to suppress mRNA from the developmental genes *nanog* and *no-tail* in zebrafish. Interestingly, the CRISPR/Cas7-11 system demonstrated an in vivo efficiency of 60.1% without collateral activity [54]. Most importantly, when targeting the green fluorescent protein (GFP), the authors observed no developmental defects, 28S rRNA cleavage, or transcriptomic alterations in zebrafish embryos [54], supporting the safety of the CRISPR/Cas7-11 system.

2.2.2. Cas13

Similar to Cas7-11, Cas13 is also an ssRNA-cutting endonuclease. Unlike Cas7-11, Cas13 exhibits collateral activity that mediates unintended, non-targeted RNA cutting [55,56]. The CRISPR/Cas13 system was recently tested by Kumar et al. (2024) [57] as a GT for retinal neovascularization by targeting VEGF mRNA in cell-derived retinal organoids and humanized VEGF transgenic mice. In the mouse model of proliferative retinopathy, the intravitreal delivery of the CRISPR/Cas13 system packaged into an AAV2 vector resulted in a ~50% reduction in human VEGF expression, along with a reduced leakage area, lower leaky vessel density, and fewer branch points in treated mice, suggesting an amelioration of disease progression [57]. No decrease in mouse VEGF was observed, supporting the high specificity of the system. Despite these promising achievements, the CRISPR/Cas13 system induced dysregulation of 23 unidentified genes, as detected by transcriptomic analysis [57], consistent with the reported collateral effect of Cas13 [55]. Novel variants of Cas13, such as CRISPR-DjCas13d, are under development to maintain the high specificity of Cas13 while eliminating its collateral effects [54].

2.3. CRISPR/Cas-Derived Alternatives

Initially, the CRISPR/Cas system was designed to modify the genome via Cas9-mediated DSB and later HR activation. Emerging approaches aim to edit the genome by leveraging enzymes such as deaminases and reverse transcriptase in CRISPR/Cas-based systems, which are referred to as base and prime editing, respectively (Figure 3). Similarly, the HR dependence for GE can be dismissed when working with the homology-independent targeted integration (HITI) system (Figure 3), potentially increasing the efficacy of the CRISPR/Cas systems in both dividing and non-dividing cells [58]. Finally, new DNA repair-independent targeted integration alternatives such as CRISPR-associated transposases (CAST) and those combining PE and integrases/recombinases have extended the CRISPR/Cas-based tools that can be evaluated as GT approaches (Figure 3).

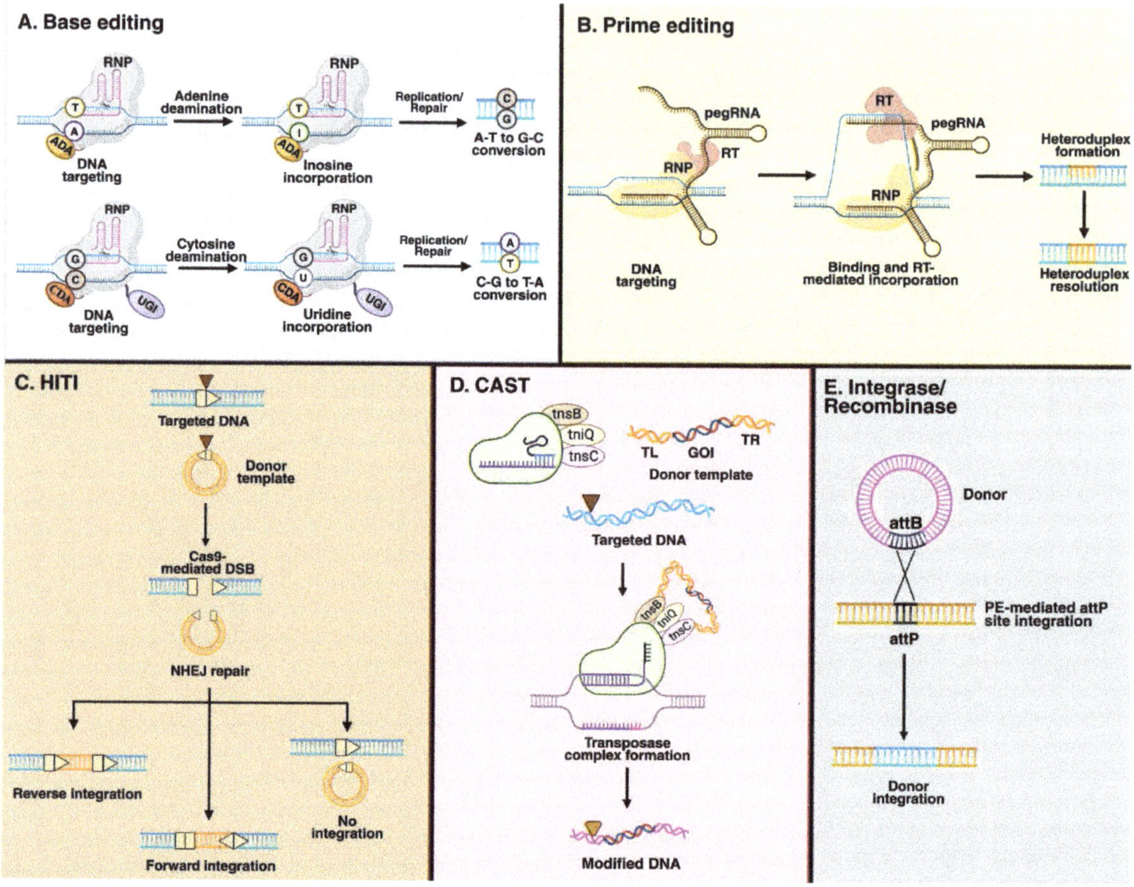

Figure 3. CRISPR/Cas9-based approaches. (**A**) BE is a genome editing tool that involves a ribonucleoprotein (RNP) complex, comprising an sgRNA and nCas9 protein. nCas9 is fused to either adenine (ADA) or cytosine (CDA) deaminase. Upon Cas9-mediated ssDNA nicking, ADA or CDA mediates the deamination of adenine and cytosine. While adenine deamination leads to inosine formation (upper), cytosine deamination leads to uracil (bottom). Later replication or reparation mechanisms allow the final conversion from I to G and U to T. (**B**) The PE system retains the RNP complex comprising a pegRNA and nCas9. nCas9 is fused to a reverse transcriptase (RT). The pegRNA is a large sgRNA containing a primer binding site (PBS) and an RT template. While the PBS leads to targeted sequence hybridization, the RT uses the RT template to synthesize the new gene information. The cell's repair mechanism later corrects the mismatch between edited and non-edited DNA strands. (**C**) HITI is intended to integrate foreign DNA within host DNA using the NHEJ pathway. Both the target DNA and the donor template should contain sgRNA and PAM sequences to undergo Cas9-mediated cutting. Upon Cas9-mediated DSB generation, the NHEJ leads to donor insertion in forward or reverse directions. If forward direction occurs, the sgRNA and PAM sequences are abolished, impeding Cas9 cutting. Conversely, if a donor template is inserted in the reverse direction, the presence of the sgRNA and PAM sequences will lead to Cas9 cutting. Repair without donor template insertion will induce additional Cas9-mediated cleavage. (**D**) The CRISPR-associated transposases (CAST) system enables the insertion of a gene of interest (GOI) through the activity of Tn7-like transposons. CAST requires the presence of an RNP comprised of Cas12k and a sgRNA to recognize and bind the targeted DNA, along with a donor template flanked by transposon left (TL) and right (TR) sequences. Cas12 brings the Tn7-like genes to the target site. Later, DNA recognition

by the RNP complex leads to the formation of a transposase complex, which ultimately enables the excision and insertion of the donor template, approximately 60 bp from the PAM sequence (GTN), within the target DNA [9,59]. (**E**) Gene insertion can also be achieved through the use of integrases/recombinases, which utilize the specific recombination sites attB and attP. In this approach, the presence of the attP sequence in the target DNA is achieved through the use of PE. Consequently, the presence of attB sites in the donor template later leads to the recombination of the GOI within the targeted DNA, resulting in a knock-in without inducing DSBs [9,59]. This figure was created with Biorender.com.

2.3.1. Base Editing

The base editors (BEs) are an alternative to the CRISPR/Cas system, based on the fusion of a Cas9 enzyme carrying mutations in its catalytic domains, RuvC and HNH [60]. Those mutations maintain PAM recognition and DNA targeting while abolishing Cas9-mediated DSB within the DNA. nCas9 is also suitable for base editing. Fused to Cas9, a base-modifying enzyme such as cytosine or adenine deaminase enables the system to convert cytosine to uracil and adenine to hypoxanthine (inosine), respectively [60,61]. While uracil is converted into thymine, hypoxanthine is converted into guanine during DNA replication and repair. Primary described BEs can only catalyze nucleotide transitions, purine–purine (A–G), and pyrimidine–pyrimidine (C–T), which limits their use for correcting transversions; nevertheless, new advances make correcting transversions via glycosylase base editors and adenine transversion editors possible [60]. Recently, a new generation of BEs has been developed, including Cas12a, which can mediate base editing in mammalian cells [62]. Cas12a can process multiple sgRNAs from a single array transcript [63], thus allowing multi-gene targeting. This could be an interesting alternative when several mutations affect a gene or different genes are involved.

2.3.2. Prime Editing (PE)

PE is a CRISPR/Cas alternative that leads to insertions, deletions, and substitutions without introducing DSBs in the DNA. PE relies on the nCas9-mediated nicking and activity of the reverse transcriptase (RT) [64]. Contrary to the classical CRISPR/Cas9 system, the sgRNA is a prime editing guide RNA (pegRNA). While the sgRNA is a single, short RNA sequence, a pegRNA is a longer RNA containing a reverse-transcription template and a primer binding site at the 3′ end [65]. The RT uses the template placed on the pegRNA to synthesize new DNA, which will later be introduced into the cell's DNA. Since its first description, PE has undergone extensive modifications aimed at improving RT efficiency (PE2, PE6, and PEmax), resolving mismatches (P3, P4, and P5), refining sizing, optimizing codons (PEmax), and increasing pegRNA stability (PE7) [65–68]. Like BEs, PE has also incorporated Cas12, enabling the system to target multiple genes in human cells [69].

2.3.3. Homology-Independent Targeted Integration (HITI)

HITI is a CRISPR/Cas9-based system that enables the knock-in of foreign DNA without relying on HR, unlike the classical CRISPR/Cas9 system [70]. In HITI, the donor template contains the desired insertion sequence, flanked by the sgRNA and PAM sequences, which are recognized by the Cas9 protein. Consequently, Cas9 will recognize and cut both the targeted DNA and the donor template, enabling the NHEJ repair mechanism [58]. Upon NHEJ repair, forward-inserted sequences will abolish the sgRNA/PAM sequence, resulting in a permanent insertion [58,70]. In contrast, reverse orientation restores the previous Cas9-mediated sgRNA/PAM cut sequences, allowing Cas9 to recut until a forward orientation is achieved.

3. CRISPR/Cas9: Ex Vivo Approaches in LSDs

The CRISPR/Cas9 system has been successfully evaluated in some LSDs using ex vivo approaches, including MPS I, MPS IVA, Gaucher, Krabbe, and Fabby. Table 2 summarizes the main findings for these ex vivo CRISPR/Cas9-based strategies.

Table 2. Ex vivo CRISPR/Cas9-based approaches in LSD models.

Model	CRISPR/Cas System Tested	Cell Type	Locus	IE (%)	OT (%)	Eng. (%)	Stemness	Ref.
MPS I	CRISPR/Cas9	iPSCs	IDUA	2.7	NR	NA	Preserved	[71]
MPS I	CRISPR/Cas9	HSPCs	CCR5	54	<0.5	21.5	Preserved	[72]
MPS I	CRISPR/Cas9	HSPCs	CCR5	TBI:5.7 BU:20	NR	* TBI:45 * BU:73	NR	[20]
MPS IVA	CRISPR/nCas9	HSPCs	AAVS1	10	ND	NA	Preserved	[73]
Gaucher	CRISPR/Cas9	HSPC	CCR5	29.9	NR	23.2	Preserved	[74]
Gaucher	CRISPR/Cas9	iPSCs	GBA	NR	ND	NA	Preserved	[75]
Krabbe	CRISPR/Cas9	NSC	IL2RG	2.8	<0.7	NA	Preserved	[76]
Krabbe	CRISPR/Cas9	NSC	CCR5	2.4	NR	NA	NR	
Fabry	CRISPR/Cas9	iPSCs	GLA	NR	NR	NA	Preserved	[77]
Fabry	CRISPR/Cas9	iPSCs	GLA	NR	ND	NA	Preserved	[78]

BM, bone marrow; BU, busulfan; Eng., engraftment in BM (CD45+); HSPCs, hematopoietic stem progenitor cells; IE, insertion efficiency; iPSCs, induced pluripotent stem cells; NA, not applicable; ND, not detected; NR, not reported; NSC, neural stem cells; OT, off-target; TBI, total body irradiation. * Median (min, max). Note: While most of the cell models correspond to human-derived stem cells, results from the study [71] were derived from mouse-derived stem cells.

3.1. Mucopolysaccharidosis (MPS) I

MPS I (OMIM 607014) is a multi-systemic LSD caused by an alpha-L-iduronidase (IDUA; EC 3.2.1.76) deficiency that leads to lysosomal accumulation of dermatan sulfate (DS) and heparan sulfate (HS) [79,80]. MPS I is a central nervous system (CNS)-affecting LSD, also producing skeletal dysplasia along with cardiac and ocular manifestations [80]. Initial proof-of-concept studies using CRISPR/Cas9 and iPSCs were conducted by Miki et al. (2019) [71]. In this study, mouse embryonic fibroblast-derived iPSCs were established from an MPS I mouse model and edited with the CRISPR/Cas9 system to remove a Neo^R gene inserted into exon VI of the *Idua* gene. The authors successfully observed the recovery of IDUA expression in CRISPR/Cas9-edited iPSCs, with no difference from WT iPSCs [71]. Although no significant changes were detected in *Oct4*, *Nanog*, and *Sox2* genes in teratoma formation assays using CRISPR/Cas9-edited iPSCs compared to WT iPSCs [71], the authors discussed the potential tumorigenicity of the iPSCs and CRISPR/Cas9 off-targeting as critical concerns before implementing this approach in clinical practice.

In 2019, Goméz-Ospina et al. demonstrated the suitability of an ex vivo CRISPR/Cas9-based GT in an MPS I mouse model. By electroporating human CD34+ cells with a CRISPR/Cas9 system targeting the human CCR5 locus, the authors successfully achieved supraphysiological IDUA expression levels in human CD34+ cells without compromising their stemness and in vivo engraftment properties [72]. Upon transplantation of edited CD34+ cells into immunodeficient MPS I mice, a significant increase in the IDUA activity in serum, liver, spleen, and brain was observed. A consistent GAG reduction and recovery of bone thickness of the zygomatic, skull, and femur were observed. Likewise, CRISPR/Cas9-modified CD34+ cells decreased the isolectin B4 and glial fibrillary acidic protein (GFAP) staining in mouse brains, suggesting that gliosis is ameliorated [72]. The molecular validation of this CRISPR/Cas9 system revealed an indel occurrence rate of less than 0.5% across 62 predicted off-target sites, supporting the safety of the sgRNA used in human CD34+ cells.

Sub-lethal irradiation and busulfan-mediated myeloablation are standard protocols for conditioning before HSCT [15,20,81], and their effectiveness in enhancing the engraftment of CRISPR/Cas9-modified human CD34+ cells was assessed by Poletto et al. (2022) in MPS I mice. Interestingly, the authors reported that busulfan-mediated myeloablation significantly improved the engraftment of CRISPR/Cas9-modified human CD34+ cells in BM, long-term homing in BM-derived visceral cells [20], and the greater migration of BM-derived cells in the brain compared to the irradiation protocol. Consistently, an improvement in IDUA expression and GAG clearance was observed in busulfan-treated mice compared to irradiation [20], highlighting the importance of preconditioning regimens before transplantation of CRISPR/Cas9-modified human CD34+ cells.

3.2. Mucopolysaccharidosis (MPS) IVA

MPS IVA (OMIM 253000) is a systemic LSD caused by a deficiency of the N-acetylgalactosamine-6-sulfatase (GALNS; EC 3.1.6.4) enzyme. GALNS leads to the lysosomal degradation of keratan sulfate (KS) and chondroitin 6-sulfate (C6S) [80,82]. MPS IVA patients are characterized by skeletal dysplasia, cardiopulmonary, and ocular complications [83]. Recently, Herreño-Pachón et al. (2025) reported, for the first time, the use of a novel CRISPR/nCas9-based GT to modify human CD34+ cells [73]. In this study, the authors designed paired sgRNAs targeting the human AAVS1 locus to knock in an expression cassette carrying the GALNS cDNA [73]. Upon electroporation of the CRISPR/nCas9 system into human CD34+ cells, the authors found that these cells retained their stemness, including the ability to proliferate and differentiate [73]. Moreover, upon co-culture of CRISPR/nCas9-modified CD34+ cells with human MPS IVA fibroblasts, a significant increase in GALNS activity was observed in MPS IVA fibroblasts, suggesting that human CD34+ cells-derived GALNS enzyme was successfully uptaken by human MPS IVA fibroblasts. Most importantly, the authors reported a significant decrease in lysosomal mass and mono-KS, supporting the intracellular sorting of GALNS into the lysosome [73]. The authors also found a recovery in global and mitochondrial-dependent oxidative stress. This study also revealed a significant increase in mitochondrial mass and a pro-apoptotic profile in untreated MPS IVA fibroblasts, which was efficiently restored to WT levels upon co-culture with CRISPR/nCas9-modified CD34+ cells [73], paving the way for a novel ex vivo CRISPR/nCas9-based GT.

3.3. Gaucher

Gaucher disease (GD) is an LSD that belongs to the sphingolipidoses and results from the deficiency of β-glucocerebrosidase (GCase; EC 3.2.1.45). GCase degrades glucosylceramide (GlcCer) and glucosylsphingosine (GlcSph) within the lysosome [84,85]. GD subtypes include nonneuronopathic (GD1; OMIM 230800), acute neuronopathic (GD2; OMIM 230900), and chronic neuronopathic (GD3; OMIM 231000). GlcCer and GlcSph accumulate primarily within mononuclear phagocyte cells, which are identified in the BM, spleen, and liver as wrinkled, paper-like cells termed Gaucher cells [85]. An ex vivo CRISPR/Cas9-based strategy was tested by Scharenberg et al. (2020) as a potential alternative for treating GD [74]. By targeting human CCR5, the authors successfully integrated an expression cassette carrying GCase cDNA, driven by the lineage-specific CD68 promoter, to restrict its expression to the monocyte/macrophage lineage [74]. The use of the spleen focus-forming virus (SFFV) promoter to drive GCase expression in human CD34+ cells induced toxic effects, underscoring the importance of careful promoter selection [74]. Upon CRISPR/Cas9 GE, the human CD34+ cells could differentiate into CD14+/Cd11b+ cells, suggesting successful human monocyte/macrophage differentiation. CRISPR/Cas9-edited CD34+ cells did not fail to differentiate into all hematopoietic lineages [74], supporting the maintenance of stemness upon knock-in. Most importantly, transplantation into immunodeficient mice demonstrated higher engraftment and self-renewal capacity upon serial transplantation [74], further supporting the preservation of stemness.

Given that GlcCer accumulation primarily occurs in the monocyte/macrophage lineage, it is not surprising that monocyte/macrophage functions, such as migration, phagocytosis, and activation, are impaired [86,87]. A recent study conducted by Ramalingam et al. (2023) [75] investigated the feasibility of using the CRISPR/Cas9 system to correct the L444P (1448T→C) mutation in GD2 iPSCs by targeting this mutation at the *GBA* gene. iPSCs-derived macrophages resulted in GCase-expressing cells and the restoration of motility and phagocytosis of zymozan [75]. It has been observed that GD protects against *Mycobacterium tuberculosis* infection [88]. Interestingly, the recovery of GCase expression increased the susceptibility of iPSC-derived macrophages to *M. tuberculosis* infection [75], revealing new considerations when attempting CRISPR/Cas9-mediated GE for GD.

3.4. Krabbe

Krabbe disease (OMIM 245200) is an LSD caused by a deficiency of galactocerebrosidase (GALC; 3.2.1.46). This deficiency leads to the lysosomal accumulation of galactosylceramide (GalCer) in Schwann and oligodendroglial cells, thereby affecting the central and peripheral nervous systems [89]. Although CD34+ cells and iPSCs are the most common sources tested for ex vivo CRISPR/Cas9-based GT, a study by Dever et al. (2019) [76] showed the feasibility of neural stem cells (NSCs) as a potential alternative for treating Krabbe disease. In this study, three loci, IL2RG, HBB, and CCR5, were targeted via CRISPR/Cas9 to insert an expression cassette carrying a functional copy of GALC in NSCs. The highest HR was achieved when targeting IL2RG (~6%), while the lowest HR was observed with CCR5 [76]. Upon subventricularly transplanting CRISPR/Cas9-edited NSCs into immunodeficient neonatal mice, the authors noticed engraftment in the cortex and corpus callosum, along with migration from the subventricular zone to the olfactory bulb and the rostral migratory stream [76], suggesting that CRISPR/Cas9-edited NSCs preserve their biological properties in vivo. Significantly, the co-culture of Krabbe fibroblasts and GALC-expressing NSCs resulted in Krabbe fibroblasts' mannose 6-phosphate-dependent GALC uptake [76]. Although functional assessment of phenotype recovery in Krabbe fibroblasts was not performed upon co-culture, the intracellular GALC increase is expected to exert a therapeutic effect. Upcoming experiments in Krabbe animal models, such as the Twitcher mouse model, are still to be conducted to validate this novel ex vivo CRISPR/Cas9-based GT preclinically.

3.5. Fabry

Fabry disease (FD; OMIM 301500), an LSD caused by α-galactosidase A (GLA; EC 3.2.1.22) deficiency, is characterized by the accumulation of globotriaosylceramide (Gb3) and Lyso-GB3 in the kidneys, heart, brain, and peripheral nervous system [90]. A recent study published by Choi et al. (2023) [77] used the CRISPR/Cas9 system to correct the mutation 1268fs*1 (c.803_806del) in FD iPSCs. While GLA enzyme activity was significantly recovered upon GE, CRISPR/Cas9-edited iPSCs retained pluripotency-associated marker expression, a normal karyotype, and displayed differentiation into three germ layers, with no detectable off-target effects [77], thereby developing an ex vivo CRISPR/Cas9 proof-of-concept to correct the deletion of the *GLA* gene. Similar outcomes were recently reported by Karl-Schöller et al. (2025) in FD iPSCs for the mutation c.1069C>T in the *GLA* gene [78]. While the CRISPR/Cas9 system led to the restoration of the WT sequence, the authors also observed the clearance of Gb3 deposits, along with the preservation of pluripotency markers, further supporting the therapeutic effect of the CRISPR/Cas9 for FD. The transplantation of these CRISPR/Cas9-edited iPSCs in animal models of Fabry disease still needs to be carried out to test their therapeutic efficacy in vivo.

4. Beyond Cas9 Enzymes and HR: An Overview in Non-LSD Models

Most ex vivo CRISPR/Cas-based GT approaches in LSDs utilize Cas9 and HR as the primary molecular mechanisms to knock in expression cassettes at safe harbors, thereby expressing the missing lysosomal enzyme and mediating cross-correction. Nevertheless, some non-LSD studies have demonstrated the feasibility of Cas proteins beyond Cas9, achieving transgene expression and/or do not depend on DBS/HR. This section describes recent non-LSD studies evaluating non-Cas9 proteins and non-classical HR-based knock-in approaches in ex vivo CRISPR/Cas-based alternatives. Table 3 compiles ex vivo CRISPR/Cas approaches in non-LSD models.

Table 3. Ex vivo CRISPR/Cas-based approaches in non-LSDs.

Model	CRISPR/Cas System Tested	Cell Type	Locus	IE (%)	OT (%)	Eng. (%)	Stemness	Ref.
CAR-T	CRISPR/wtCas12	T cells	TCR	30	NR	NA	NA	[91]
	CRISPR/cCas12			52.7				
CAR-T	CRISPR/MAD7	T cells	AAVS1	85	ND	NA	NA	[92]
			* AAVS1	30				
SCD	ABE	HSPCs	HBB	80	<2	70	Preserved	[93]
SCD	PE3max	HSPCs	HBB	41	<1	97	Preserved	[94]
X-SCID	HITI	HSPCs	IL2RG	18.7	ND	13.2	Preserved	[95]
LAD-1	HITI	HSPCs	ITGB2	12	ND	21	Preserved	[96]

ABE, adenine base editing; Eng., engraftment in BM (CD45+); HITI, homology-independent targeted integration; HSPCs; hematopoietic stem progenitor cells; IE, insertion efficiency; LAD-1, leukocyte adhesion deficiency type 1; NA, not applicable; ND, not detected; NR, not reported; OT, off-target; PE3max, prime editing; SCD, sickle cell disease; X-SCID, X-linked severe combined immunodeficiency. * AAVS1 + multi-gene targeting at AAVS1 and CTLA4, PDCD1, TIGIT, CD247, PTPN6, or PTPN11. Note: While most of the cell models correspond to human-derived stem cells, results from the study [95] were derived from mouse-derived stem cells.

4.1. CAR-T Cells: Implementing Cas12a

Chimeric antigen receptor (CAR)-T cell therapy is an innovative ex vivo treatment option for cancer, particularly hematological malignancies [97]. Patients' T cells are isolated via leukapheresis and engineered with a CAR construct to express synthetic receptors on the surface of the T cell. These receptors will recognize and suppress specific malignancy-antigen-expressing tumor cells [98,99]. Even though the Food and Drug Administration (FDA) has recently approved several CAR-T-cell alternatives for B-cell lymphoma, B-cell acute lymphoblastic leukemia (B-ALL), mantle cell lymphoma, and follicular lymphoma [100], cytokine release syndrome (CRS) and immune effector-cell associated neurotoxicity syndrome (ICANS) are significant concerns that still need to be addressed [101]. To overcome this, some studies have utilized the CRISPR/Cas system to edit CAR-T cells, aiming to enhance their function, develop universal CAR-T cells, and improve their safety.

Early studies by Ling et al. (2021) [91] demonstrated the efficiency of the CRISPR/Cas12a system in enhancing the knock-in of an anti-CD19-containing expression cassette at the endogenous T-cell receptor (TCR) locus in primary T cells. While wtCas12 showed about 30% knock-in efficiency, a Cas12 covalently conjugated to the 5′ terminus of crRNA (cCas12) enhanced the knock-in efficiency up to 52.7% [91], suggesting that this chemical modification increases the Cas12 stability. Major concerns in CAR-T cell preparations arise from patients' poor quality and quantity of T cells [102]; thus, creating universal CAR-T cells is now a promising research field. As Cas12 enables multi-gene targeting, the study conducted by Ling et al. (2021) [91] also evaluated the simultaneous CRISPR/Cas12-mediated knock-out of TCR, β2 microglobulin (β2M), PD1, and CTLA4, along with the anti-CD19-containing expression cassette knock-in. Interestingly, cCas12 showed up to 7.4-fold GE four-gene knock-out efficacy improvement compared to wtCas12, while cCas12 led to 37% GE anti-CD19-containing expression cassette knock-in, which was higher than that observed with wtCas12 (28%) [91]. These interesting results show the potential of ex vivo CRISPR/Cas12-mediated multi-gene targeting.

The AAVS1 locus is a well-known safe harbor for inserting large expression cassettes [103]. We have extensively tested the AAVS1 locus in several LSDs using human fibroblasts [104–106], and we recently evaluated its feasibility for inserting a GALNS-containing expression cassette in CD34+ cells [73]. In a study by Mohr et al. (2023) [92], the AAVS1 locus was used for inserting a GFP-containing expression cassette under the control of the promoters JET (195 bp), PGK (511 bp), EF-1α (1195 bp), and CAG (1723 bp), which differ in size, using the CRISPR/MAD7 system in Jurkat cells. MAD7 is a modified Cas12 released by Inscripta in 2017 that requires a crRNA and identifies AT-rich genome regions, resulting in staggered DNA cutting [107]. In this study, the authors reported up to 30% CRISPR/MAD7-mediated GE HR efficiency, regardless of the promoter evaluated [92], suggesting that insert size did not influence knock-in efficiency. Later experiments assessed the insertion of a CAR expression cassette at the AAVS1 locus in primary T cells using CRISPR/MAD7-mediated GE [92]. Although a high GE efficiency of up to 85% was observed in primary T cells at 13 days after GE, it was dramatically reduced (~30%) upon CRISPR/MAD7-mediated multi-gene knock-out targeting of CTLA4, PDCD1, TIGIT, CD247, PTPN6, or PTPN11 [92], which are common targets for enhancing and invigorating CAR-T cells [108].

4.2. Hematological Disorders: Base and Prime Editing

Ex vivo BE and PE have been tested in several hematological disorders to modify hematopoietic stem progenitor cells (HSPCs), which can differentiate into all hematological progenitors and virtually reach any tissue in the body. The first FDA-approved CRISPR/Cas9-based GT (Casgevy) was developed for sickle cell disease (SCD) and was approved in December 2023 [109]. Casgevy is a GT based on disrupting the *BCL11A* gene via Cas9-mediated DSB, which ultimately favors the expression of γ-globin to compensate for the absence of functional β-globin affected by mutations on the *HBB* gene [110].

An ABE was evaluated by Newby et al. (2021) [93] to correct a point mutation in the β-globin gene (*HBB*) in HSPCs from patients with SCD. The ex vivo delivery of an mRNA ABE resulted in the 80% conversion of the pathogenic HBB^S (CAC) into a non-pathogenic HBB^G (CGC) known as Makassar. HSCT in immunodeficient mice achieved up to 68% HBB^G as well as a 5-fold decrease in BM reticulocytes following hypoxia-induced sickling [93]. Moreover, Makassar globin represented up to 79% of β-globin after 16 weeks post-transplantation into humanized mice. Transplanted mice also exhibited normalization of hematological parameters and a reduction in splenomegaly compared to untreated mice [93]. Contrary to what was observed with some ex vivo CRISPR/Cas9 GTs [111], ABE did not activate p53 or large deletions in BE HSPCs [93]. Similar findings have been achieved in β-thalassemia patient-derived HSPCs [112], with a new generation of ABEs termed ABE8e [113]. Importantly, using mRNA to deliver ABEs leads to higher off-target effects than the ribonucleoprotein complex delivery [112], which is commonly associated with the persistence of the BE within the cell.

Although studies such as those conducted by Newby et al. (2021) reported using BE to modify HBB^S to HBB^G [93], the transversion causing HBB^S (A>T) cannot be directly reverted through BE [60]. Instead, PE is suitable for modifying this type of mutation [66], overcoming the limitations of BE. In 2023, Everette et al. [94] showed the correction of HBB^S (A>T) to WT HBB^A with a frequency of up to 41% in HSPCs from patients with SCD when using PE3max. Interestingly, HBB^A was found in ~42% of erythroblasts and reticulocytes at 17 weeks post-transplantation, while HSPC-derived erythrocytes resulted in 28–43% normal hemoglobin [94]. Although PE leads to the recovery of the transversion causing HBB^S to WT HBB^A, the GE efficiency observed by Everette et al. (2023) [94] was significantly lower (43%) than that reported by Newby et al. (2021) [93] for BE (79%) when modifying HBB^S by HBB^G. Regardless of the efficacy, both BE and PE offer novel alternatives to modify any mutation with lower risks than the classical CRISPR/Cas9 system.

4.3. HITI: Advancing in Proof-of-Concept Assays

Current ex vivo HITI-based GE using ex vivo assays has tested the efficacy of the HITI system for inserting genes at specific genomic regions. In 2021, Bloomer et al. [96] evaluated the insertion of GFP at the first non-coding exon of the *ITGB2* gene. Mutations within the *ITGB2* gene, which encodes for the β2 integrin subunit CD18, produce leukocyte adhesion deficiency type 1 (LAD-1), an inherited immunodeficiency. In this study, the authors found that while 63% of insertions occurred in a forward direction, approximately 11% of bulk CD34+ cells exhibited GFP positivity from day 14 to 28 post GE, suggesting stable gene integration [96]. No significant changes in differentiation potential from HITI-edited CD34+ were noticed when compared with untreated CD34+ cells. Most interestingly, the transplantation of HITI-edited CD34+ cells in immunodeficient mice resulted in up to 21% of GPF-positive CD45+ cells, supporting the engraftment of HITI-edited CD34+ cells [96]. The therapeutic effect of HITI-edited CD34+ cells on LAD-1 still needs to be addressed.

Similar outcomes were reported by Byambaa et al. (2021) when using the HITI system to insert an Il2rg exon 2–8 cDNA-containing cassette into the intron 1 of the *Il2rg* gene in mouse HSCs [95]. Disruption of the *Il2rg* gene produces X-linked severe combined immunodeficiency (X-SCID) [114]. Up to 18.7% of knock-in efficacy and stemness preservation were reported upon HITI treatment. When transplanting HITI-edited mouse HSCs in an X-SCID mouse model, 3 out of 13 recipients showed an increase in white blood counts and T- and B-cell development after 8 weeks post transplantation compared to untreated mice [95]. In those three mice, CD45+ CD3+ T cells resulted in the expression of IL2rg in peripheral blood, BM, and thymus, supporting the therapeutic effect of the ex vivo HITI-based CRISPR/Cas9 system [95].

5. Conclusions and Future Perspectives

Undoubtedly, the CRISPR/Cas systems have opened novel alternatives for treating genetic diseases, including LSDs. While it is clear that CRISPR/Cas9 is the most common approach used in ex vivo strategies, implementing non-Cas9 proteins could become an innovative alternative. The CAR-T technology has demonstrated that higher GE efficiencies can be achieved with Cas12a than with Cas9 at several safe harbors, including the AAVS1 locus. The AAVS1 locus has been evaluated in several LSDs, and recently, we tested it for editing CD34+ cells using a CRISPR/nCas9-based GT approach for treating MPS IVA [73]. Further experiments comparing the efficacy of the Cas12a, wtCas9, and nCas9 should be performed to establish the higher GE strategy. Most interestingly, Cas12a can mediate multi-gene targeting, offering a new avenue for attempting gene insertion at multiple loci simultaneously (i.e., CCR5, AAS1, Album intron 1, etc.), which could lead to higher lysosomal enzyme expression than that observed when targeting a single locus.

On the other hand, given that Cas9-mediated DSB can induce unwanted DNA cutting, implementing DSB-free strategies such as PE could offer a promising alternative to correct mutations in LSDs, as PE can correct both transitions and transversions. Indeed, an advanced PE variant, PE3max, has shown comparable outcomes for stemness preservation and genome editing efficiency to those observed with CRISPR/Cas9 in HSPCs [94], making it feasible to substitute the classical CRISPR/Cas9 system. Notably, it is essential to highlight that hundreds of mutations can affect genes involved in lysosomal function, limiting the universal use of a single PE-based GT. Nonetheless, some mutations are commonly expressed, thus enabling the use of PE to treat frequent mutations.

Finally, because primary CD34+ HSPCs can undergo quiescent phenotypes in which NHEJ is the primary DSB repair mechanism [96,115], the HITI system could offer a new avenue for editing HSPCs. This premise has been tested in some proof-of-concept experiments; therefore, upcoming preclinical studies are necessary to establish the real therapeutic effect of an ex vivo HITI-based GT for treating LSDs. Likewise, HITI relies on Cas9-mediated DSB, raising concerns about its potential off-target effect. Consequently, low off-target scoring sgRNA and high-fidelity Cas9 enzymes should always be attempted, along with rigorous next-generation sequencing, to validate the absence of unwanted Cas9 cutting while preserving the higher efficiency of HITI in quiescent primary cells.

Author Contributions: Conceptualization, A.F.L.; writing—original draft, A.F.L., L.E.P. and H.P.; writing—review and editing, A.F.L. and H.P.; funding acquisition, A.F.L. and H.P. All authors have read and agreed to the published version of the manuscript.

Funding: A.F.L. and H.P. were supported by Universidad Icesi and Fundación Valle de Lili (Contract ID: 2025-ART-001).

Institutional Review Board Statement: Not applicable.

Informed Consent Statement: Not applicable.

Data Availability Statement: No new data were created or analyzed in this study.

Conflicts of Interest: The authors declare no conflicts of interest.

Abbreviations

The following abbreviations are used in this manuscript:

BE	Base editing
BM	Bone marrow
DSB	Double-strand break
GAG	Glycosaminoglycan
GE	Genome editing
GT	Gene therapy
HR	Homologous recombination
HSCT	Hematopoietic stem cell transplantation
HSPCs	Hematopoietic stem progenitor cells
iPSCs	Induced pluripotent stem cells
NHEJ	Non-homologous end joining
PE	Prime editing
WT	Wild type

References

1. Leal, A.F.; Espejo-Mojica, A.J.; Sánchez, O.F.; Ramírez, C.M.; Reyes, L.H.; Cruz, J.C.; Alméciga-Díaz, C.J. Lysosomal storage diseases: Current therapies and future alternatives. *J. Mol. Med.* **2020**, *98*, 931–946. [CrossRef] [PubMed]
2. Fernández-Pereira, C.; Millán-Tejado, B.S.; Gallardo-Gómez, M.; Pérez-Márquez, T.; Alves-Villar, M.; Melcón-Crespo, C.; Fernández-Martín, J.; Ortolano, S. Therapeutic Approaches in Lysosomal Storage Diseases. *Biomolecules* **2021**, *11*, 1775. [CrossRef] [PubMed]
3. Parenti, G.; Medina, D.L.; Ballabio, A. The rapidly evolving view of lysosomal storage diseases. *EMBO Mol. Med.* **2021**, *13*, e12836. [CrossRef] [PubMed]
4. Leal, A.F.; Benincore-Flórez, E.; Solano-Galarza, D.; Jaramillo, R.G.G.; Echeverri-Peña, O.Y.; Suarez, D.A.; Alméciga-Díaz, C.J.; Espejo-Mojica, A.J. GM2 Gangliosidoses: Clinical Features, Pathophysiological Aspects, and Current Therapies. *Int. J. Mol. Sci.* **2020**, *21*, 6213. [CrossRef] [PubMed]
5. Ertl, H.C.J. Immunogenicity and toxicity of AAV gene therapy. *Front. Immunol.* **2022**, *13*, 975803. [CrossRef] [PubMed]
6. Pupo, A.; Fernández, A.; Low, S.H.; François, A.; Suárez-Amarán, L.; Samulski, R.J. AAV vectors: The Rubik's cube of human gene therapy. *Mol. Ther.* **2022**, *30*, 3515–3541. [CrossRef] [PubMed]
7. Zhao, Z.; Anselmo, A.C.; Mitragotri, S. Viral vector-based gene therapies in the clinic. *Bioeng. Transl. Med.* **2022**, *7*, e10258. [CrossRef] [PubMed]
8. Leal, A.F.; Herreno-Pachón, A.M.; Benincore-Flórez, E.; Karunathilaka, A.; Tomatsu, S. Current Strategies for Increasing Knock-In Efficiency in CRISPR/Cas9-Based Approaches. *Int. J. Mol. Sci.* **2024**, *25*, 2456. [CrossRef] [PubMed]
9. Chen, X.; Du, J.; Yun, S.; Xue, C.; Yao, Y.; Rao, S. Recent advances in CRISPR-Cas9-based genome insertion technologies. *Mol. Ther. Nucleic Acids* **2024**, *35*, 102138. [CrossRef] [PubMed]
10. Leal, A.F.; Celik, B.; Fnu, N.; Khan, S.; Tomatsu, S.; Alméciga-Díaz, C.J. Iron oxide-coupled CRISPR-nCas9-based genome editing assessment in mucopolysaccharidosis IVA mice. *Mol. Ther. Methods Clin. Dev.* **2023**, *31*, 101153. [CrossRef] [PubMed]
11. Schuh, R.S.; Gonzalez, E.A.; Tavares, A.M.V.; Seolin, B.G.; De Elias, L., Sr.; Vera, L.N.P.; Kubaski, F.; Poletto, E.; Giugliani, R.; Teixeira, H.F.; et al. Neonatal nonviral gene editing with the CRISPR/Cas9 system improves some cardiovascular, respiratory, and bone disease features of the mucopolysaccharidosis I phenotype in mice. *Gene Ther.* **2020**, *27*, 74–84. [CrossRef] [PubMed]
12. Schuh, R.S.; Poletto, É.; Pasqualim, G.; Tavares, A.M.V.; Meyer, F.S.; Gonzalez, E.A.; Giugliani, R.; Matte, U.; Teixeira, H.F.; Baldo, G. In vivo genome editing of mucopolysaccharidosis I mice using the CRISPR/Cas9 system. *J. Control. Release* **2018**, *288*, 23–33. [CrossRef] [PubMed]
13. Charlesworth, C.T.; Deshpande, P.S.; Dever, D.P.; Camarena, J.; Lemgart, V.T.; Cromer, M.K.; Vakulskas, C.A.; Collingwood, M.A.; Zhang, L.; Bode, N.M.; et al. Identification of preexisting adaptive immunity to Cas9 proteins in humans. *Nat. Med.* **2019**, *25*, 249–254. [CrossRef] [PubMed]
14. Simhadri, V.L.; McGill, J.; McMahon, S.; Wang, J.; Jiang, H.; Sauna, Z.E. Prevalence of Pre-existing Antibodies to CRISPR-Associated Nuclease Cas9 in the USA Population. *Mol. Ther. Methods Clin. Dev.* **2018**, *10*, 105–112. [CrossRef] [PubMed]
15. Poletto, E.; Silva, A.O.; Weinlich, R.; Martin, P.K.M.; Torres, D.C.; Giugliani, R.; Baldo, G. Ex vivo gene therapy for lysosomal storage disorders: Future perspectives. *Expert. Opin. Biol. Ther.* **2023**, *23*, 353–364. [CrossRef] [PubMed]
16. Pará, C.; Bose, P.; Pshezhetsky, A.V. Neuropathophysiology of Lysosomal Storage Diseases: Synaptic Dysfunction as a Starting Point for Disease Progression. *J. Clin. Med.* **2020**, *9*, 616. [CrossRef] [PubMed]
17. Ellison, S.; Parker, H.; Bigger, B. Advances in therapies for neurological lysosomal storage disorders. *J. Inherit. Metab. Dis.* **2023**, *46*, 874–905. [CrossRef] [PubMed]
18. Capotondo, A.; Milazzo, R.; Politi, L.S.; Quattrini, A.; Palini, A.; Plati, T.; Merella, S.; Nonis, A.; di Serio, C.; Montini, E.; et al. Brain conditioning is instrumental for successful microglia reconstitution following hematopoietic stem cell transplantation. *Proc. Natl. Acad. Sci. USA* **2012**, *109*, 15018–15023. [CrossRef] [PubMed]
19. Archie, S.R.; Al Shoyaib, A.; Cucullo, L. Blood-Brain Barrier Dysfunction in CNS Disorders and Putative Therapeutic Targets: An Overview. *Pharmaceutics* **2021**, *13*, 1779. [CrossRef] [PubMed]
20. Poletto, E.; Colella, P.; Vera, L.N.P.; Khan, S.; Tomatsu, S.; Baldo, G.; Gomez-Ospina, N. Improved engraftment and therapeutic efficacy by human genome-edited hematopoietic stem cells with Busulfan-based myeloablation. *Mol. Ther. Methods Clin. Dev.* **2022**, *25*, 392–409. [CrossRef] [PubMed]
21. Taylor, M.; Khan, S.; Stapleton, M.; Wang, J.; Chen, J.; Wynn, R.; Yabe, H.; Chinen, Y.; Boelens, J.J.; Mason, R.W.; et al. Hematopoietic Stem Cell Transplantation for Mucopolysaccharidoses: Past, Present, and Future. *Biol. Blood Marrow Transpl.* **2019**, *25*, e226–e246. [CrossRef] [PubMed]
22. Yalcin, K.; Uygun, V.; Hismi, B.O.; Celen, S.; Ozturkmen, S.; Zhumatayev, S.; Daloglu, H.; Karasu, G.; Yesilipek, A. Hematopoietic stem cell transplantation in children with mucopolysaccharidosis IVA: Single center experience. *Bone Marrow Transplant.* **2024**, *60*, 47–51. [CrossRef] [PubMed]
23. Rossini, L.; Durante, C.; Marzollo, A.; Biffi, A. New Indications for Hematopoietic Stem Cell Gene Therapy in Lysosomal Storage Disorders. *Front. Oncol.* **2022**, *12*, 885639. [CrossRef] [PubMed]

24. Malard, F.; Holler, E.; Sandmaier, B.M.; Huang, H.; Mohty, M. Acute graft-versus-host disease. *Nat. Rev. Dis. Primers* **2023**, *9*, 27. [CrossRef] [PubMed]
25. Leal, A.F.; Fnu, N.; Benincore-Flórez, E.; Herreño-Pachón, A.M.; Echeverri-Peña, O.Y.; Alméciga-Díaz, C.J.; Tomatsu, S. The landscape of CRISPR/Cas9 for inborn errors of metabolism. *Mol. Genet. Metab.* **2022**, *138*, 106968. [CrossRef] [PubMed]
26. Jinek, M.; Chylinski, K.; Fonfara, I.; Hauer, M.; Doudna, J.A.; Charpentier, E. A programmable dual-RNA-guided DNA endonuclease in adaptive bacterial immunity. *Science* **2012**, *337*, 816–821. [CrossRef] [PubMed]
27. Hryhorowicz, M.; Lipiński, D.; Zeyland, J. Evolution of CRISPR/Cas Systems for Precise Genome Editing. *Int. J. Mol. Sci.* **2023**, *24*, 14233. [CrossRef] [PubMed]
28. Wang, J.Y.; Doudna, J.A. CRISPR technology: A decade of genome editing is only the beginning. *Science* **2023**, *379*, eadd8643. [CrossRef] [PubMed]
29. Huo, Y.; Nam, K.H.; Ding, F.; Lee, H.; Wu, L.; Xiao, Y.; Farchione, M.D., Jr.; Zhou, S.; Rajashankar, K.; Kurinov, I.; et al. Structures of CRISPR Cas3 offer mechanistic insights into Cascade-activated DNA unwinding and degradation. *Nat. Struct. Mol. Biol.* **2014**, *21*, 771–777. [CrossRef] [PubMed]
30. Kim, D.Y.; Lee, S.Y.; Ha, H.J.; Park, H.H. Structural basis of Cas3 activation in type I-C CRISPR-Cas system. *Nucleic Acids Res.* **2024**, *52*, 10563–10574. [CrossRef] [PubMed]
31. Yang, L.; Li, H.; Han, Y.; Song, Y.; Wei, M.; Fang, M.; Sun, Y. CRISPR/Cas9 Gene Editing System Can Alter Gene Expression and Induce DNA Damage Accumulation. *Genes* **2023**, *14*, 806. [CrossRef] [PubMed]
32. Guo, C.; Ma, X.; Gao, F.; Guo, Y. Off-target effects in CRISPR/Cas9 gene editing. *Front. Bioeng. Biotechnol.* **2023**, *11*, 1143157. [CrossRef] [PubMed]
33. Csörgő, B.; León, L.M.; Chau-Ly, I.J.; Vasquez-Rifo, A.; Berry, J.D.; Mahendra, C.; Crawford, E.D.; Lewis, J.D.; Bondy-Denomy, J. A compact Cascade–Cas3 system for targeted genome engineering. *Nat. Methods* **2020**, *17*, 1183–1190. [CrossRef] [PubMed]
34. Morisaka, H.; Yoshimi, K.; Okuzaki, Y.; Gee, P.; Kunihiro, Y.; Sonpho, E.; Xu, H.; Sasakawa, N.; Naito, Y.; Nakada, S.; et al. CRISPR-Cas3 induces broad and unidirectional genome editing in human cells. *Nat. Commun.* **2019**, *10*, 5302. [CrossRef] [PubMed]
35. Vaillend, C.; Aoki, Y.; Mercuri, E.; Hendriksen, J.; Tetorou, K.; Goyenvalle, A.; Muntoni, F. Duchenne muscular dystrophy: Recent insights in brain related comorbidities. *Nat. Commun.* **2025**, *16*, 1298. [CrossRef] [PubMed]
36. Meliawati, M.; Schilling, C.; Schmid, J. Recent advances of Cas12a applications in bacteria. *Appl. Microbiol. Biotechnol.* **2021**, *105*, 2981–2990. [CrossRef] [PubMed]
37. Wu, J.; Gao, P.; Shi, Y.; Zhang, C.; Tong, X.; Fan, H.; Zhou, X.; Zhang, Y.; Yin, H. Characterization of a thermostable Cas12a ortholog. *Cell Insight* **2023**, *2*, 100126. [CrossRef] [PubMed]
38. Ji, S.; Wang, X.; Wang, Y.; Sun, Y.; Su, Y.; Lv, X.; Song, X. Advances in Cas12a-Based Amplification-Free Nucleic Acid Detection. *CRISPR J.* **2023**, *6*, 405–418. [CrossRef] [PubMed]
39. Wang, S.; Shen, X.; Chen, G.; Zhang, W.; Tan, B. Application and development of CRISPR-Cas12a methods for the molecular diagnosis of cancer: A review. *Anal. Chim. Acta* **2024**, *1341*, 343603. [CrossRef] [PubMed]
40. Santos, L.; Mention, K.; Cavusoglu-Doran, K.; Sanz, D.J.; Bacalhau, M.; Lopes-Pacheco, M.; Harrison, P.T.; Farinha, C.M. Comparison of Cas9 and Cas12a CRISPR editing methods to correct the W1282X-CFTR mutation. *J. Cyst. Fibros.* **2022**, *21*, 181–187. [CrossRef] [PubMed]
41. Liu, J.-J.; Orlova, N.; Oakes, B.L.; Ma, E.; Spinner, H.B.; Baney, K.L.M.; Chuck, J.; Tan, D.; Knott, G.J.; Harrington, L.B.; et al. CasX enzymes comprise a distinct family of RNA-guided genome editors. *Nature* **2019**, *566*, 218–223. [CrossRef] [PubMed]
42. Li, D.; Zhang, S.; Lin, S.; Xing, W.; Yang, Y.; Zhu, F.; Su, D.; Chen, C.; Liu, J.-J.G. Cas12e orthologs evolve variable structural elements to facilitate dsDNA cleavage. *Nat. Commun.* **2024**, *15*, 10727. [CrossRef] [PubMed]
43. Wu, W.Y.; Adiego-Pérez, B.; van der Oost, J. Biology and applications of CRISPR–Cas12 and transposon-associated homologs. *Nat. Biotechnol.* **2024**, *42*, 1807–1821. [CrossRef] [PubMed]
44. Pausch, P.; Al-Shayeb, B.; Bisom-Rapp, E.; Tsuchida, C.A.; Li, Z.; Cress, B.F.; Knott, G.J.; Jacobsen, S.E.; Banfield, J.F.; Doudna, J.A. CRISPR-CasΦ from huge phages is a hypercompact genome editor. *Science* **2020**, *369*, 333–337. [CrossRef] [PubMed]
45. Xiao, R.; Li, Z.; Wang, S.; Han, R.; Chang, L. Structural basis for substrate recognition and cleavage by the dimerization-dependent CRISPR–Cas12f nuclease. *Nucleic Acids Res.* **2021**, *49*, 4120–4128. [CrossRef] [PubMed]
46. Takeda, S.N.; Nakagawa, R.; Okazaki, S.; Hirano, H.; Kobayashi, K.; Kusakizako, T.; Nishizawa, T.; Yamashita, K.; Nishimasu, H.; Nureki, O. Structure of the miniature type V-F CRISPR-Cas effector enzyme. *Mol. Cell* **2021**, *81*, 558–570.e3. [CrossRef] [PubMed]
47. Cui, T.; Cai, B.; Tian, Y.; Liu, X.; Liang, C.; Gao, Q.; Li, B.; Ding, Y.; Li, R.; Zhou, Q.; et al. Therapeutic In Vivo Gene Editing Achieved by a Hypercompact CRISPR-Cas12f1 System Delivered with All-in-One Adeno-Associated Virus. *Adv. Sci.* **2024**, *11*, e2308095. [CrossRef] [PubMed]
48. Yang, J.; Li, X.; He, Q.; Wang, X.; Tang, J.; Wang, T.; Zhang, Y.; Yu, F.; Zhang, S.; Liu, Z.; et al. Structural basis for the activity of the type VII CRISPR–Cas system. *Nature* **2024**, *633*, 465–472. [CrossRef] [PubMed]

49. Zhou, B.; Yang, R.; Sohail, M.; Kong, X.; Zhang, X.; Fu, N.; Li, B. CRISPR/Cas14 provides a promising platform in facile and versatile aptasensing with improved sensitivity. *Talanta* **2023**, *254*, 124120. [CrossRef] [PubMed]
50. Savage, D.F. Cas14: Big Advances from Small CRISPR Proteins. *Biochemistry* **2019**, *58*, 1024–1025. [CrossRef] [PubMed]
51. Goswami, H.N.; Rai, J.; Das, A.; Li, H.; Institute of Molecular Biophysics, Florida State University, United States; Department of Chemistry and Biochemistry, Florida State University, United States. Molecular mechanism of active Cas7-11 in processing CRISPR RNA and interfering target RNA. *eLife* **2022**, *11*, 81678. [CrossRef] [PubMed]
52. Özcan, A.; Krajeski, R.; Ioannidi, E.; Lee, B.; Gardner, A.; Makarova, K.S.; Koonin, E.V.; Abudayyeh, O.O.; Gootenberg, J.S. Programmable RNA targeting with the single-protein CRISPR effector Cas7-11. *Nature* **2021**, *597*, 720–725. [CrossRef] [PubMed]
53. Lin, C.P.; Li, H.; Brogan, D.J.; Wang, T.; Akbari, O.S.; A Komives, E. CRISPR RNA binding drives structural ordering that primes Cas7-11 for target cleavage. *Nucleic Acids Res.* **2025**, *53*, gkaf271. [CrossRef] [PubMed]
54. Moreno-Sánchez, I.; Hernández-Huertas, L.; Nahón-Cano, D.; Martínez-García, P.M.; Treichel, A.J.; Gómez-Marin, C.; Tomás-Gallardo, L.; Pescador, G.d.S.; Kushawah, G.; Egidy, R.; et al. Enhanced RNA-targeting CRISPR-Cas technology in zebrafish. *Nat. Commun.* **2025**, *16*, 2591. [CrossRef] [PubMed]
55. Yoon, P.H.; Zhang, Z.; Loi, K.J.; Adler, B.A.; Lahiri, A.; Vohra, K.; Shi, H.; Rabelo, D.B.; Trinidad, M.; Boger, R.S.; et al. Struc-ture-guided discovery of ancestral CRISPR-Cas13 ribonucleases. *Science* **2024**, *385*, 538–543. [CrossRef] [PubMed]
56. Yang, H.; Patel, D.J. Structures, mechanisms and applications of RNA-centric CRISPR-Cas13. *Nat. Chem. Biol.* **2024**, *20*, 673–688. [CrossRef] [PubMed]
57. Kumar, S.; Hsiao, Y.-W.; Wong, V.H.Y.; Aubin, D.; Wang, J.-H.; Lisowski, L.; Rakoczy, E.P.; Li, F.; Alarcon-Martinez, L.; Gonzalez-Cordero, A.; et al. Characterization of RNA editing and gene therapy with a compact CRISPR-Cas13 in the retina. *Proc. Natl. Acad. Sci. USA* **2024**, *121*, e2408345121. [CrossRef] [PubMed]
58. Suzuki, K.; Belmonte, J.C.I. In vivo genome editing via the HITI method as a tool for gene therapy. *J. Hum. Genet.* **2017**, *63*, 157–164. [CrossRef] [PubMed]
59. Wang, X.; Xu, G.; Johnson, W.A.; Qu, Y.; Yin, D.; Ramkissoon, N.; Xiang, H.; Cong, L. Long sequence insertion via CRISPR/Cas gene-editing with transposase, recombinase, and integrase. *Curr. Opin. Biomed. Eng.* **2023**, *28*, 100491. [CrossRef] [PubMed]
60. Zhang, C.; Xu, J.; Wu, Y.; Xu, C.; Xu, P. Base Editors-Mediated Gene Therapy in Hematopoietic Stem Cells for Hematologic Diseases. *Stem Cell Rev. Rep.* **2024**, *20*, 1387–1405. [CrossRef] [PubMed]
61. Liang, Y.; Chen, F.; Wang, K.; Lai, L. Base editors: Development and applications in biomedicine. *Front. Med.* **2023**, *17*, 359–387. [CrossRef] [PubMed]
62. Chen, F.; Lian, M.; Ma, B.; Gou, S.; Luo, X.; Yang, K.; Shi, H.; Xie, J.; Ge, W.; Ouyang, Z.; et al. Multiplexed base editing through Cas12a variant-mediated cytosine and adenine base editors. *Commun. Biol.* **2022**, *5*, 1163. [CrossRef] [PubMed]
63. Fonfara, I.; Richter, H.; Bratovič, M.; Le Rhun, A.; Charpentier, E. The CRISPR-associated DNA-cleaving enzyme Cpf1 also processes precursor CRISPR RNA. *Nature* **2016**, *532*, 517–521. [CrossRef] [PubMed]
64. Anzalone, A.V.; Randolph, P.B.; Davis, J.R.; Sousa, A.A.; Koblan, L.W.; Levy, J.M.; Chen, P.J.; Wilson, C.; Newby, G.A.; Raguram, A.; et al. Search-and-replace genome editing without double-strand breaks or donor DNA. *Nature* **2019**, *576*, 149–157. [CrossRef] [PubMed]
65. Nelson, J.W.; Randolph, P.B.; Shen, S.P.; Everette, K.A.; Chen, P.J.; Anzalone, A.V.; An, M.; Newby, G.A.; Chen, J.C.; Hsu, A.; et al. Engineered pegRNAs improve prime editing efficiency. *Nat. Biotechnol.* **2022**, *40*, 402–410. [CrossRef] [PubMed]
66. Liu, Z.; Guo, D.; Wang, D.; Zhou, J.; Chen, Q.; Lai, J. Prime editing: A gene precision editing tool from inception to present. *FASEB J.* **2024**, *38*, e70148. [CrossRef] [PubMed]
67. Anzalone, A.V.; Gao, X.D.; Podracky, C.J.; Nelson, A.T.; Koblan, L.W.; Raguram, A.; Levy, J.M.; Mercer, J.A.M.; Liu, D.R. Programmable deletion, replacement, integration and inversion of large DNA sequences with twin prime editing. *Nat. Biotechnol.* **2022**, *40*, 731–740. [CrossRef] [PubMed]
68. Doman, J.L.; Pandey, S.; Neugebauer, M.E.; An, M.; Davis, J.R.; Randolph, P.B.; McElroy, A.; Gao, X.D.; Raguram, A.; Richter, M.F.; et al. Phage-assisted evolution and protein engineering yield compact, efficient prime editors. *Cell* **2023**, *186*, 3983–4002.e26. [CrossRef] [PubMed]
69. Liang, R.; He, Z.; Zhao, K.T.; Zhu, H.; Hu, J.; Liu, G.; Gao, Q.; Liu, M.; Zhang, R.; Qiu, J.-L.; et al. Prime editing using CRISPR-Cas12a and circular RNAs in human cells. *Nat. Biotechnol.* **2024**, *42*, 1867–1875. [CrossRef] [PubMed]
70. Meng, X.; Jia, R.; Zhao, X.; Zhang, F.; Chen, S.; Yu, S.; Liu, X.; Dou, H.; Feng, X.; Zhang, J.; et al. In vivo genome editing via CRISPR/Cas9 mediated homology-independent targeted integration. *Nature* **2016**, *540*, 144–149.
71. Miki, T.; Vazquez, L.; Yanuaria, L.; Lopez, O.; Garcia, I.M.; Ohashi, K.; Rodriguez, N.S. Induced Pluripotent Stem Cell Derivation and Ex Vivo Gene Correction Using a Mucopolysaccharidosis Type 1 Disease Mouse Model. *Stem Cells Int.* **2019**, *2019*, 6978303. [CrossRef] [PubMed]
72. Gomez-Ospina, N.; Scharenberg, S.G.; Mostrel, N.; Bak, R.O.; Mantri, S.; Quadros, R.M.; Gurumurthy, C.B.; Lee, C.; Bao, G.; Suarez, C.J.; et al. Human genome-edited hematopoietic stem cells phenotypically correct Mucopolysaccharidosis type I. *Nat. Commun.* **2019**, *10*, 4045. [CrossRef] [PubMed]

73. Herreño-Pachón, A.M.; Leal, A.F.; Khan, S.; Alméciga-Díaz, C.J.; Tomatsu, S. CRISPR/nCas9-Edited CD34+ Cells Rescue Mucopolysaccharidosis IVA Fibroblasts Phenotype. *Int. J. Mol. Sci.* **2025**, *26*, 4334. [CrossRef] [PubMed]
74. Scharenberg, S.G.; Poletto, E.; Lucot, K.L.; Colella, P.; Sheikali, A.; Montine, T.J.; Porteus, M.H.; Gomez-Ospina, N. Engineering monocyte/macrophage-specific glucocerebrosidase expression in human hematopoietic stem cells using genome editing. *Nat. Commun.* **2020**, *11*, 3327. [CrossRef] [PubMed]
75. Ramalingam, S.; Kumar, A.; Krug, S.; Mohan, H.; Rao, D.N.; Bishai, W.R.; Chandrasegaran, S. CRISPR Correction of the GBA Mutation in Human-Induced Pluripotent Stem Cells Restores Normal Function to Gaucher Macrophages and Increases Their Susceptibility to *Mycobacterium tuberculosis*. *J. Infect. Dis.* **2023**, *228*, 777–782. [CrossRef] [PubMed]
76. Dever, D.P.; Scharenberg, S.G.; Camarena, J.; Kildebeck, E.J.; Clark, J.T.; Martin, R.M.; Bak, R.O.; Tang, Y.; Dohse, M.; Birgmeier, J.A.; et al. CRISPR/Cas9 Genome Engineering in Engraftable Human Brain-Derived Neural Stem Cells. *iScience* **2019**, *15*, 524–535. [CrossRef] [PubMed]
77. Choi, J.B.; Seo, D.; Do, H.-S.; Han, Y.-M. Generation of a CRISPR/Cas9-corrected-hiPSC line (DDLABi001-A) from Fabry disease (FD)-derived iPSCs having α-galactosidase (GLA) gene mutation (c.803_806del). *Stem Cell Res.* **2023**, *66*, 103001. [CrossRef] [PubMed]
78. Karl-Schöller, F.; Breyer, M.; Klopocki, E.; Üçeyler, N. Generation of a gene-corrected human isogenic iPSC line from a patient with Fabry disease carrying the GLA variant c.1069C>T using CRISPR/Cas9-mediated homology directed repair. *Stem Cell Res.* **2025**, *86*, 103711. [CrossRef] [PubMed]
79. Rintz, E.; Banacki, M.; Ziemian, M.; Kobus, B.; Wegrzyn, G. Causes of death in mucopolysaccharidoses. *Mol. Genet. Metab.* **2024**, *142*, 108507. [CrossRef] [PubMed]
80. Leal, A.F.; Benincore-Flórez, E.; Rintz, E.; Herreño-Pachón, A.M.; Celik, B.; Ago, Y.; Alméciga-Díaz, C.J.; Tomatsu, S. Mucopolysaccharidoses: Cellular Consequences of Glycosaminoglycans Accumulation and Potential Targets. *Int. J. Mol. Sci.* **2022**, *24*, 477. [CrossRef] [PubMed]
81. Consiglieri, G.; Bernardo, M.E.; Brunetti-Pierri, N.; Aiuti, A. Ex Vivo and In Vivo Gene Therapy for Mucopolysaccharidoses: State of the Art. *Hematol. Oncol. Clin. N. Am.* **2022**, *36*, 865–878. [CrossRef] [PubMed]
82. Leal, A.F.; Alméciga-Díaz, C.J.; Tomatsu, S. Mucopolysaccharidosis IVA: Current Disease Models and Drawbacks. *Int. J. Mol. Sci.* **2023**, *24*, 16148. [CrossRef] [PubMed]
83. Sawamoto, K.; González, J.V.Á.; Piechnik, M.; Otero, F.J.; Couce, M.L.; Suzuki, Y.; Tomatsu, S. Mucopolysaccharidosis IVA: Diagnosis, Treatment, and Management. *Int. J. Mol. Sci.* **2020**, *21*, 1517. [CrossRef] [PubMed]
84. Özdemir, G.N.; Gündüz, E. Gaucher Disease for Hematologists. *Turk. J. Haematol.* **2022**, *39*, 136–139. [CrossRef] [PubMed]
85. Daykin, E.C.; Ryan, E.; Sidransky, E. Diagnosing neuronopathic Gaucher disease: New considerations and challenges in assigning Gaucher phenotypes. *Mol. Genet. Metab.* **2021**, *132*, 49–58. [CrossRef] [PubMed]
86. Aflaki, E.; Stubblefield, B.K.; Maniwang, E.; Lopez, G.; Moaven, N.; Goldin, E.; Marugan, J.; Patnaik, S.; Dutra, A.; Southall, N.; et al. Macrophage Models of Gaucher Disease for Evaluating Disease Pathogenesis and Candidate Drugs. *Sci. Transl. Med.* **2014**, *6*, 240ra73. [CrossRef] [PubMed]
87. Şoroğlu, C.V.; Berkay, E.G. Old disease-New reflections: Gaucher, immunity, and inflammation. *J. Cell Mol. Med.* **2024**, *28*, e70087. [CrossRef] [PubMed]
88. Fan, J.; Hale, V.L.; Lelieveld, L.T.; Whitworth, L.J.; Busch-Nentwich, E.M.; Troll, M.; Edelstein, P.H.; Cox, T.M.; Roca, F.J.; Aerts, J.M.F.G.; et al. Gaucher disease protects against tuberculosis. *Proc. Natl. Acad. Sci. USA* **2023**, *120*, 2217673120. [CrossRef] [PubMed]
89. Maghazachi, A. Globoid Cell Leukodystrophy (Krabbe Disease): An Update. *ImmunoTargets Ther.* **2023**, *12*, 105–111. [CrossRef] [PubMed]
90. Soloway, S.; Lister, D. Fabry's Disease. *N. Engl. J. Med.* **2024**, *391*, 1038. [CrossRef] [PubMed]
91. Ling, X.; Chang, L.; Chen, H.; Gao, X.; Yin, J.; Zuo, Y.; Huang, Y.; Zhang, B.; Hu, J.; Liu, T. Improving the efficiency of CRISPR-Cas12a-based genome editing with site-specific covalent Cas12a-crRNA conjugates. *Mol. Cell* **2021**, *81*, 4747–4756.e7. [CrossRef] [PubMed]
92. Mohr, M.; Damas, N.; Gudmand-Høyer, J.; Zeeberg, K.; Jedrzejczyk, D.; Vlassis, A.; Morera-Gómez, M.; Pereira-Schoning, S.; Puš, U.; Oliver-Almirall, A.; et al. The CRISPR-Cas12a Platform for Accurate Genome Editing, Gene Disruption, and Efficient Transgene Integration in Human Immune Cells. *ACS Synth. Biol.* **2023**, *12*, 375–389. [CrossRef] [PubMed]
93. Newby, G.A.; Yen, J.S.; Woodard, K.J.; Mayuranathan, T.; Lazzarotto, C.R.; Li, Y.; Sheppard-Tillman, H.; Porter, S.N.; Yao, Y.; Mayberry, K.; et al. Base editing of haematopoietic stem cells rescues sickle cell disease in mice. *Nature* **2021**, *595*, 295–302. [CrossRef] [PubMed]
94. Everette, K.A.; Newby, G.A.; Levine, R.M.; Mayberry, K.; Jang, Y.; Mayuranathan, T.; Nimmagadda, N.; Dempsey, E.; Li, Y.; Bhoopalan, S.V.; et al. Ex vivo prime editing of patient haematopoietic stem cells rescues sickle-cell disease phenotypes after engraftment in mice. *Nat. Biomed. Eng.* **2023**, *7*, 616–628. [CrossRef] [PubMed]

95. Byambaa, S.; Uosaki, H.; Ohmori, T.; Hara, H.; Endo, H.; Nureki, O.; Hanazono, Y. Non-viral ex vivo genome-editing in mouse bona fide hematopoietic stem cells with CRISPR/Cas9. *Mol. Ther. Methods Clin. Dev.* **2021**, *20*, 451–462. [CrossRef] [PubMed]
96. Bloomer, H.; Smith, R.H.; Hakami, W.; Larochelle, A. Genome editing in human hematopoietic stem and progenitor cells via CRISPR-Cas9-mediated homology-independent targeted integration. *Mol. Ther.* **2021**, *29*, 1611–1624. [CrossRef] [PubMed]
97. Tao, R.; Han, X.; Bai, X.; Yu, J.; Ma, Y.; Chen, W.; Zhang, D.; Li, Z. Revolutionizing cancer treatment: Enhancing CAR-T cell therapy with CRISPR/Cas9 gene editing technology. *Front. Immunol.* **2024**, *15*, 1354825. [CrossRef] [PubMed]
98. Sterner, R.C.; Sterner, R.M. CAR-T cell therapy: Current limitations and potential strategies. *Blood Cancer J.* **2021**, *11*, 69. [CrossRef] [PubMed]
99. Ceja, M.A.; Khericha, M.; Harris, C.M.; Puig-Saus, C.; Chen, Y.Y. CAR-T cell manufacturing: Major process parameters and next-generation strategies. *J. Exp. Med.* **2024**, *221*, 20230903. [CrossRef] [PubMed]
100. Brudno, J.N.; Maus, M.V.; Hinrichs, C.S. CAR T Cells and T-Cell Therapies for Cancer: A Translational Science Review. *JAMA* **2024**, *332*, 1924–1935. [CrossRef] [PubMed]
101. Chohan, K.L.; Siegler, E.L.; Kenderian, S.S. CAR-T Cell Therapy: The Efficacy and Toxicity Balance. *Curr. Hematol. Malign-Rep.* **2023**, *18*, 9–18. [CrossRef] [PubMed]
102. Depil, S.; Duchateau, P.; Grupp, S.A.; Mufti, G.; Poirot, L. 'Off-the-shelf' allogeneic CAR T cells: Development and challenges. *Nat. Rev. Drug Discov.* **2020**, *19*, 185–199. [CrossRef] [PubMed]
103. Gu, J.; Rollo, B.; Sumer, H.; Cromer, B. Targeting the AAVS1 Site by CRISPR/Cas9 with an Inducible Transgene Cassette for the Neuronal Differentiation of Human Pluripotent Stem Cells. *Methods Mol. Biol.* **2022**, *2495*, 99–114. [PubMed]
104. Leal, A.F.; Alméciga-Díaz, C.J. Efficient CRISPR/Cas9 nickase-mediated genome editing in an in vitro model of mucopolysaccharidosis IVA. *Gene Ther.* **2022**, *30*, 107–114. [CrossRef] [PubMed]
105. Leal, A.F.; Cifuentes, J.; Torres, C.E.; Suárez, D.; Quezada, V.; Gómez, S.C.; Cruz, J.C.; Reyes, L.H.; Espejo-Mojica, A.J.; Alméciga-Díaz, C.J. Delivery and assessment of a CRISPR/nCas9-based genome editing system on in vitro models of mucopolysaccharidoses IVA assisted by magnetite-based nanoparticles. *Sci. Rep.* **2022**, *12*, 15045. [CrossRef] [PubMed]
106. Leal, A.F.; Cifuentes, J.; Quezada, V.; Benincore-Flórez, E.; Cruz, J.C.; Reyes, L.H.; Espejo-Mojica, A.J.; Alméciga-Díaz, C.J. CRISPR/nCas9-Based Genome Editing on GM2 Gangliosidoses Fibroblasts via Non-Viral Vectors. *Int. J. Mol. Sci.* **2022**, *23*, 10672. [CrossRef] [PubMed]
107. Hozumi, S.; Chen, Y.-C.; Takemoto, T.; Sawatsubashi, S. Cas12a and MAD7, genome editing tools for breeding. *Breed. Sci.* **2024**, *74*, 22–31. [CrossRef] [PubMed]
108. Agarwal, S.; Aznar, M.A.; Rech, A.J.; Good, C.R.; Kuramitsu, S.; Da, T.; Gohil, M.; Chen, L.; Hong, S.-J.A.; Ravikumar, P.; et al. Deletion of the inhibitory co-receptor CTLA-4 enhances and invigorates chimeric antigen receptor T cells. *Immunity* **2023**, *56*, 2388–2407.e9. [CrossRef] [PubMed]
109. Singh, A.M.; Irfan, H.M.; Fatima, E.M.; Nazir, Z.M.; Verma, A.M.; Akilimali, A. Revolutionary breakthrough: FDA approves CASGEVY, the first CRISPR/Cas9 gene therapy for sickle cell disease. *Ann. Med. Surg.* **2024**, *86*, 4555–4559. [CrossRef] [PubMed]
110. Kerwash, E.; Sajic, M.; Rantell, K.R.; McBlane, J.W.; Johnston, J.D.; Niewiarowska, A.; Butler, A.S.; Cole, S. Regulatory Assessment of Casgevy for the Treatment of Transfusion-Dependent β-Thalassemia and Sickle Cell Disease with Recurrent Vaso-Occlusive Crises. *Curr. Issues Mol. Biol.* **2024**, *46*, 8209–8225. [CrossRef] [PubMed]
111. Singh, A.; Babu, S.; Phan, M.; Yuan, S. CRISPR/Cas9-Based Protocol for Precise Genome Editing in Induced Pluripotent Stem Cells. *Bio-Protocol.* **2024**, *14*, e5141. [CrossRef] [PubMed]
112. Liao, J.; Chen, S.; Hsiao, S.; Jiang, Y.; Yang, Y.; Zhang, Y.; Wang, X.; Lai, Y.; Bauer, D.E.; Wu, Y. Therapeutic adenine base editing of human hematopoietic stem cells. *Nat. Commun.* **2023**, *14*, 207. [CrossRef] [PubMed]
113. Richter, M.F.; Zhao, K.T.; Eton, E.; Lapinaite, A.; Newby, G.A.; Thuronyi, B.W.; Wilson, C.; Koblan, L.W.; Zeng, J.; Bauer, D.E.; et al. Phage-assisted evolution of an adenine base editor with improved Cas domain compatibility and activity. *Nat. Biotechnol.* **2020**, *38*, 883–891. [CrossRef] [PubMed]
114. Pai, S.-Y.; Thrasher, A.J. Gene therapy for X-linked severe combined immunodeficiency: Historical outcomes and current status. *J. Allergy Clin. Immunol.* **2020**, *146*, 258–261. [CrossRef] [PubMed]
115. Mohrin, M.; Bourke, E.; Alexander, D.; Warr, M.R.; Barry-Holson, K.; Le Beau, M.M.; Morrison, C.G.; Passegué, E. Hematopoietic Stem Cell Quiescence Promotes Error-Prone DNA Repair and Mutagenesis. *Cell Stem Cell* **2010**, *7*, 174–185. [CrossRef] [PubMed]

cells

Review

Towards a Cure for Diamond–Blackfan Anemia: Views on Gene Therapy

Matilde Vale [1], Jan Prochazka [1,2] and Radislav Sedlacek [1,2,*]

[1] Laboratory of Transgenic Models of Diseases, Institute of Molecular Genetics of the Czech Academy of Sciences, v.v.i, 252 50 Vestec, Czech Republic; ana-matilde.vale@img.cas.cz (M.V.); jan.prochazka@img.cas.cz (J.P.)
[2] Czech Centre for Phenogenomics, Institute of Molecular Genetics of the Czech Academy of Sciences, v.v.i, 252 50 Vestec, Czech Republic
* Correspondence: radislav.sedlacek@img.cas.cz

Abstract: Diamond–Blackfan anemia (DBA) is a rare genetic disorder affecting the bone marrow's ability to produce red blood cells, leading to severe anemia and various physical abnormalities. Approximately 75% of DBA cases involve heterozygous mutations in ribosomal protein (RP) genes, classifying it as a ribosomopathy, with RPS19 being the most frequently mutated gene. Non-RP mutations, such as in GATA1, have also been identified. Current treatments include glucocorticosteroids, blood transfusions, and hematopoietic stem cell transplantation (HSCT), with HSCT being the only curative option, albeit with challenges like donor availability and immunological complications. Gene therapy, particularly using lentiviral vectors and CRISPR/Cas9 technology, emerges as a promising alternative. This review explores the potential of gene therapy, focusing on lentiviral vectors and CRISPR/Cas9 technology in combination with non-integrating lentiviral vectors, as a curative solution for DBA. It highlights the transformative advancements in the treatment landscape of DBA, offering hope for individuals affected by this condition.

Keywords: Diamond–Blackfan anemia; ribosomopathy; ribosomal protein genes; rare genetic disorder; hematopoietic stem cell transplantation; gene therapy; lentiviral vector; non-integrating lentiviral vector; CRISPR/Cas9

1. Introduction

Diamond–Blackfan anemia (DBA) is a rare inherited bone marrow failure syndrome (IBMFS) characterized by erythroid hypoplasia, primarily affecting infants [1]. This condition, estimated to occur in 5–7 cases per million live births [1], is considered one of the emerging group of disorders known as ribosomopathies [2], which arise from defects in ribosome biogenesis and function. Approximately 75% of DBA cases involve heterozygous mutations in ribosomal protein (RP) genes [1]. In fact, the initial discovery of genetic mutations in DBA was attributed to mutations in the RPS19 gene, which encodes one of the proteins in the 40S small ribosomal subunit [3]. Among the 81 RP-encoding genes, mutations have been identified in 19 of them, with RPS19 (25%), RPL5 (7%), RPS26 (6.6%), and RPL11 (5%) being the most frequently mutated in DBA [2]. Recent advancements have identified mutations in GATA1, a key erythroid transcription factor, as the first non-RP mutations in DBA patients. This discovery followed the identification of other non-RP gene mutations in the RPS26 chaperone protein TSR2 [3].

The current therapeutic strategies for DBA include glucocorticosteroids (GC), blood transfusions, and hematopoietic stem cell transplantation (HSCT), each with its own set of limitations. Glucocorticosteroids, despite being commonly used, may lose effectiveness over time, particularly in patients who become non-responsive to long-term treatment. Moreover, long-term or high-dose therapies with GCs can lead to a range of adverse effects, including osteoporosis, skin atrophy, diabetes, abdominal obesity, glaucoma, cataracts, avascular necrosis and infection, growth retardation, and hypertension [4]. Blood transfusions serve as a vital supportive measure to alleviate symptoms and manage anemia in DBA patients, but it is important to notice that there is a toxicity associated with iron overload [5]. HSCT is the only curative treatment for DBA and although it can be an option for patients with steroid resistance and transfusion dependency, it presents challenges as finding suitable donors and the risk of immunological complications [6]. Amidst these challenges, gene therapy emerges as a promising tool for treating DBA.

Gene therapy holds significant importance in the treatment of DBA due to the limitations of current therapies. Recent advances in gene therapy, particularly the use of lentiviral vectors, show promise for treating DBA. These therapies aim to correct the genetic defects causing DBA by introducing functional copies of the mutated genes into the patient's cells. As research in this field progresses, there is growing potential for gene therapy to correct the underlying genetic mutations associated with DBA, using techniques such as the CRISPR/Cas9 editing tool [1,3].

In this review, we aim to explore the potential of gene therapy based on CRISPR/Cas9 technology, particularly in combination with non-integrating lentiviral vectors, as a curative solution for DBA. We will delve into how these innovative approaches hold the key to restoring normal hematopoiesis, thereby offering transformative advancements in the treatment landscape of DBA. Continued research and refinement of gene therapy strategies can unlock this potential and dive into a new era of hope for individuals affected by DBA.

2. Clinical Presentation of DBA and Diagnosis

In individuals with DBA, the hematological profile typically shows macrocytic or occasionally normocytic anemia along with reticulocytopenia. Patients usually present normal neutrophil and platelet counts, and the bone marrow appears normal in terms of cellularity but has a deficiency in erythroid precursors [7]. Symptoms of DBA often surface in infancy, with 95% of cases being diagnosed before 2 years of age and 99% before 5 years of age. These symptoms include anemia-related signs such as pallor, fatigue, and feeding difficulties [8]. Although DBA is primarily a hematological disorder, patients also exhibit a spectrum of physical abnormalities. Common features among DBA patients include delayed growth, short stature, and a distinct facial appearance known as Cathie facies, which is characterized by a cute snub nose and wide-spaced eyes. Triphalangeal thumbs, a condition known as Aase syndrome, are also common, and are often accompanied by craniofacial malformations, cleft palate, cardiac defects, and urogenital malformations (Figure 1) [7,9]. DBA patients also face an elevated risk of developing various cancers, including hematological malignancies and solid tumors, such as colon carcinoma and osteosarcomas [1].

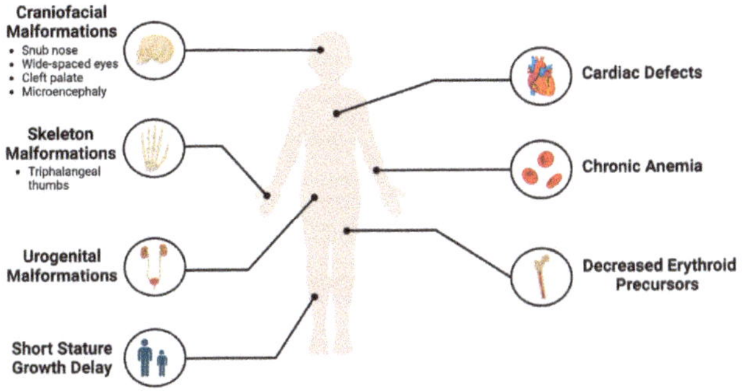

Figure 1. Clinical manifestations of DBA. Created with BioRender.com (accessed on 18 April 2024).

In addition to the classic hematological profile of DBA patients, a significant number of non-classic cases have been identified, requiring alternative diagnostic approaches beyond traditional methods, such as complete blood count, reticulocyte count, and bone marrow aspiration and biopsy. Diagnosis typically involves assessing fetal hemoglobin (HbF) levels and erythrocyte adenosine deaminase (eADA) activity, as these are considered biomarkers of DBA [7,10]. When clinical suspicion arises, mutation analysis for known DBA genes is conducted to confirm the diagnosis [7]. These diagnostic strategies allow for comprehensive evaluation and accurate identification of DBA, facilitating appropriate management and care for affected individuals.

3. Molecular Mechanism of DBA

The intricate molecular pathways underlying DBA remain incompletely elucidated, motivating ongoing scientific efforts to unravel the relationship between mutations, mostly in RP genes, and the resultant anomalies in ribosome assembly and biogenesis, ultimately culminating in impaired erythropoiesis.

Research has shed light on one aspect of this complexity, revealing that haploinsufficiency in certain RP genes leads to the stabilization of p53, leading to cell cycle arrest and apoptosis (Figure 2a) [1]. Remarkably, studies using zebrafish and patient samples have shown that mutations in RP genes are related to the activation of p53 and target genes [11–13]. Additionally, unbalanced globin/heme synthesis emerges as another critical aspect of DBA pathogenesis. Reports indicate that primary DBA cells exhibit imbalanced globin and heme synthesis, resulting in the accumulation of reactive oxygen species (ROS) within early erythroid precursors. This accumulation significantly contributes to the impairment of erythropoiesis (Figure 2b) [14]. Translational dysfunction also emerges as a key player in DBA pathology (Figure 2c). This occurs when ribosomal stress, induced by mutations in RP genes, leads to issues in protein synthesis. It is possible that ribosome dysfunction can affect mRNA production and that certain specific cells or tissues may be more vulnerable to ribosome dysfunction [15]. Notably, in patients with RP mutations, the mRNA for GATA1, a master hematopoietic transcription factor, is poorly translated, further exacerbating the impaired erythroid defect characteristic of DBA, which might be due to the fact that this mRNA has a higher threshold for initiation in comparison to other mRNAs [16]. Moreover, emerging evidence suggests that inflammatory signaling pathways may also contribute to the pathology of DBA [17] as Iskander et al. found elevated levels of IFN-γ and TNF-α in bone marrow plasma, known instigators of stress erythropoiesis (Figure 2d).

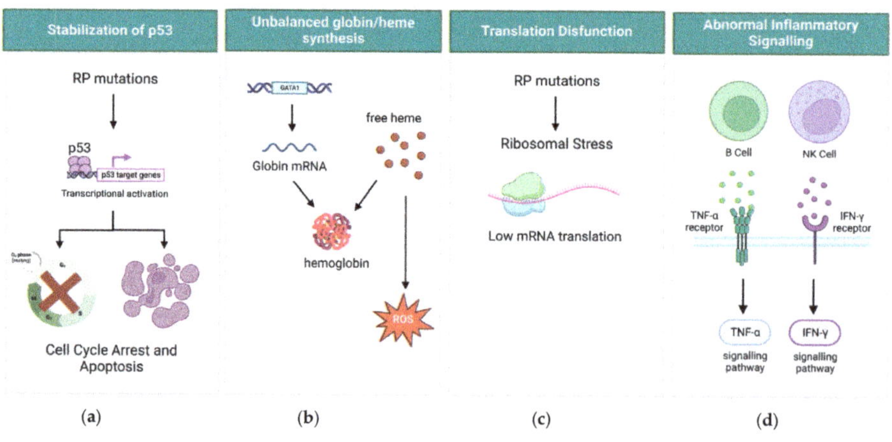

Figure 2. Molecular mechanism of DBA. (**a**) RP mutations lead to activation of p53, cell cycle arrest, and apoptosis; (**b**) Unbalanced globin/heme synthesis leads to accumulation of ROS in erythroid precursors; (**c**) Translation dysfunction caused by RP mutations; (**d**) Abnormal inflammatory signaling caused by RP mutations. Created with BioRender.com (accessed on 22 May 2024).

In conclusion, the multifaceted nature DBA presents a complex puzzle for researchers. The intricate interplay between mutations in RP genes, disruptions in ribosome assembly, and subsequent impairment of erythropoiesis underscores the need for continued investigation.

4. Existing Treatment Options for DBA

The primary therapeutic options for anemia in DBA are the use of glucocorticosteroids, red blood cell transfusions, and hematopoietic stem cell transplantation (HSCT) (Figure 3). Vlachos and Muir et al. provide a comprehensive guide on how to treat DBA [7]. However, in this review we will emphasize mostly the limitations of these treatments and why it is necessary to develop gene therapy.

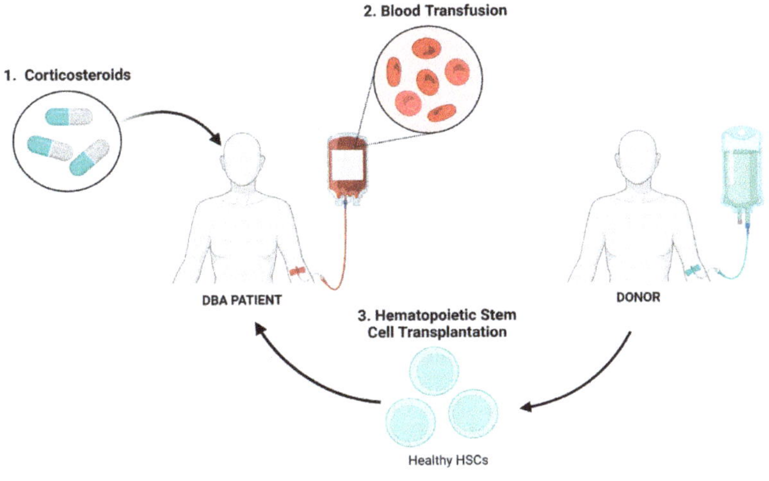

Figure 3. Existing treatment options for DBA patients. Created with BioRender.com (accessed on 18 April 2024).

Glucocorticosteroids (GCs) serve as the primary treatment for DBA, yet their precise mechanism of action in DBA remains unclear [1]. They seem to exert a non-specific anti-apoptotic effect on erythroid progenitor cells [5,18]. Initially, approximately 80% of patients show a positive response to steroid therapy. However, approximately half of these individuals discontinue treatment due to either a loss of response or severe side effects [1,19]. These adverse effects may include growth retardation, increased risk of heart disease, osteoporosis, and severe infections [1]. Only approximately 20% of patients are able to fully discontinue steroid treatment without experiencing a relapse of anemia, achieving a state referred to as "remission" [6]. Given the profound impact of GCs on growth, physical, and neurocognitive development, the initiation of steroid administration in infants is carefully delayed, if feasible, and is maintained with chronic transfusion therapy until the child reaches one year of age [7]. For patients who do not respond to corticosteroids, blood transfusions are administered as an alternative treatment. However, a significant drawback of this approach is the potential toxicity associated with iron overload. Consequently, patients require intensive chelation therapy to mitigate the risks posed by excessive iron accumulation [1,5].

HSCT stands as the sole curative option for DBA typically recommended when resistance to corticosteroid therapy and dependence on transfusions occur [1,7]. HLA-matched sibling HSCT has demonstrated significant success rates, particularly in patients younger than 9 years old. However, each potential sibling donor undergoes thorough screening for DBA mutations, even in the absence of hematological or physical DBA manifestations [7].

Despite the efficacy of this approach, the availability of HLA-matched donors is not always guaranteed. HSCT carries several drawbacks, including the risk of graft-versus-host disease (GvHD), adverse effects stemming from preconditioning, the possibility of undetected mutations in silent carriers, and the necessity for immunosuppressive therapy post-transplantation [1,5,19].

5. Gene Therapy for DBA—From Research Now to Clinic in the Future

Utilizing autologous HSCT with genetically modified hematopoietic stem and progenitor cells (HSPCs) presents a potential solution to address the limitations associated with allogeneic HSCT. This innovative approach could circumvent challenges such as the scarcity of suitable donors, the risk of GvHD, the potential for graft rejection, and the possibility of donors being silent carriers of DBA mutations.

5.1. Lentiviral Vectors as a Potential Gene Therapy Approach for DBA

RP-mutations are the primary cause of DBA. Consequently, gene therapy aimed at enabling the expression of the functional RP gene represents a potential solution for DBA patients. Lentiviral vectors (LVs) have emerged as effective delivery tools for hematopoietic stem and progenitor cells (HSPCs) [20]. When pseudotyped with the vesicular stomatitis virus G protein (VSV-G), LVs demonstrate their versatility by transducing a wide array of cells [21]. Their large genetic capacity (up to 10 kb) and the ability to transduce both dividing and non-dividing cells make them exceptional tools for gene therapy [22]. Traditional LVs integrate the viral genome into the host's genome, ensuring stable expression of the gene of interest [22]. The general strategy involves developing lentiviral vectors that encode the various functional RP genes mutated in DBA patients. The effectiveness of any LV-based gene therapy hinges on the successful high-level transduction of patient HSPCs that are capable of long-term hematopoietic repopulation [23]. Upon integration into the patients' HSPCs, these cells would begin producing functional ribosomal proteins, facilitating normal erythropoiesis. However, LVs also carry oncogenic potential, as integration can occur at multiple sites, potentially leading to the disruption of normal gene function, activation of oncogenes or inactivation of tumor suppressor genes [24]. Notably, there have been several adverse events observed in clinical trials attributed to insertional mutagenesis, wherein the integration of the vector may disrupt normal genomic function or even activate oncogenes, potentially leading to adverse outcomes [24,25].

The RPS19 gene stands out as the most frequently mutated gene among individuals diagnosed with DBA, affecting approximately 25% of patients [26]. Consequently, research in this field has been directed towards elucidating methods to restore the protein encoded by this gene, aiming to reverse the hematological abnormalities associated with DBA. In a pivotal study conducted by Hamaguchi et al., the potential of gene transfer techniques to address RPS19-related pathology was demonstrated [27]. Specifically, the researchers utilized lentiviral vectors to introduce the RPS19 gene into hematopoietic progenitors from RPS19-deficient DBA patients. Remarkably, this intervention resulted in notable improvements in $CD34^+$ cell proliferation, as well as in erythroid development. Further supporting the feasibility of gene transfer as a therapeutic strategy for DBA, additional studies have corroborated these findings. For instance, Jaako et al. used transgenic mice containing a RPS19-targeting shRNA under a doxycycline-responsive promoter for lentiviral-based gene therapy. They transduced uninduced BM cells from heterozygous (D/+) and homozygous (D/D) RPS19-targeting shRNA DBA mice with lentivirus containing RPS19 cDNA and transduced cells were transplanted into wild type mice. Following engraftment, the mice were administered doxycycline to downregulate the endogenous RPS19 and induce the disease. They concluded that enforced expression of RPS19 cures anemia and prevents fatal bone marrow failure in RPS19-deficient mice. Additionally, they observed that cells corrected with the RPS19 gene displayed sustained improvement in pan-hematopoietic function over time, contrasting with untreated cells, and showed no adverse effects attributable to the gene transfer process [28]. Following this study, Debnath et al. engineered lentiviral vectors capable of expressing the RPS19 gene under the control of the human elongation factor 1α short (EFS) promoter, a clinically relevant promoter [29]. To evaluate the efficacy of this vector, they transfected c-Kit-enriched BM cells from both control and heterozygous RPS19 shRNA mice in the presence of doxycycline, and subsequently injected these cells into lethally irradiated wild type mice. Their results revealed that recipients transduced with EFS-RPS19 shRNA BM exhibited near normal blood cellularity, indicating that enforced expression of RPS19 driven by the EFS promoter can effectively treat severe anemia and bone marrow failure in RPS19-deficient mice. However, it is worth noting that this model does not mimic the haploinsufficiency seen in DBA patients, which is caused by mutations in the RPS19 gene. More recently, this group designed a clinically applicable self-inactivating (SIN) lentiviral vector containing the human RPS19 driven by the human EFS promoter for the clinical development of gene therapy for RPS19-deficient DBA patients. Their study showcased that this vector effectively rescues the anemia and lethal BM failure phenotype observed in the mouse models of RPS19-deficient DBA, with low risk of mutagenesis and a highly polyclonal insertion site pattern. Additionally, they observed the restauration of impaired erythroid differentiation in human RPS19-deficient CD34+ cord blood cells treated with this vector, underscoring its potential for clinical translation and therapeutic benefit in DBA patients [30].

Furthermore, the exploration of using lentiviral vectors to express GATA1 for the promotion of red blood cell production is under investigation. This approach offers significant advantages, particularly in targeting the majority of DBA mutations rather than a specific one. In vitro studies have shown that overexpression of GATA1 in hematopoietic stem and progenitor cells (HSPCs) from DBA patients can rescue erythroid differentiation defects [1,31]. While gene therapy presents an attractive strategy for curing DBA, the traditional approach of overexpressing a functional copy of a mutated gene is not the most efficient. This is because it would necessitate the development and validation of numerous gene therapy vectors, each containing a copy of one of the mutated DBA genes. Instead, a unified gene therapy strategy is being proposed, which involves the developmentally regulated and highly restricted expression of GATA1. This strategy is anticipated to be curative for most, if not all, DBA patients, regardless of the specific mutation causing the disease [31].

While extensive research has been conducted on the utilization of LVs for gene therapy in DBA, these efforts have not yet been translated into clinical trials. Nevertheless, the successful implementation of LVs in gene therapy for various genetic blood disorders, including sickle cell disease (SCD), β-thalassemia, and Fanconi anemia (FA), underscores the potential of LV-based gene therapy as a promising avenue for treating DBA effectively (Table 1).

Table 1. An overview of clinical trials utilizing LVs for gene therapy of genetic blood disorders.

Disease	Clinical Trial ID	Intervention/Treatment	Ref.
Sickle Cell Disease (SCD)	NCT02186418	CD34+ cells transduced with gamma-globin lentiviral vector.	[32]
	NCT03282656	Single infusion of autologous bone marrow derived CD34+ HSC cells transduced with the lentiviral vector containing a short-hairpin RNA targeting BCL11a.	[33]
	NCT05353647	Autologous transplantation of CD34+ HSC cells transduced with the lentiviral vector containing a shRNA targeting BCL11a.	[34]
	NCT02247843	Autologous transplantation of peripheral blood CD34+ cells transduced ex vivo by the Lenti/G-βAS3-FB lentiviral vector to express an anti-sickling (βAS3) gene.	[35]
	NCT03964792	Transplantation of an autologous CD34+ enriched cell fraction that contains CD34+ cells transduced ex vivo with the GLOBE1 lentiviral vector expressing the βAS3 globin gene (GLOBE1 βAS3 modified autologous CD34+ cells).	[36]

Table 1. *Cont.*

Disease	Clinical Trial ID	Intervention/Treatment	Ref.
SCD and β-Thalassemia	NCT02151526	Administration of LentiGlobin BB305 drug product to participants with either transfusion dependent β-thalassemia (TDT) or sickle cell disease (SCD).	[37]
β-Thalassemia	NCT03276455	Autologous transplantation of HSCs transduced with lentiviral vector encoding for beta-globin gene.	[38]
	NCT02453477	Autologous transplantation of HSCs genetically modified with GLOBE lentiviral vector encoding for the human beta-globin gene.	[39]
	NCT06219239	Autologous transplantation of HSCs transduced with lentiviral vector encoding βA-T87Q-globin gene.	[40]
	NCT05745532	Autologous transplantation of HSCs transduced with LentiHBBT87Q system to restore β-globin expression.	[41]
	NCT06280378	Autologous transplantation of CD34+ stem cells transduced ex vivo with a lentiviral vector encoding βA-T87Q-globin.	[42]
	NCT01639690	Autologous transplantation of CD34+ cells transduced with TNS9.3.55 lentiviral vector encoding the normal human ß-globin gene.	[43]
	NCT05762510	Autologous transplantation of CD34+ HSCs transduced with LentiRed lentiviral vector.	[44]
	NCT05757245	Autologous HSCT using GMCN-508A drug product (autologous CD34+ HSCs transduced with GMCN-508A lentiviral vector encoding the human α-globin gene).	[45]
	NCT05015920	Transplantation of autologous CD34+ stem cells transduced with a lentiviral vector encoding βA-T87Q-globin.	[46]
Fanconi Anemia (FA)	NCT01331018	Transplantation of autologous patient blood stem cells that have been corrected in the laboratory by introduction of the normal FANCA gene.	[47]
	NCT04437771	Transplantation of autologous CD34+ cells transduced with lentiviral vector carrying the FANCA gene.	[48]

5.2. CRISPR/Cas9 Non-Integrating Lentiviral-Based Gene Therapy

Another strategy relies on non-integrating lentiviruses (NILVs) offering two primary methodologies for generating non-integrating lentiviral vectors: (i) introducing mutations in the viral integrase protein, and (ii) inhibiting the recognition of viral DNA by this enzyme through mutations in the sites [49,50]. Gurumoorthy et al. summarize the point mutations that have been used to develop NILVs [50]. By employing NILVs, the viral genome can persist in the host cell as an episome, rather than integrating into the host genomic DNA [51]. This mechanism creates an opportunity for the application of CRISPR/Cas9 technologies. Once the desired DNA editing event occurs, the Cas9 protein and the guide RNA (gRNA) are no longer required for ongoing transgene expression. Therefore, they can be removed from the cell, minimizing the risk of off-target effects and undesirable side effects associated with their continued presence [49]. The difference between the traditional integrating LVs and NILVs is highlighted in Table 2.

Table 2. Main characteristics, applications, and limitations of integrating LVs and NILVs.

Aspect	Integrating Lentiviral Vectors (ILVs)	Non-Integrating Lentiviral Vectors (NILVs)
Integration	Integrates the transgene into the host genome [22,50]	Does not integrate the transgene into the host genome [52] Expresses the transgene from episomal DNA in non-dividing cells or transiently in dividing cells [49]
Expression	Stable integration of the transgene into the host genome [49,50]	Enables transient expression or sustained episomal expression [50]
Safety	Higher risk of insertional mutagenesis and malignant transformation [50]	Reduced risk of insertional mutagenesis and malignant transformation [50]
Applications	Gene therapy for long-term gene expression [50,52] Recombinant protein production [50] Vaccination [50] Cell imaging [50]	Gene therapy for mutation correction [50,52] Cytotoxic cancer therapies [49,50] Cellular differentiation [49] Vaccination [49,52] Immunotherapies [49,50]
Limitations	Insertional mutagenesis [24] Oncogenic potential [24]	Residual integration risks [50] Transient expression is not suitable for all applications

CRISPR/Cas9 technology has revolutionized the field of gene editing, offering a highly precise and efficient method for modifying the genome. This breakthrough has paved the way for innovative approaches in addressing genetic disorders. Alongside these benefits come significant ethical and safety concerns. One of the primary concerns is the risk of off-target effects, wherein the CRISPR/Cas9 mechanism may unintentionally alter DNA sequences at unintended sites, resulting in unintended mutations and the activation of oncogenes [53]. Oncogenic changes created by these mechanisms will be carried by the target cell and its progeny, adding another layer of complexity to the potential long-term impacts of gene therapy technologies [24]. Additionally, the induction of double-strand breaks (DSBs) leads to genomic instability and can consequently lead to more accumulation of mutations. DSBs are primarily repaired by the error-prone non-homologous end joining (NHEJ) pathway, which can lead to small insertions and deletions (INDEL mutations) [54]. Such occurrences could potentially give rise to unforeseen health issues or even the emergence of novel diseases.

One of the key advancements in the field of gene therapy is the development of base and prime editors, which are tools that can be used to correct mutations in DNA. While CRISPR/Cas9 genome editing conventionally triggers DSBs at specific DNA target sites, potentially resulting in genomic instability and off-target effects, base editors (BE) and prime editors (PE) use Cas9 nickases (dCas9), which are variants of the Cas9 that have been engineered to induce nicks in the DNA strand instead of cleaving it [55]. Table 3 lists the main differences between the traditional CRISPR/Cas9 system, base editors, and prime editors.

Table 3. Comparison between CRISPR/Cas9 system, base editors, and prime editor. CBE: cytosine base editor; ABE: adenine base editor; PE: prime editor.

Tools	Components	Applicability	Advantages	Drawbacks	Ref.
CRISPR/Cas9	Cas9, sgRNA, and donor DNA (for HDR)	Point mutations Large DNA insertions and deletions Gene knock-out	Versatility in gene insertion, deletion, and modification	DSB induction Off-target effects Low efficiency for HDR	[56]
CBE	dCa9-cytosine deaminase, and sgRNA	Transitions mutations ($C \rightarrow T$, $G \rightarrow A$, $A \rightarrow G$, $T \rightarrow C$)	Induction of SSBs	Requires precise positioning of editing window Off-target DNA and RNA editing Bystander edits Only capable of four transition mutations	[56,57]
ABE	dCas9-adenine deaminase and sgRNA				

Table 3. *Cont.*

Tools	Components	Applicability	Advantages	Drawbacks	Ref.
PE	dCas9(H840A)-M-MLV RT and pegRNA	Point mutations Small deletions and insertions	Induction of SSBs Allows for all precise modifications	Genomic scope constraints Variable efficiency in different cell types Delivery challenges due to large size of PE	[56,58]

In the context of base editors, dCas9 is combined with a deaminase enzyme, enabling the alteration of a single DNA through a process called deamination [59]. Given that approximately half of all known pathogenic genetic variants are single nucleotide variants (SNVs), base editing represents a promising approach for treating a wide range of genetic diseases [60]. There are two classes of DNA base-editors: cytosine base-editors (CBE) and adenine base-editors (ABE) [55]. CBEs convert cytosine bases to uracil which is then recognized by the cell's replication machinery as thymine, leading to a C-G to T-A substitution in the DNA sequence. This mechanism is particularly useful for correcting disease-causing mutations that involve C-G to T-A changes. However, the efficiency of CBE in human cells has been limited due to the cellular repair pathway that reverts the uracil to cytosine, known as base excision repair (BER). To overcome this, researchers have developed improved versions of CBE, such as BE2 and BE3, which incorporate strategies to inhibit BER, thereby enhancing the editing efficiency and specificity of CBE [55]. ABEs, on the other hand, convert adenine to inosine, which is then recognized as guanine during DNA replication, leading to the substitution of A-T to G-C in the DNA [55], which represent the most common pathogenic SNVs reported in ClinVar database [60]. However, base editors are not without limitations. Besides being restricted to making a maximum of four substitutions, the limitations include the requirement for precise positioning of the base editor edit window, and off-target DNA and RNA editing [57].

The prime editor system is a sophisticated genetic editing tool that allows for precise modifications of DNA, including point mutations, insertions, and deletions of small fragments [58]. This system is composed of two main components: a reverse transcriptase (RT) and a Cas9 nickase fused together. The RT component is guided by a prime editing guide RNA (pegRNA), which contains a primer binding site (PBS) and a template for the reverse transcription. This process is facilitated by dCas9, which exposes the 3′ end of the DNA strand, allowing the RT to bind to the PBS on the pegRNA and synthesize the new DNA strand with the desired edit. The edited DNA strand then has two overhangs: one unmodified 5′ flap and one purposed 3′ flap. The 3′ flap, which contains the desired edit, is retained, while the 5′ flap is cleaved away. The cellular DNA repair system then integrates the edited DNA strand into the genome, replacing the original sequence with the modified version [55,58]. To enhance efficiency, an improved version, PE3, was developed. This version incorporates a second nicking guide RNA (ngRNA) to induce a nick in the non-edited strand. This process leads to the corrected strand being used as a template for correction of the nicked, resulting in incorporation of the desired change in both strands. However, this approach also increases the number of insertions and deletions (INDELs). To mitigate this issue, the PE3b version was introduced. This version utilizes a ngRNA that specifically recognizes the non-edited strand after the edit has occurred, thereby enhancing the safety of the editing process [56].

Although the PE system offers enhanced flexibility in targeting compared to other genome editing methods, such as BE, it faces several practical limitations. These include genomic scope constraints, variable efficiency across cell types, and delivery challenges due to its original large size [58]. To address the genomic scope constraints, researchers have focused on developing PAM-relaxed Cas9 enzymes that are compatible with various protospacer adjacent motif (PAM) sequences, beyond the traditional SpCas9 which requires the NGG PAM [58]. The NGG motif represents a particular three-nucleotide sequence

(NGG), where N can represent any base. This sequence is what the SpCas9 enzyme identifies and attaches to, enabling it to sever the DNA at the desired location [61,62].

Additionally, researchers have focused on developing miniaturized Cas9 versions to potentially reduce the size of the PE system and facilitate its delivery [58]. The size limitation is of particular importance when developing NILVs that carry the prime editor system. As mentioned before, a typical lentiviral vector can carry up to 10 kb of insert fragments [22]. For prime editing, the lentiviral vector must include the PE components, including the dCas9 (~4.2 kb) and the RT (~2.0 kb), with a total size of approximately 6.2 kb [58]. While NILVs can offer advantages in gene therapy applications, the capacity of the lentiviral vector may exceed when including both PE components and the pegRNA, along with promoters, which further states the need to use miniaturized Cas9 versions. An alternative approach involves the delivery of the Cas9 protein along with NILVs that carry the gRNA. This method has been successfully demonstrated by Uchida et al., who developed a Cas9 delivery system using NILVs that encode both a gRNA and a donor template for correction of the sickle cell disease mutation [63]. This system erases the possibility of exceeding the lentiviral vector size capacity and could be adapted for prime editing.

While non-integrating lentivirus still requires further development to reach clinical trials, the successful application of gene therapy using the CRISPR/Cas9 tool in treating genetic blood disorders like SCD and β-thalassemia has sparked optimism for correcting mutations associated with DBA through gene therapy (Table 4).

Table 4. An overview of clinical trials utilizing CRISPR/Cas9 technology for gene therapy of genetic blood disorders.

Disease	Clinical Trial ID	Intervention/Treatment	Ref
Sickle Cell Disease (SCD)	NCT06287099	Autologous CRISPR/Cas9 modified CD34+ hHSPCs (BRL-101)	[64]
	NCT04819841	Gene correction in autologous CD34+ HSCs (HbS to HbA) to treat severe SCD	[65]
	NCT03745287	Autologous CRISPR/Cas9 modified CD34+ hHSPCs using CTX001	[66]
	NCT05951205	Single dose of CTX001 in subjects with severe SCD with βS/βC genotype	[67]
	NCT04774536	Transplantation of CRISPR/Cas9 corrected HSCs (CRISPR_SCD001) in patients with severe SCD	[68]
	NCT05329649	Administration of a single dose of CTX001 in pediatric subjects with severe SCD	[69]
SCD and β-Thalassemia	NCT05477563	Single dose of autologous CRISPR/Cas9 modified CD34+ hHSPCs (CTX001) in subjects with transfusion-dependent β-Thalassemia or severe SCD	[70]
	NCT04208529	A long-term follow-up study of subjects with β-thalassemia or SCD treated with autologous CRISPR/Cas9 modified HSCs (CTX001)	[71]
β-Thalassemia	NCT03655678	Autologous CRISPR/Cas9 modified CD34+ hHSPCs using CTX001 in subjects with transfusion-dependent β-Thalassemia	[72]
	NCT04925206	Autologous CRISPR/Cas9 modified CD34+ hHSPCs using ET-01 in subjects with transfusion-dependent β-Thalassemia	[73]
	NCT05577312	Autologous CRISPR/Cas9 modified CD34+ hHSPCs (BRL-101)	[74]
	NCT05356195	Autologous CRISPR/Cas9 modified CD34+ hHSPCs (CTX001) in pediatric subjects with transfusion-dependent β-Thalassemia	[75]
	NCT03728322	Gene correction of HBB in patient-specific iHSCs using CRISPR/Cas9	[76]

This breakthrough paves the way for exploring similar approaches to target and rectify DBA mutations, offering hope for more effective treatment options and potentially even a cure for this rare hematological disorder [77,78]. Additionally, recent research highlights the effective use of prime editing to correct the HBB gene in hematopoietic stem cells (HSCs) of mice with sickle cell disease (SCD). This innovative approach utilized a HDAd vector to deliver the prime editing machinery directly into the bloodstream of the mice, showcasing the potential for in vivo gene correction as a promising therapeutic strategy for genetic blood disorders like SCD [79].

Furthermore, beyond the numerous ongoing and completed clinical trials, FDA has approved two gene therapies for the treatment of SCD: Casgevy and Lyfgenia. Casgevy employs a novel genome editing technique that modifies a particular gene to restore the production of fetal hemoglobin, thereby mitigating the abnormal red blood cells typical of SCD. On the other hand, Lyfgenia employs a lentiviral vector to deliver a healthy hemoglobin-producing gene to patients, aiming to correct the underlying genetic defect causing SCD [80,81]. Additionally, there is also an FDA-approved gene therapy for β-Thalassemia. Zynteglo utilizes a replication-incompetent lentiviral vector to deliver a modified β-globin gene to the patient's own hematopoietic stem cells (HSCs). This approach allows to produce functional adult hemoglobin, addressing the underlying genetic cause of β-thalassemia by correcting the α/β-globin imbalance [81,82].

Despite the promising advancements in gene therapy, significant challenges persist, particularly in the areas of long-term patient follow-up, cost, safety, efficacy, and manufacturing. Ensuring the long-term safety of gene therapy products necessitates extensive follow-up beyond the active clinical trial period to monitor for delayed adverse effects [83]. Additionally, HSCT continues to present significant challenges, including its high cost, inherent safety concerns, variability in efficacy, and manufacturing difficulties. A key challenge lies in the production of therapeutic agents at high titers and with consistent quality [84].

Despite these challenges, advancements in gene therapy, particularly using lentivirus and CRISPR/Cas9 tools, have demonstrated significant potential in treating various blood diseases. These innovative technologies offer a promising future for DBA treatment by providing precise and targeted corrections to genetic defects (Figure 4). These advancements not only offer a safer and more effective alternative to traditional treatments like allogenic HSCT but also hold the promise of a long-term cure for DBA and other monogenic diseases. Further research and clinical trials are necessary to fully realize the potential of these gene therapy approaches and to address the unique challenges posed by DBA, such as the involvement of multiple genes and the unknown causative mutation in some patients.

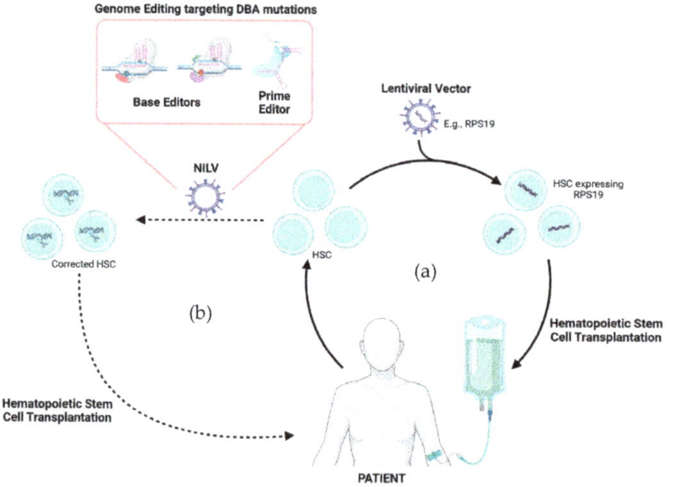

Figure 4. Gene therapy as a therapeutic alternative for DBA treatment. (**a**) Traditional gene therapy employes integrating lentiviral vectors carrying the functional gene (e.g., RPS19), which are delivered to the patients HSCs ex vivo. (**b**) A more novel approach includes correcting the mutation ex vivo using non-integrating lentiviral vectors (NILVs) that carry CRISPR/Cas9 tools specific for the DBA mutation. Created with BioRender.com (accessed on 18 April 2024).

6. Conclusions

The utilization of autologous HSCT combined with genetically modified HSPCs presents a promising alternative to overcome the limitations associated with allogenic HSCT and represents a promising leap forward in addressing the challenges associated with DBA treatment. Through targeted research efforts focused on restoring the function of the RPS19 gene, frequently mutated in DBA patients, and the development of innovative gene therapy techniques such as base and prime editing using NILVs, significant progress can be made towards effective therapeutic interventions. However, it is important to note that autologous HSCT is associated with high costs, which can be a limiting factor for its widespread adoption. Despite these challenges, the potential benefits of this approach including improved outcomes, reduced risks, and the potential for long-term remission, make it a compelling option for treating genetic blood diseases like DBA.

Author Contributions: M.V.: writing—original draft preparation, conceptualization, J.P.: supervision, writing—review and editing, conceptualization, R.S.: funding acquisition, writing—review and editing, conceptualization, project administration. All authors have read and agreed to the published version of the manuscript.

Funding: This work was supported by the Czech Academy of Science (RVO: 68378050), and the Ministry of Education, Youth and Sports (MEYS) (LM202303), and by European Union's Horizon Europe under the Marie Skłodowska-Curie grant agreement "Gene Therapy of Rare Diseases (GetRadi)", no101072427. The project was also supported by the Strategy AV21 Research program "Towards Precision Medicine and Gene Therapy".

Conflicts of Interest: The authors declare no conflicts of interest.

References

1. Liu, Y.; Karlsson, S. Perspectives of current understanding and therapeutics of Diamond-Blackfan anemia. *Leukemia* **2024**, *38*, 1–9. [CrossRef] [PubMed]
2. Kang, J.; Brajanovski, N.; Chan, K.T.; Xuan, J.; Pearson, R.B.; Sanij, E. Ribosomal proteins and human diseases: Molecular mechanisms and targeted therapy. *Signal Transduct. Target. Ther.* **2021**, *6*, 323. [CrossRef] [PubMed]
3. Ulirsch, J.C.; Verboon, J.M.; Kazerounian, S.; Guo, M.H.; Yuan, D.; Ludwig, L.S.; Handsaker, R.E.; Abdulhay, N.J.; Fiorini, C.; Genovese, G.; et al. The Genetic Landscape of Diamond-Blackfan Anemia. *Am. J. Hum. Genet.* **2018**, *103*, 930–947. [CrossRef] [PubMed]
4. Maceckova, Z.; Kubickova, A.; De Sanctis, J.B.; Hajduch, M. Effect of Glucocorticosteroids in Diamond-Blackfan Anaemia: Maybe Not as Elusive as It Seems. *Int. J. Mol. Sci.* **2022**, *23*, 1886. [CrossRef] [PubMed]
5. Bartels, M.; Bierings, M. How I manage children with Diamond-Blackfan anaemia. *Br. J. Haematol.* **2019**, *184*, 123–133. [CrossRef] [PubMed]
6. Strahm, B.; Loewecke, F.; Niemeyer, C.M.; Albert, M.; Ansari, M.; Bader, P.; Bertrand, Y.; Burkhardt, B.; Da Costa, L.M.; Ferster, A.; et al. Favorable outcomes of hematopoietic stem cell transplantation in children and adolescents with Diamond-Blackfan anemia. *Blood Adv.* **2020**, *4*, 1760–1769. [CrossRef] [PubMed]
7. Vlachos, A.; Muir, E. How I treat Diamond-Blackfan anemia. *Blood* **2010**, *116*, 3715–3723. [CrossRef] [PubMed]
8. Da Costa, L.; Leblanc, T.; Mohandas, N. Diamond-Blackfan Anemia. *Blood* **2020**, *136*, 1262–1273. [CrossRef] [PubMed]
9. Lipton, J.M.; Ellis, S.R. Diamond-Blackfan anemia: Diagnosis, treatment, and molecular pathogenesis. *Hematol. Oncol. Clin. North. Am.* **2009**, *23*, 261–282. [CrossRef]
10. Jahan, D.; Al Hasan, M.M.; Haque, M. Diamond-Blackfan anemia with mutation in RPS19: A case report and an overview of published pieces of literature. *J. Pharm. Bioallied Sci.* **2020**, *12*, 163–170. [CrossRef]
11. Danilova, N.; Sakamoto, K.M.; Lin, S. Ribosomal protein S19 deficiency in zebrafish leads to developmental abnormalities and defective erythropoiesis through activation of p53 protein family. *Blood* **2008**, *112*, 5228–5237. [CrossRef] [PubMed]
12. Chakraborty, A.; Uechi, T.; Higa, S.; Torihara, H.; Kenmochi, N. Loss of ribosomal protein L11 affects zebrafish embryonic development through a p53-dependent apoptotic response. *PLoS ONE* **2009**, *4*, e4152. [CrossRef] [PubMed]
13. Moniz, H.; Gastou, M.; Leblanc, T.; Hurtaud, C.; Cretien, A.; Lecluse, Y.; Raslova, H.; Larghero, J.; Croisille, L.; Faubladier, M.; et al. Primary hematopoietic cells from DBA patients with mutations in RPL11 and RPS19 genes exhibit distinct erythroid phenotype in vitro. *Cell Death Dis.* **2012**, *3*, e356. [CrossRef] [PubMed]
14. Rio, S.; Gastou, M.; Karboul, N.; Derman, R.; Suriyun, T.; Manceau, H.; Leblanc, T.; El Benna, J.; Schmitt, C.; Azouzi, S.; et al. Regulation of globin-heme balance in Diamond-Blackfan anemia by HSP70/GATA1. *Blood* **2019**, *133*, 1358–1370. [CrossRef] [PubMed]
15. Mills, E.W.; Green, R. Ribosomopathies: There's strength in numbers. *Science* **2017**, *358*, eaan2755. [CrossRef] [PubMed]

16. Ludwig, L.S.; Gazda, H.T.; Eng, J.C.; Eichhorn, S.W.; Thiru, P.; Ghazvinian, R.; George, T.I.; Gotlib, J.R.; Beggs, A.H.; Sieff, C.A.; et al. Altered translation of GATA1 in Diamond-Blackfan anemia. *Nat. Med.* **2014**, *20*, 748–753. [CrossRef] [PubMed]
17. Iskander, D.; Wang, G.; Heuston, E.F.; Christodoulidou, C.; Psaila, B.; Ponnusamy, K.; Ren, H.; Mokhtari, Z.; Robinson, M.; Chaidos, A.; et al. Single-cell profiling of human bone marrow progenitors reveals mechanisms of failing erythropoiesis in Diamond-Blackfan anemia. *Sci. Transl. Med.* **2021**, *13*, eabf0113. [CrossRef]
18. Narla, A.; Vlachos, A.; Nathan, D.G. Diamond Blackfan anemia treatment: Past, present, and future. *Semin. Hematol.* **2011**, *48*, 117–123. [CrossRef] [PubMed]
19. Bhoopalan, S.V.; Suryaprakash, S.; Sharma, A.; Wlodarski, M.W. Hematopoietic cell transplantation and gene therapy for Diamond-Blackfan anemia: State of the art and science. *Front. Oncol.* **2023**, *13*, 1236038. [CrossRef]
20. Sutton, R.E.; Reitsma, M.J.; Uchida, N.; Brown, P.O. Transduction of Human Progenitor Hematopoietic Stem Cells by Human Immunodeficiency Virus Type 1-Based Vectors Is Cell Cycle Dependent. *J. Virol.* **1999**, *73*, 3649–3660. [CrossRef]
21. Peluffo, H.; Foster, E.; Ahmed, S.G.; Lago, N.; Hutson, T.H.; Moon, L.; Yip, P.; Wanisch, K.; Caraballo-Miralles, V.; Olmos, G.; et al. Efficient gene expression from integration-deficient lentiviral vectors in the spinal cord. *Gene Ther.* **2013**, *20*, 645–657. [CrossRef] [PubMed]
22. Kalidasan, V.; Ng, W.H.; Ishola, O.A.; Ravichantar, N.; Tan, J.J.; Das, K.T. A guide in lentiviral vector production for hard-to-transfect cells, using cardiac-derived c-kit expressing cells as a model system. *Sci. Rep.* **2021**, *11*, 19265. [CrossRef] [PubMed]
23. Jang, Y.; Kim, Y.S.; Wielgosz, M.M.; Ferrara, F.; Ma, Z.; Condori, J.; Palmer, L.E.; Zhao, X.; Kang, G.; Rawlings, D.J.; et al. Optimizing lentiviral vector transduction of hematopoietic stem cells for gene therapy. *Gene Ther.* **2020**, *27*, 545–556. [CrossRef] [PubMed]
24. Schlimgen, R.; Howard, J.; Wooley, D.; Thompson, M.; Baden, L.R.; Yang, O.O.; Christiani, D.C.; Mostoslavsky, G.; Diamond, D.V.; Duane, E.G.; et al. Risks Associated with Lentiviral Vector Exposures and Prevention Strategies. *J. Occup. Env. Med.* **2016**, *58*, 1159–1166. [CrossRef] [PubMed]
25. Howe, S.J.; Mansour, M.R.; Schwarzwaelder, K.; Bartholomae, C.; Hubank, M.; Kempski, H.; Brugman, M.H.; Pike-Overzet, K.; Chatters, S.J.; de Ridder, D.; et al. Insertional mutagenesis combined with acquired somatic mutations causes leukemogenesis following gene therapy of SCID-X1 patients. *J. Clin. Investig.* **2008**, *118*, 3143–3150. [CrossRef] [PubMed]
26. Flygare, J.; Aspesi, A.; Bailey, J.C.; Miyake, K.; Caffrey, J.M.; Karlsson, S.; Ellis, S.R. Human RPS19, the gene mutated in Diamond-Blackfan anemia, encodes a ribosomal protein required for the maturation of 40S ribosomal subunits. *Blood* **2006**, *109*, 980–986. [CrossRef] [PubMed]
27. Hamaguchi, I.; Flygare, J.; Nishiura, H.; Brun, A.C.; Ooka, A.; Kiefer, T.; Ma, Z.; Dahl, N.; Richter, J.; Karlsson, S. Proliferation deficiency of multipotent hematopoietic progenitors in ribosomal protein S19 (RPS19)-deficient diamond-Blackfan anemia improves following RPS19 gene transfer. *Mol. Ther.* **2003**, *7*, 613–622. [CrossRef] [PubMed]
28. Jaako, P.; Debnath, S.; Olsson, K.; Modlich, U.; Rothe, M.; Schambach, A.; Flygare, J.; Karlsson, S. Gene therapy cures the anemia and lethal bone marrow failure in a mouse model of RPS19-deficient Diamond-Blackfan anemia. *Haematologica* **2014**, *99*, 1792–1798. [CrossRef] [PubMed]
29. Debnath, S.; Jaako, P.; Siva, K.; Rothe, M.; Chen, J.; Dahl, M.; Gaspar, H.B.; Flygare, J.; Schambach, A.; Karlsson, S. Lentiviral Vectors with Cellular Promoters Correct Anemia and Lethal Bone Marrow Failure in a Mouse Model for Diamond-Blackfan Anemia. *Mol. Ther.* **2017**, *25*, 1805–1814. [CrossRef]
30. Liu, Y.; Dahl, M.; Debnath, S.; Rothe, M.; Smith, E.M.; Grahn, T.H.M.; Warsi, S.; Chen, J.; Flygare, J.; Schambach, A.; et al. Successful gene therapy of Diamond-Blackfan anemia in a mouse model and human CD34(+) cord blood hematopoietic stem cells using a clinically applicable lentiviral vector. *Haematologica* **2022**, *107*, 446–456. [CrossRef]
31. Voit, R.A.; Liao, X.; Cohen, B.; Armant, M.; Kamal, E.; Huang, M.-M.; Clarke, W.; Williams, D.A.; Shimamura, A.; Sankaran, V.G. Regulated Expression of GATA1 As a Gene Therapy Cure for Diamond-Blackfan Anemia. *Blood* **2022**, *140*, 986–987. [CrossRef]
32. Gene Transfer For Patients with Sickle Cell Disease Using A Gamma Globin Lentivirus Vector: An Open-Label Phase 1/2 Pilot Study, NCT02186418. Available online: https://clinicaltrials.gov/study/NCT02186418?cond=Sickle%20Cell%20Disease&term=lentivirus%20gene%20therapy%20&rank=1 (accessed on 17 April 2024).
33. Pilot and Feasibility Study of Hematopoietic Stem Cell Gene Transfer for Sickle Cell Disease, NCT03282656. Available online: https://clinicaltrials.gov/study/NCT03282656?cond=Sickle%20Cell%20Disease&term=lentivirus%20gene%20therapy%20&rank=2 (accessed on 17 April 2024).
34. A Multi-Center, Phase 2 Gene Transfer Study Inducing Fetal Hemoglobin in Sickle Cell (GRASP, BMT CTN 2001), NCT05353647. Available online: https://clinicaltrials.gov/study/NCT05353647?cond=Sickle%20Cell%20Disease&term=lentivirus%20gene%20therapy%20&rank=3 (accessed on 17 April 2024).
35. Clinical Research Study of Autologous Stem Cell Transplantation for Sickle Cell Disease (SCD) Using Peripheral Blood CD34+ Cells Modified with the Lenti/G-βAS3-FB Lentiviral Vector, NCT02247843. Available online: https://clinicaltrials.gov/study/NCT02247843?cond=Sickle%20Cell%20Disease&term=lentivirus%20gene%20therapy%20&rank=4 (accessed on 17 April 2024).
36. A Phase 1/2 Open Label Study Evaluating the Safety and Efficacy of Gene Therapy of the Sickle Cell Disease by Transplantation of an Autologous CD34+ Enriched Cell Fraction That Contains CD34+ Cells Transduced Ex Vivo with the GLOBE1 Lentiviral Vector Expressing the βAS3 Globin Gene (GLOBE1 βAS3 Modified Autologous CD34+ Cells) in Patients with Sickle Cell Disease (SCD), NCT03964792. Available online: https://clinicaltrials.gov/study/NCT03964792?cond=Sickle%20Cell%20Disease&term=lentivirus%20gene%20therapy%20&rank=5 (accessed on 17 April 2024).

37. A Phase 1/2 Open Label Study Evaluating the Safety and Efficacy of Gene Therapy of the β-Hemoglobinopathies (Sickle Cell Anemia and β-Thalassemia Major) by Transplantation of Autologous CD34+ Stem Cells Transduced Ex Vivo with a Lentiviral β-A-T87Q Globin Vector (LentiGlobin BB305 Drug Product), NCT02151526. Available online: https://clinicaltrials.gov/study/NCT02151526?cond=Sickle%20Cell%20Disease&term=lentivirus%20gene%20therapy%20&rank=6 (accessed on 17 April 2024).
38. Evaluation of Safety and Efficacy of Transplantation of Autologous Hematopoietic Stem Cell Genetically Modified in Beta-Thalassemia Major, NCT03276455. Available online: https://clinicaltrials.gov/study/NCT03276455?cond=Thalassemia&term=lentivirus%20gene%20therapy%20&rank=3 (accessed on 17 April 2024).
39. A Phase I/II Study Evaluating Safety and Efficacy of Autologous Hematopoietic Stem Cells Genetically Modified with GLOBE Lentiviral Vector Encoding for the Human Beta-Globin Gene for the Treatment of Patients Affected by Transfusion Dependent Beta-Thalassemia, NCT02453477. Available online: https://clinicaltrials.gov/study/NCT02453477?cond=Thalassemia&term=lentivirus%20gene%20therapy%20&rank=4 (accessed on 17 April 2024).
40. Evaluation of the Safety and Efficacy of KL003 Gene Therapy in Patients with Transfusion-Dependent β-Thalassemia with No Conditioning Regimen, NCT06219239. Available online: https://clinicaltrials.gov/study/NCT06219239?cond=Thalassemia&term=lentivirus%20gene%20therapy%20&rank=6 (accessed on 17 April 2024).
41. Safety and Efficacy Evaluation of β-globin Restored Autologous Hematopoietic Stem Cells in β-Thalassemia Major Patients, NCT05745532. Available online: https://clinicaltrials.gov/study/NCT05745532?cond=Thalassemia&term=lentivirus%20gene%20therapy%20&rank=7 (accessed on 17 April 2024).
42. A Phase I/II Clinical Study Evaluating the Safety and Efficacy of KL003 Cell Injection in Transfusion-Dependent β-Thalassemia, NCT06280378. Available online: https://clinicaltrials.gov/study/NCT06280378?cond=Thalassemia&term=lentivirus%20gene%20therapy%20&rank=8 (accessed on 17 April 2024).
43. A Phase I Clinical Trial for the Treatment of ß-Thalassemia Major with Autologous CD34+ Hematopoietic Progenitor Cells Transduced with TNS9.3.55 a Lentiviral Vector Encoding the Normal Human ß-Globin Gene, NCT01639690. Available online: https://clinicaltrials.gov/study/NCT01639690?cond=Thalassemia&term=lentivirus%20gene%20therapy%20&rank=9 (accessed on 17 April 2024).
44. An Open Label Study Evaluating the Safety and Efficacy of Gene Therapy for Transfusion-Dependent β-Thalassemia by Transplantation of Autologous CD34+ Stem Cells Transduced Ex Vivo with a LentiRed Lentiviral Vector (GMCN-508B Drug Product, Also Called LentiRed), NCT05762510. Available online: https://clinicaltrials.gov/study/NCT05762510?cond=Thalassemia&term=lentivirus%20gene%20therapy%20&rank=11 (accessed on 17 April 2024).
45. A Phase 1 Open Label Study Evaluating the Safety and Efficacy of Gene Therapy in Subjects with Transfusion-Dependent α-Thalassemia by Transplantation of Autologous CD34+ Cells Transduced Ex Vivo with a Lentiviral Vector (GMCN-508A Drug Product), NCT05757245. Available online: https://clinicaltrials.gov/study/NCT05757245?cond=Thalassemia&term=lentivirus%20gene%20therapy%20&rank=12 (accessed on 17 April 2024).
46. A Phase 1 Open Label Study Evaluating the Safety and Efficacy of Gene Therapy in Subjects with β-Thalassemia Major by Transplantation of Autologous CD34+Stem Cells Transduced with a Lentiviral Vector Encoding βA-T87Q-Globin, NCT05015920. Available online: https://clinicaltrials.gov/study/NCT05015920?cond=Thalassemia&term=lentivirus%20gene%20therapy%20&rank=13 (accessed on 17 April 2024).
47. Gene Transfer for Patients with Fanconi Anemia Complementation Group A (FANCA), NCT01331018. Available online: https://clinicaltrials.gov/study/NCT01331018?cond=Fanconi%20Anemia&term=lentivirus%20gene%20therapy%20&rank=1 (accessed on 17 April 2024).
48. Long-Term Follow-up: Phase I/II Clinical Study to Evaluate the Safety and Efficacy of the Infusion of Autologous CD34+ Cells Transduced with a Lentiviral Vector Carrying the FANCA Gene in Patients with Fanconi Anaemia Subtype A: FANCOLEN-I, NCT04437771. Available online: https://clinicaltrials.gov/study/NCT04437771?cond=Fanconi%20Anemia&term=lentivirus%20gene%20therapy%20&rank=2 (accessed on 17 April 2024).
49. Luis, A. The Old and the New: Prospects for Non-Integrating Lentiviral Vector Technology. *Viruses* **2020**, *12*, 1103. [CrossRef] [PubMed]
50. Gurumoorthy, N.; Nordin, F.; Tye, G.J.; Wan Kamarul Zaman, W.S.; Ng, M.H. Non-Integrating Lentiviral Vectors in Clinical Applications: A Glance Through. *Biomedicines* **2022**, *10*, 107. [CrossRef] [PubMed]
51. Milone, M.C.; O'Doherty, U. Clinical use of lentiviral vectors. *Leukemia* **2018**, *32*, 1529–1541. [CrossRef]
52. Shaw, A.; Cornetta, K. Design and Potential of Non-Integrating Lentiviral Vectors. *Biomedicines* **2014**, *2*, 14–35. [CrossRef] [PubMed]
53. Wienert, B.; Cromer, M.K. CRISPR nuclease off-target activity and mitigation strategies. *Front. Genome Ed.* **2022**, *4*, 1050507. [CrossRef] [PubMed]
54. Hunt, J.M.T.; Samson, C.A.; Rand, A.D.; Sheppard, H.M. Unintended CRISPR-Cas9 editing outcomes: A review of the detection and prevalence of structural variants generated by gene-editing in human cells. *Hum. Genet.* **2023**, *142*, 705–720. [CrossRef] [PubMed]
55. Kantor, A.; McClements, M.E.; MacLaren, R.E. CRISPR-Cas9 DNA Base-Editing and Prime-Editing. *Int. J. Mol. Sci.* **2020**, *21*, 6240. [CrossRef] [PubMed]
56. Hryhorowicz, M.; Lipinski, D.; Zeyland, J. Evolution of CRISPR/Cas Systems for Precise Genome Editing. *Int. J. Mol. Sci.* **2023**, *24*, 4233. [CrossRef]

57. Jeong, Y.K.; Song, B.; Bae, S. Current Status and Challenges of DNA Base Editing Tools. *Mol. Ther.* **2020**, *28*, 1938–1952. [CrossRef]
58. Lu, C.; Kuang, J.; Shao, T.; Xie, S.; Li, M.; Zhu, L.; Zhu, L. Prime Editing: An All-Rounder for Genome Editing. *Int. J. Mol. Sci.* **2022**, *23*, 9862. [CrossRef] [PubMed]
59. Mishra, R.; Joshi, R.K.; Zhao, K. Base editing in crops: Current advances, limitations and future implications. *Plant Biotechnol. J.* **2020**, *18*, 20–31. [CrossRef] [PubMed]
60. Porto, E.M.; Komor, A.C.; Slaymaker, I.M.; Yeo, G.W. Base editing: Advances and therapeutic opportunities. *Nat. Rev. Drug Discov.* **2020**, *19*, 839–859. [CrossRef] [PubMed]
61. Collias, D.; Beisel, C.L. CRISPR technologies and the search for the PAM-free nuclease. *Nat. Commun.* **2021**, *12*, 555. [CrossRef] [PubMed]
62. Aldag, P.; Welzel, F.; Jakob, L.; Schmidbauer, A.; Rutkauskas, M.; Fettes, F.; Grohmann, D.; Seidel, R. Probing the stability of the SpCas9-DNA complex after cleavage. *Nucleic Acids Res.* **2021**, *49*, 12411–12421. [CrossRef] [PubMed]
63. Uchida, N.; Drysdale, C.M.; Nassehi, T.; Gamer, J.; Yapundich, M.; DiNicola, J.; Shibata, Y.; Hinds, M.; Gudmundsdottir, B.; Haro-Mora, J.J.; et al. Cas9 protein delivery non-integrating lentiviral vectors for gene correction in sickle cell disease. *Mol. Ther. Methods Clin. Dev.* **2021**, *21*, 121–132. [CrossRef] [PubMed]
64. Clinical Study on the Safety and Efficacy of a Single Intravenous Dose of CRISPR/Cas9-Edited Autologous CD34+ Hematopoietic Stem/Progenitor Cells (BRL-101) in the Treatment of Sickle Cell Disease, NCT06287099. Available online: https://clinicaltrials.gov/study/NCT06287099?cond=Sickle%20Cell%20Disease&intr=CRISPR/Cas9&rank=1 (accessed on 17 April 2024).
65. A Phase I/II Study of Nula-cel in Autologous CD34+ Hematopoietic Stem Cells to Convert HbS to HbA for Treating Severe Sickle Cell Disease, NCT04819841. Available online: https://clinicaltrials.gov/study/NCT04819841?cond=Sickle%20Cell%20Disease&intr=CRISPR/Cas9&rank=4 (accessed on 17 April 2024).
66. A Phase 1/2/3 Study to Evaluate the Safety and Efficacy of a Single Dose of Autologous CRISPR-Cas9 Modified CD34+ Human Hematopoietic Stem and Progenitor Cells (CTX001) in Subjects with Severe Sickle Cell Disease, NCT03745287. Available online: https://clinicaltrials.gov/study/NCT03745287?cond=Sickle%20Cell%20Disease&intr=CRISPR/Cas9&rank=5 (accessed on 17 April 2024).
67. A Phase 3 Study to Evaluate Efficacy and Safety of a Single Dose of Exa-cel in Subjects with Severe Sickle Cell Disease, βS/βC Genotype, NCT05951205. Available online: https://clinicaltrials.gov/study/NCT05951205?cond=Sickle%20Cell%20Disease&intr=CRISPR/Cas9&rank=7 (accessed on 17 April 2024).
68. Transplantation of CRISPRCas9 Corrected Hematopoietic Stem Cells (CRISPR_SCD001) in Patients with Severe Sickle Cell Disease, NCT04774536. Available online: https://clinicaltrials.gov/study/NCT04774536?cond=Sickle%20Cell%20Disease&intr=CRISPR/Cas9&rank=8 (accessed on 17 April 2024).
69. A Phase 3 Study to Evaluate the Safety and Efficacy of a Single Dose of CTX001 in Pediatric Subjects with Severe Sickle Cell Disease, NCT05329649. Available online: https://clinicaltrials.gov/study/NCT05329649?cond=Sickle%20Cell%20Disease&intr=CRISPR/Cas9&rank=10 (accessed on 17 April 2024).
70. A Phase 3b Study to Evaluate Efficacy and Safety of a Single Dose of Autologous CRISPR Cas9 Modified CD34+ Human Hematopoietic Stem and Progenitor Cells (CTX001) in Subjects with Transfusion-Dependent β-Thalassemia or Severe Sickle Cell Disease, NCT05477563. Available online: https://clinicaltrials.gov/study/NCT05477563?cond=Thalassemia&intr=CRISPR/Cas9&rank=2 (accessed on 17 April 2024).
71. A Long-term Follow-up Study of Subjects with β-Thalassemia or Sickle Cell Disease Treated with Autologous CRISPR-Cas9 Modified Hematopoietic Stem Cells (CTX001), NCT04208529. Available online: https://clinicaltrials.gov/study/NCT04208529?cond=Thalassemia&intr=CRISPR/Cas9&rank=7 (accessed on 17 April 2024).
72. A Phase 1/2/3 Study of the Safety and Efficacy of a Single Dose of Autologous CRISPR-Cas9 Modified CD34+ Human Hematopoietic Stem and Progenitor Cells (hHSPCs) in Subjects with Transfusion-Dependent β-Thalassemia, NCT03655678. Available online: https://clinicaltrials.gov/study/NCT03655678?cond=Thalassemia&intr=CRISPR/Cas9&rank=1 (accessed on 17 April 2024).
73. A Multicenter, Open Label Phase 1 Study to Evaluate the Safety and Efficacy of a Single Dose of Autologous CRISPR-Cas9 Modified CD34+ Human Hematopoietic Stem and Progenitor Cells (hHSPCs) in Subjects with Transfusion Dependent β-Thalassaemia, NCT04925206. Available online: https://clinicaltrials.gov/study/NCT04925206?cond=Thalassemia&intr=CRISPR/Cas9&rank=3 (accessed on 17 April 2024).
74. A Phase 1/2 Clinical Study to Evaluate the Safety and Efficacy of Single Dose Intravenous Infusion of BRL-101 in Subjects with Transfusion-Dependent β-Thalassemia, NCT05577312. Available online: https://clinicaltrials.gov/study/NCT05577312?cond=Thalassemia&intr=CRISPR/Cas9&rank=4 (accessed on 17 April 2024).
75. A Phase 3 Study to Evaluate the Safety and Efficacy of a Single Dose of CTX001 in Pediatric Subjects with Transfusion-Dependent β-Thalassemia, NCT05356195. Available online: https://clinicaltrials.gov/study/NCT05356195?cond=Thalassemia&intr=CRISPR/Cas9&rank=5 (accessed on 17 April 2024).
76. A Safety and Efficacy Study of a Single Center, Open-Label, Single Arm About the Gene Correction of HBB in Patient-Specific iHSCs Using CRISPR/Cas9 That Intervent Subjests with β-Thalassemia Mutations, NCT03728322. Available online: https://clinicaltrials.gov/study/NCT03728322?cond=Thalassemia&intr=CRISPR/Cas9&rank=6 (accessed on 17 April 2024).

77. Frangoul, H.; Altshuler, D.; Cappellini, M.D.; Chen, Y.S.; Domm, J.; Eustace, B.K.; Foell, J.; de la Fuente, J.; Grupp, S.; Handgretinger, R.; et al. CRISPR-Cas9 Gene Editing for Sickle Cell Disease and beta-Thalassemia. *N. Engl. J. Med.* **2021**, *384*, 252–260. [CrossRef]
78. Newby, G.A.; Yen, J.S.; Woodard, K.J.; Mayuranathan, T.; Lazzarotto, C.R.; Li, Y.; Sheppard-Tillman, H.; Porter, S.N.; Yao, Y.; Mayberry, K.; et al. Base editing of haematopoietic stem cells rescues sickle cell disease in mice. *Nature* **2021**, *595*, 295–302. [CrossRef] [PubMed]
79. Li, C.; Georgakopoulou, A.; Newby, G.A.; Chen, P.J.; Everette, K.A.; Paschoudi, K.; Vlachaki, E.; Gil, S.; Anderson, A.K.; Koob, T.; et al. In vivo HSC prime editing rescues sickle cell disease in a mouse model. *Blood* **2023**, *141*, 2085–2099. [CrossRef]
80. FDA. FDA Approves First Gene Therapies to Treat Patients with Sickle Cell Disease. Available online: https://www.fda.gov/news-events/press-announcements/fda-approves-first-gene-therapies-treat-patients-sickle-cell-disease (accessed on 14 May 2024).
81. Mohammadian Gol, T.; Zahedipour, F.; Trosien, P.; Urena-Bailen, G.; Kim, M.; Antony, J.S.; Mezger, M. Gene therapy in pediatrics—Clinical studies and approved drugs (as of 2023). *Life Sci.* **2024**, *348*, 122685. [CrossRef]
82. ZYNTEGLO. ZYNTEGLO™ Mechanism of Action. Available online: https://www.zynteglohcp.com/mechanism-of-action (accessed on 14 May 2024).
83. FDA. Long Term Follow-up After Administration of Human Gene Therapy Products. Available online: https://www.fda.gov/regulatory-information/search-fda-guidance-documents/long-term-follow-after-administration-human-gene-therapy-products (accessed on 14 May 2024).
84. Lee, N.K.; Chang, J.W. Manufacturing Cell and Gene Therapies: Challenges in Clinical Translation. *Ann. Lab. Med.* **2024**, *44*, 314–323. [CrossRef] [PubMed]

MDPI AG
Grosspeteranlage 5
4052 Basel
Switzerland
Tel.: +41 61 683 77 34

Cells Editorial Office
E-mail: cells@mdpi.com
www.mdpi.com/journal/cells

www.ingramcontent.com/pod-product-compliance
Lightning Source LLC
Chambersburg PA
CBHW061444210926
51850CB00008B/1612
9783725874460